Coral Reefs of the World

Volume 11

Series editors:

Bernhard M. Riegl, Nova Southeastern University, Dania Beach, USA
Richard E. Dodge, Nova Southeastern University, Dania Beach, USA

Coral Reefs of the World is a series presenting the status of knowledge of the world's coral reefs authored by leading scientists. The volumes are organized according to political or regional oceanographic boundaries. Emphasis is put on providing authoritative overviews of biology and geology, explaining the origins and peculiarities of coral reefs in each region. The information is so organized that it is up to date and can be used as a general reference and entry-point for further study. The series will cover all recent and many of the fossil coral reefs of the world.

Prospective authors and/or editors should consult the Series Editors B.M. Riegl and R.E. Dodge for more details. Any comments or suggestions for future volumes are welcomed:

Dr. Bernhard M. Riegl/Dr. Richard E. Dodge Nova Southeastern University Dania Beach, FL 33004 USA

e-mail: rieglb@nova.edu and dodge@nova.edu

More information about this series at http://www.springer.com/series/7539

Christian R. Voolstra · Michael L. Berumen
Editors

Coral Reefs of the Red Sea

Editors
Christian R. Voolstra
Red Sea Research Center
KAUST University
Thuwal, Saudi Arabia

Michael L. Berumen
Red Sea Research Center
KAUST University
Thuwal, Saudi Arabia

ISSN 2213-719X   ISSN 2213-7203   (electronic)
Coral Reefs of the World
ISBN 978-3-030-05800-5   ISBN 978-3-030-05802-9   (eBook)
https://doi.org/10.1007/978-3-030-05802-9

Library of Congress Control Number: 2019933737

© Springer Nature Switzerland AG 2019
This work is subject to copyright. All rights are reserved by the Publisher, whether the whole or part of the material is concerned, specifically the rights of translation, reprinting, reuse of illustrations, recitation, broadcasting, reproduction on microfilms or in any other physical way, and transmission or information storage and retrieval, electronic adaptation, computer software, or by similar or dissimilar methodology now known or hereafter developed.
The use of general descriptive names, registered names, trademarks, service marks, etc. in this publication does not imply, even in the absence of a specific statement, that such names are exempt from the relevant protective laws and regulations and therefore free for general use.
The publisher, the authors, and the editors are safe to assume that the advice and information in this book are believed to be true and accurate at the date of publication. Neither the publisher nor the authors or the editors give a warranty, express or implied, with respect to the material contained herein or for any errors or omissions that may have been made. The publisher remains neutral with regard to jurisdictional claims in published maps and institutional affiliations.

This Springer imprint is published by the registered company Springer Nature Switzerland AG.
The registered company address is: Gewerbestrasse 11, 6330 Cham, Switzerland

A shallow coral reef in the Saudi Arabian central Red Sea near Thuwal village. The nutrient-poor and clear waters of the Red Sea benefit reef-building corals, which construct some of the world's largest and diverse coral reef habitats. The success of reef-building corals depends on the symbiosis with unicellular algae of the family Symbiodiniaceae, which is an obligatory but temperature-sensitive association. The Red Sea is one of the warmest regions, where reef-building corals form and maintain coral reef ecosystems. The Red Sea has emerged as an important region for coral reef research, since its corals may contribute to a better understanding of the adaptation capacity of reef-building corals to global warming. (picture credit: Anna Roik 2014)

Preface

The Red Sea is the northernmost tropical sea and the only enclosed coral sea in the world. Although known to many for its spectacular and truly wondrous coral reefs, the Red Sea is becoming increasingly acknowledged as a unique place to study corals. Some of the most temperature-resilient corals live in the Red Sea, and the Northern Red Sea might constitute a refuge for corals for the decades to come – making it the largest laboratory in the world to study coral adaptation and a beacon of hope in such desperate times, when we lost an estimated 30% of global reef cover to climate change-driven coral bleaching.

Much of the research considering the role of the Red Sea as a model for Future Oceans and what it can teach us about the effects of climate change is not even a decade old, and there really is no good resource that comprehensively and concisely summarizes the current state of research around coral reefs in the Red Sea.

Primary as a resource for my own, I thus became motivated to summarize the current state of research and knowledge of Red Sea coral reefs in the form of a pocket guide and reference handbook that holds the important information in a quickly accessible manner.

The book starts with an overview over the strong environmental gradient across the latitudinal spread of the Red Sea and how this characterizes and structures the habitats in the Red Sea. It then details the physicochemical environment that enables reef building in this ocean basin. After a review of coral physiology under the backdrop of these unique environmental settings, we explore reef-associated bacteria and the diversity of microalgal photosymbionts in the family Symbiodiniaceae. It is the symbioses between these microalgae and their coral animal hosts that provide the foundation of coral reef ecosystems and give rise to the unparalleled diversity of sponge, coral, and reef fish, which are highlighted in the concluding chapters.

I would like to thank my coeditor and long-term colleague Michael L. Berumen for his contribution to this book. We both came as founding faculty to the King Abdullah University of Science and Technology (KAUST) in Saudi Arabia in 2009, when the prospect of making a difference and studying the unique coral reefs of the Saudi Arabian Red Sea drew us into this adventure – an adventure that keeps giving and let's us discover new and remarkable things about marine life every day.

May this book prove as useful of a resource to you, as it is for us; and may this book, in addition to the science, also convey the wonders of Red Sea coral reefs – a truly unique and precious resource that deserves and requests our attention, respect, and protection.

Thuwal, Saudi Arabia Christian R. Voolstra
October 1, 2018

Contents

1. **The Red Sea: Environmental Gradients Shape a Natural Laboratory in a Nascent Ocean** 1
   Michael L. Berumen, Christian R. Voolstra, Daniele Daffonchio, Susana Agusti, Manuel Aranda, Xabier Irigoien, Burton H. Jones, Xosé Anxelu G. Morán, and Carlos M. Duarte

2. **Environmental Setting for Reef Building in the Red Sea** 11
   James Churchill, Kristen Davis, Eyal Wurgaft, and Yonathan Shaked

3. **Ecophysiology of Reef-Building Corals in the Red Sea** 33
   Maren Ziegler, Anna Roik, Till Röthig, Christian Wild, Nils Rädecker, Jessica Bouwmeester, and Christian R. Voolstra

4. **Microbial Communities of Red Sea Coral Reefs** 53
   Matthew J. Neave, Amy Apprill, Greta Aeby, Sou Miyake, and Christian R. Voolstra

5. **Symbiodiniaceae Diversity in Red Sea Coral Reefs & Coral Bleaching** 69
   Maren Ziegler, Chatchanit Arif, and Christian R. Voolstra

6. **Sponges of the Red Sea** 91
   Michael K. Wooster, Oliver Voigt, Dirk Erpenbeck, Gert Wörheide, and Michael L. Berumen

7. **Corals of the Red Sea** 123
   Michael L. Berumen, Roberto Arrigoni, Jessica Bouwmeester, Tullia I. Terraneo, and Francesca Benzoni

8. **Fishes and Connectivity of Red Sea Coral Reefs** 157
   Michael L. Berumen, May B. Roberts, Tane H. Sinclair-Taylor, Joseph D. DiBattista, Pablo Saenz-Agudelo, Stamatina Isari, Song He, Maha T. Khalil, Royale S. Hardenstine, Matthew D. Tietbohl, Mark A. Priest, Alexander Kattan, and Darren J. Coker

The Red Sea: Environmental Gradients Shape a Natural Laboratory in a Nascent Ocean

Michael L. Berumen, Christian R. Voolstra, Daniele Daffonchio, Susana Agusti, Manuel Aranda, Xabier Irigoien, Burton H. Jones, Xosé Anxelu G. Morán, and Carlos M. Duarte

Abstract

This chapter introduces the environmental gradients that characterize the broader Red Sea habitat. The Red Sea is formed by an actively spreading rift and notably has only one natural connection to the Indian Ocean – a narrow, shallow opening known as the Strait of Bab al Mandab. The resultant isolation undoubtedly plays a key role in shaping the environmental gradients, species endemism, and distinct evolutionary trajectory observed within the Red Sea. While this young ocean is known to be among the saltiest and warmest seas on Earth, there are important spatial and temporal gradients that likely influence the biological communities residing in its waters.

Keywords

Red Sea · Physical environment · Ecosystems · Environmental gradients · Coral reefs · Brine pools · Seagrass meadows · Biogeography

1.1 Introduction

The Red Sea, a narrow, marginal sea of the Indian Ocean nearly 2000 km long and 200–300 km wide, offers many opportunities as a 'natural laboratory' due to the different gradients, often reaching extreme values, that occur within this unique body of water. While the Red Sea is recognized as one of the warmest and saltiest seas on the planet, these traits are not uniform but change strongly along its main axis, extending from 12.5°N in the south to 30°N at Suez in the north. The lack of freshwater input (very low regional rainfall, no riverine input, etc.) and the warm desert climate of the region lead to high evaporation rates. This contributes to the pattern of salinity increasing with distance from the Bab al Mandab, where inputs of Indian Ocean water from the Gulf of Aden enter the Red Sea, creating an inverse estuarine circulation (Sofianos and Johns 2002). Temperature generally shows an opposite pattern in that the highest sea surface temperature (SST) occurs in the south and decreases northward (Chaidez et al. 2017; Ngugi et al. 2012). Inorganic nutrients also exhibit strong spatial gradients in the upper 200 m of the water column. Dissolved inorganic nitrogen (DIN) concentrations in excess of 15 µmol N L^{-1} occur in the south and at depth, while they approach nanomolar concentrations in the surface layer of the central and northern Red Sea (Churchill et al. 2014; Kürten et al. 2016; Wafar et al. 2016b). Surface chlorophyll concentration inferred from remote sensing techniques generally shows similar spatial variability, declining from south to north, while the southern Red Sea typically exhibits consistently higher values year-round (Kheireddine et al. 2017; Li et al. 2017; Raitsos et al. 2013). A vertical gradient is also present. Water temperature decreases gradually within the upper layers (approximately 400 m), as observed in other oceans. However, uniquely to the Red Sea, temperature stabilizes at depth and does not fall below ~21 °C even at depths below 2000 m (Roder et al. 2013). This contrasts with other seas and oceans, where water temperature below the upper layers continues to decrease with increasing depth, typically reaching 2–4 °C at depths of 1000 m and dropping to less than 2 °C in the deep basins. This unique temperature gradient is likely to influence the metabolic requirements of any organisms venturing to or residing at these depths. For example, deep-water corals are usually associated with the cold water typically found at depths below 500-600 m in the ocean, but somehow in the Red Sea they have adapted to the much warmer conditions encountered at depth and associated energetic requirements (Roder et al. 2013; Roik et al. 2015b; Röthig et al. 2017; Yum et al. 2017). Other physical gradients and processes that contribute to the uniqueness of the Red Sea include regional wind patterns, surface currents and eddies, and dust inputs (Zarokanellos et al. 2017b).

Notably, most of the aforementioned gradients also show strong seasonal signals, increasing from more tropical regimes in the south to more temperate regimes in the north. For many of these variables, reliable time-series data with sufficient spatial coverage is only recently available (Roik et al. 2016). Contemporary work has focused on understanding the nature of these gradients and explicitly addressing spatial and temporal variability. This provides an important introduction to the environment in which Red Sea coral reefs reside, which are the focus of this book.

These environmental gradients are also expected to influence other largely unknown or under-sampled biological variables such as primary productivity (Qurban et al. 2017), dissolved organic matter, microbial standing stocks/biomass, plankton community metabolism, and the genetic makeup of most of Red Sea species. It remains to be seen if the Red Sea conforms to the same biological rules as other oceans and tropical seas; some of the limitations and drivers in other systems (e.g., nutrient availability, seasonal mixing, stratification and upwelling, etc.) may function quite differently in the Red Sea. The potential flow-on effects could play a role in structuring entire ecosystems present in the Red Sea.

Despite these gradients and the relatively extreme conditions, the Red Sea is host to many diverse ecosystems along its length, including mangroves, seagrass habitats, soft-bottom sediment flats, coral reefs, and more. The roles of the prominent environmental gradients on these ecosystems are not thoroughly studied and warrant further investigation. The recent increase in interest in research institutions in the region (Mervis 2009) has led to growing efforts to examine the interactions of environmental gradients.

The Red Sea exhibits high levels of endemism and has long been recognized as a biodiversity hotspot for tropical marine fauna (DiBattista et al. 2015b). Comparisons of the unique species found in the Red Sea to broad-ranging species that co-occur in the Red Sea and species outside of the Red Sea may offer some insight to the evolutionary history of Red Sea fauna and the adaptations necessary for survival in these conditions. Since the present Red Sea may reflect future conditions in other regions of the world, its communities might provide some forecast as to how reefs and other marine ecosystems in other parts of the world will fare under climate change scenarios, particularly in the context of genetic capacities for adaptation (Aranda et al. 2016; Voolstra et al. 2017). However, it is important to note that the Red Sea is experiencing its own rapidly-changing conditions as a result of global climate change (Chaidez et al. 2017; Raitsos et al. 2011) and, unfortunately, its coral reef communities are not immune to impacts of these changes (Cantin et al. 2010; Furby et al. 2013; Hughes et al. 2018; Monroe et al. 2018; Osman et al. 2018; Riegl et al. 2012; Roik et al. 2015a).

It is in the background and in the context of these gradients that coral reef ecosystems exist within the Red Sea. In several of the Red Sea countries, the reefs represent a critical component of their respective tourism industries. In other Red Sea countries, a substantial number of people rely on reef-based fisheries (Jin et al. 2012). In some locations, these industries remain under-developed or under-utilized. However, the Kingdom of Saudi Arabia, one of the areas where tourism has remained low, has announced a major eco-tourism area in the northern Red Sea, with the unique premise that ecosystems should receive no impacts from this development, where coral reefs, together with seagrass meadows, are arguably the more vulnerable habitats. Therefore, while the Red Sea provides a window into fascinating scientific questions, there are also important practical applications for understanding the relationship between environmental conditions and the general state of coral reefs in this region.

In this chapter, we seek to address the various gradients of the Red Sea in four broad categories:

1. The physical environment
2. Nutrients and productivity
3. Gene flow and genetic diversity
4. Biogeography

Although knowledge of some of these remains imperfect, we believe that this overview provides a useful context for subsequent chapters in this volume, where many of the facets mentioned here will be addressed in detail.

1.2 The Physical Environment of the Red Sea

Much of what has been known about the physical environment of the Red Sea is derived from remotely-sensed measurements and simulations of the physical dynamics, with limited *in situ* observations (Acker et al. 2008; Raitsos et al. 2013). Temperature shows a latitudinal gradient in all seasons, with warmest temperatures in the south and cooler temperatures in the north (Fig. 1.1). However, the surface temperature range in the north is greater than in the south. The northern Red Sea can see an average SST seasonal change from a winter low of ~22 °C to a summer high of 29–30 °C. In the southern Red Sea, the SST may range from a winter low of <26 °C to a summer high of >31 °C. Salinity shows an inverse gradient, with higher salinities in the north (>40) and lowest salinities in the south (see Fig. 1.1).

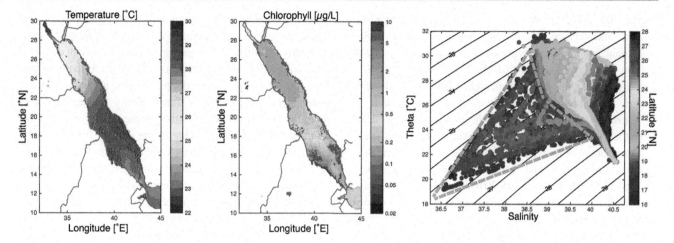

Fig. 1.1 The left and center panels show the 15-year averages for SST and chlorophyll a from the MODIS Aqua instrument (source NASA's Giovanni: http://giovanni.gsfc.nasa.gov/). The right panel shows a temperature-salinity diagram for a joint KAUST-WHOI hydrographic cruise during September–October 2011. The spatial extent of the temperature-salinity dataset is from 16°N to 28°N within the Saudi Arabian EEZ and extends vertically through the water column. Latitude in the right panel is indicated by color coding from blue in the south to red in the north. The black lines are the contours of density anomaly [kg/m^3]. Gulf of Aden intermediate water (sometimes known as Gulf of Aden intrusion water) is outlined with light blue dashed lines

Temperature gradients are dynamic and changing (Chaidez et al. 2017; Raitsos et al. 2011). Red Sea SSTs have increased rapidly in the latter half of 1990–2010, at an average rate of 0.17 ± 0.07 °C decade^{-1}, exceeding the global rate while the northern Red Sea is warming even faster, at between 0.40 and 0.45 °C decade^{-1}, rendering the northern Red Sea one of the fastest-warming areas in the world (Chaidez et al. 2017).

The along-basin change in salinity results from large annual evaporation of approximately 2 m year^{-1} (Patzert 1974; Sofianos and Johns 2002; Tragou et al. 1999). The only significant source of "fresh" water for the Red Sea is Gulf of Aden water, which enters the Red Sea with a salinity of about 36. Thus, the Red Sea functions as an "inverse" estuary where fresher water enters the system and the strong thermohaline circulation resulting from evaporation results in a warm, salty outflow that is traceable throughout much of the Indian Ocean (Beal et al. 2000; Zhai et al. 2015). This unique environmental profile is likely to have major impacts on the biological and ecological functions of the Red Sea environment, as there appear to never have been any permanent riverine inputs to the Red Sea. For example, the Nile River does not appear to have ever emptied into the Red Sea (DiBattista et al. 2015a). Rare rainfall events on the coast may provide local input of freshwater via ephemeral rivers known locally as "wadis", sometimes in such quantities as to have significant ecological impacts (e.g., freshwater-induced coral bleaching events) (Antonius 1988). Such events are rare, with average rainfall of about 100 mm year^{-1}; rainfall declines from 100 mm year^{-1} in the south to about 50 mm year^{-1} in the northern Red Sea (Almazroui et al. 2012).

It is well-established that numerous transient eddies characterize the Red Sea. Analysis of sea level anomaly (SLA) has demonstrated that the central Red Sea is especially characterized by high levels of eddy activity (Zhan et al. 2014). While long term averages from models and SLA have suggested that subregions may be dominated by either cyclonic or anti-cyclonic eddies, in the central Red Sea both types of eddies have been observed and often form in eddy pairs (Zarokanellos et al. 2017b). The duration of these eddies varies, typically ranging from a few weeks to several months. These eddies have been detected by various measurements, including SLA, remotely sensed chlorophyll, SST patterns, and in-situ measurements (Kürten et al. 2016; Raitsos et al. 2013; Sofianos Sarantis and Johns William 2007; Zhai and Bower 2013; Zhan et al. 2014); the eddies are also indicated in numerical models of Red Sea circulation (Clifford et al. 1997; Sofianos and Johns 2002; Yao et al. 2014). These eddies play an important role in modulating the depth of the nutricline, and may contribute to nutrient exchange between the open sea and coral reefs. The eddies likely play further important roles in disrupting potential boundary currents along the latitudinal gradient of the Red Sea coastline (Zarokanellos et al. 2017a). The ability of the eddies to facilitate longitudinal connections (i.e., east-west connectivity across the Red Sea) has not often been investigated, but is a potentially important feature (Raitsos et al. 2017).

The evaporation-driven thermohaline circulation results in a northward transport within the upper layer of the basin (Sofianos and Johns 2003). Modeling indicates that this current initiates along the western boundary in the south and crosses over to the eastern coast mid-basin, influenced by topographically-steered winds (Zhai et al. 2015). Observations are fewer, but support the northward transport along the eastern boundary in the northern half of the basin (Bower and

Farrar 2015). Recent hydrographic surveys utilizing ship, gliders, and surface current mapping observations provide additional support for the presence of an eastern boundary current (Zarokanellos et al. 2017b). This northward transport will also contribute to the dispersal of organisms along the axis of the basin but has yet to be carefully investigated.

There are no permanent riverine inputs to the Red Sea. Nonetheless, terrigenous inputs are likely to be a very important component of Red Sea nutrient cycles, delivered via deposition of atmospheric dust. Soils in coastal plains surrounding the Red Sea (Jish Prakash et al. 2016) are calculated to deliver 7.5 Mt of dust suspended in the atmosphere per year, corresponding to 76 Kt of iron oxides and 6 Kt of phosphorus per year, with over 65% of dust emitted from the northern region, much of which is likely to be deposited in the Red Sea (Anisimov et al. 2017). The Red Sea is exposed to ~15–20 dust storms per year (Jish Prakash et al. 2015), with atmospheric dust loads (and likely the subsequent inputs) double over the southern Red Sea compared to the northern Red Sea (Banks et al. 2017), particularly during the summer, when dust loads are highest (Osipov and Stenchikov 2018). In addition to delivering nutrients, dust cools the Red Sea, reduces the surface wind speed, and weakens both the exchange at the Bab al Mandab strait and the overturning circulation, affecting salinity distribution and heat budgets, and thereby circulation of the Red Sea (Osipov and Stenchikov 2018).

The Red Sea, geologically, has been formed by the separation of the Nubian plate from the Arabian plate. These plates continue to separate at a rate of about 16 mm per year. This activity leads to potential influence on the oceanography. At present, the activity seems to be more prominent in the south as opposed to the north (Xu et al. 2015). For example, between 2011–2013, two new islands emerged in the Zubair archipelago (in Yemeni waters) as a result of volcanic activity (Xu et al. 2015). The emergence of these two new islands would otherwise provide a unique opportunity to observe the establishment and succession of benthic communities, but the islands are unfortunately located in an area of intense political unrest. The separation of the two plates that surround the Red Sea determined the formation of deep anoxic brine pools mainly along the central axis of the basin (Backer and Schoell 1972; Pautot et al. 1984). Due to temporary isolation from fresh seawater inputs during the Miocene, the precipitation of thick evaporitic layers occurred in the ancient Red Sea. These evaporites were later dissolved when they were exposed to fresh seawater, similarly to what occurred after the Messinian Salinity Crisis in the Mediterranean Sea (Garcia-Castellanos and Villaseñor 2011; Searle and Ross 2007). The dissolution of the evaporites determined the formation of deep anoxic brine lakes, some of which are extremely sulfuric and others that are influenced by hydrothermal fluids (Schardt 2016; Swift et al. 2012).

The brine pools have a high density that limits the mixing with the overlying seawater and may include different vertically-stratified water bodies (of tens or more m depth) of increasing salinities (Bougouffa et al. 2013). At the transition between the brine and the deep seawater, a very productive water layer occurs along a sharp salinity gradient and chemocline referred to as the brine-seawater interface (Daffonchio et al. 2006). Due to the density barrier, such a chemocline (whose thickness varies between one to tens of m) entraps the organic matter sinking from the overlying water column. The different combinations of redox couples along the chemocline enable the selection of unique groups of microorganisms (Ngugi et al. 2015) that remain stratified along the salinity gradient per the availability of the suitable redox couples for their metabolisms (Borin et al. 2009). The environment surrounding the brine pools hosts complex communities of animals that exploit the carbon and nutrient resources emanating from the pool (Batang et al. 2012; Vestheim and Kaartvedt 2016).

The gradients on the surface of the brine pools represent a further source of microorganism variability in the Red Sea with potential biotechnology applications (Grotzinger et al. 2018).

1.3 Nutrients and Productivity in the Red Sea

The Red Sea, as a whole, is an oligotrophic system. However, the few nutrients that are present in the upper layers do not have an even distribution throughout the basin. This is apparent even in remotely-sensed chlorophyll-a concentrations (Raitsos et al. 2013) used as a proxy for the biomass of phytoplankton, the pelagic primary producers. The northern half of the sea typically has less chlorophyll than the southern half. There are seasonal maxima of productivity in coastal habitats (Racault et al. 2015), likely linked to irregular oceanographic features such as eddy circulation (Kürten et al. 2016; Zarokanellos et al. 2017b) or the delivery of nutrient-rich Gulf of Aden intrusion water (Churchill et al. 2014; Wafar et al. 2016a). Based on the strong differences in inorganic nutrient availability, small-sized phytoplankton (*Synechococcus* and *Prochlorococcus* cyanobacteria and picoeukaryotes) are relatively more abundant in the central and northern reaches (Kheireddine et al. 2017), although not in absolute numbers (Kürten et al. 2014). Stronger nitrogen limitation at higher latitudes would be the likely cause for the frequent presence of *Trichodesmium* in the northern reaches, including the Gulf of Aqaba (Post et al. 2002), although no evidence of latitudinal gradients in *Trichodesmium* distribution is available yet (Devassy et al. 2017).

The latitudinal gradient in phytoplankton biomass and productivity translates into other components of pelagic food webs. Thus, heterotrophic prokaryotes (bacteria and archaea) and zooplankton increase their abundance towards the south (Kürten et al. 2014). Although much less is known about other planktonic groups, the scant evidence points to larger stocks in the richer lower latitude waters, as recently reported for chaetognaths (Al-Aidaroos et al. 2017).

While total biodiversity (i.e., species richness) may not be greatly affected (Devassy et al. 2017), changes in species composition of planktonic assemblages do occur along the latitudinal axis. For instance, changes in the species composition of prokaryotic plankton have been described (Ngugi et al. 2012), including the widespread SAR11 clade (Ngugi and Stingl 2012). This pattern accompanies the trend of increasing planktonic biomass towards the south, a pattern that is consistent across the inshore-offshore gradient (Pearman et al. 2016, 2017). However, northern and southern regions did not differ significantly in the ecotype compositions of *Prochlorococcus* (Shibl et al. 2016). More conspicuous changes in functional genes (rather than taxa) have been recently reported in a study of 45 metagenomes obtained in a latitudinal transect along the eastern coast (Thompson et al. 2016). In spite of these geographical differences, the temporal variability, often overlooked in subtropical and tropical waters, seems comparable to the latitudinal one for planktonic microbial communities (Pearman et al. 2017; Silva et al. 2019).

At higher trophic levels, the latitudinal patterns of productivity may be influencing the biology of Red Sea fish populations. The southern half of the Red Sea is the preferred habitat for Red Sea whale sharks (Berumen et al. 2014). Other parts of the Red Sea may not provide sufficient food for these planktivorous sharks. The seasonal fluctuations in productivity may also reflect food availability for reef-dwelling planktivorous fishes. Another study (Robitzch et al. 2016) linked local variations in food availability to larval growth rates and metabolism in *Dascyllus* damselfishes.

1.4 Gene Flow and Genetic Diversity in the Red Sea

The presence of such clear environmental gradients within the Red Sea might be expected to lead to a corresponding genetic gradient if populations exhibit local adaptation to the potentially stressful conditions (e.g., high temperature or salinity). Even without the environmental gradients, the physical length of the Red Sea coastline may create genetically distinguishable populations at opposite ends. In terms of coral reef habitats, the Red Sea presents an interesting case with several thousands of kilometers of nearly-continuous reef habitat. It is arguably the longest fringing reef of the world, bordering over 4000 km of the eastern and western shorelines of the Red Sea and the Gulf of Aqaba. The aforementioned temporal eddies thus introduce the potential for east-west transport of larvae and, thus, genetic connectivity between both sides of the Red Sea (Raitsos et al. 2017).

To date, these topics have been addressed in several fishes and a few coral species but very rarely in other taxa. Various methods and genetic markers have been employed. Notably, not all studies are able to include samples at either extreme of the Red Sea (e.g., the Gulf of Suez or the Gulf of Aqaba in the north, or near the Bab al Mandab in the south).

Studies on benthic invertebrates have indicated very little population differentiation (Giles et al. 2015; Robitzch et al. 2015) along the Red Sea. Giles et al. (2015) analyzed samples of a reef sponge (*Stylissa carteri*) with high sampling coverage along the Saudi Arabian Red Sea coast, including samples from the Gulf of Aqaba to the Farasan Islands (at the border of Saudi Arabia and Yemen), and one site on the opposite coast (from Sudan). The authors found that the marked environmental difference of the Farasan Islands habitat explained more of the genetic variation in the samples than geographic distance. In other words, the majority of the Red Sea was relatively genetically homogeneous for this common sponge, but the distinct environment of the Farasan Islands was reflected in a genetically distinct population.

Various studies of fish genetics within the Red Sea reached similar conclusions. The majority of the Red Sea appears to be reasonably well-mixed genetically, with some exceptions found in samples from the farthest ends. For example, Froukh and Kochzius (2008) found limited connectivity between the Gulf of Aqaba and central Red Sea sites in the fourline wrasse (*Larabicus quadrilineatus*), but did not have samples from the northern Red Sea (outside of the Gulf of Aqaba). Nanninga et al. (2014) studied the two-band anemonefish (*Amphiprion bicinctus*) using spatial coverage similar to that of Giles et al. (2015) and likewise found that environmental gradients explained more variation than geographic distance. Interestingly, the location of the genetic break in the anemonefish population was not exactly the same as it was for the sponge. There could be a few potential explanations for this, such as differences in larval ecology of the two species or differences in the sensitivity to the environmental conditions. Saenz-Agudelo et al. (2015) analyzed a subset of the anemonefish samples from Nanninga et al. (2014) and added additional samples from outside the Red Sea. Using a next-generation sequencing approach, Saenz-Agudelo et al. (2015) tested explicit theories about potential barriers to connectivity within and outside of the Red Sea, confirming the within-Red Sea patterns previously mentioned. These three brief examples from fishes suggest that each species may have slightly different genetic gradients throughout the Red Sea, but the general pattern holds that populations are well-mixed throughout the majority of the Red Sea.

A recent study used long-term particle dispersion model simulations and satellite-derived biophysical observations to show that physical oceanography largely explains the genetic homogeneity through the Red Sea (Raitsos et al. 2017). The modeling results also suggest that there is a high degree of west-east connectivity, which is driven by frequent occurrences of eddies that transport surface water across the Red Sea. Wilson (2017) examined the west-east connectivity of two *Plectropomus* grouper species and found relatively high levels of gene flow between Sudanese and Saudi Arabian populations, which supports the predictions of the models from Raitsos et al. (2017). For more information about connectivity and fishes please refer to Chap. 8 in this book.

1.5 Biogeography of Red Sea Organisms

In some well-studied groups, the Red Sea is well-known for hosting a relatively high proportion of endemic species. This is true, for example, among conspicuous reef fishes and corals, and is also true in other taxonomic groups where sufficient data is available (DiBattista et al. 2015b). This latter caveat is important – there are very few groups for which large-scale, systematic surveys of the faunal communities and their distributions within the Red Sea have been completed. Establishing an organism's status as endemic requires some reasonable confidence that the organism does not occur in neighboring areas. For many of the waters around the Arabian Peninsula and the northwestern Indian Ocean, this is a non-trivial undertaking. The evolutionary history of the Red Sea, combined with its unique environmental properties, has likely played a role in the observed level of endemism. Several case studies, however, reveal that there are multiple scenarios of evolutionary connectivity (DiBattista et al. 2013). Some species are recent invaders (i.e., colonized the Red Sea from the Indian Ocean within the last 20,000–25,000 years) while others seem to have persisted inside the Red Sea with minimal genetic connection to the Indian Ocean for hundreds of thousands of years (DiBattista et al. 2013). There does not appear to be, therefore, a single mechanism responsible for the endemism found in the Red Sea.

A global assessment of marine ecoregions identified two putative ecoregions within the Red Sea, demarcated at approximately 20°N (Spalding et al. 2007). However, biogeographic patterns have been explored in detail only for some of the biological assemblages. Corals and reef communities have probably received the most attention. The coastal, fringing communities show notable gradients along the Saudi Arabian coast (Sheppard and Sheppard 1991), with diversity highest in the northern and central Red Sea, declining in the southern, more turbid region. This is in contrast to surveys of the offshore communities, which found relatively homogenous compositions of fish and benthic assemblages (Roberts et al. 2016). (It is important to note that the latter study did not include the communities on the extreme north or south ends of the Red Sea.) Similar results were found when surveying offshore coral communities (Sawall et al. 2014). It is worth mentioning that the Sheppard and Sheppard (1991) surveys occurred ~20 years prior to Sawall et al. (2014) and (Roberts et al. 2016), so the assemblages may have experienced changes or homogenization over this timeframe (Riegl et al. 2012).

In overall assessments of the reef fish communities, Roberts et al. (2016) did not find any trends in endemism along the Red Sea's latitudinal gradient. When analyzing the distribution patterns of more specific taxonomic groups (e.g., families), however, some species appear to have restricted ranges within the Red Sea. For example, some species of butterflyfishes occur primarily in the northern and central regions of the Red Sea (Roberts et al. 1992). Each of these cases provides an interesting opportunity to examine factors that might control distribution of a given species, but this remains understudied.

While latitudinal gradients in assemblage composition throughout most of the offshore Red Sea reefs may be difficult to detect, stronger cross-shelf patterns may exist. At a scale of only 10s of km, (Khalil et al. 2017) found differences in both benthic and fish assemblages on reefs in the central Saudi Arabian Red Sea. Such cross-shelf gradients have been documented in other parts of the world (Malcolm et al. 2010), but have rarely been assessed in the Red Sea. Some species appear to be inshore specialists, while others only occur in offshore habitats. It is arguable that there are stronger environmental gradients between offshore and inshore habitats than there are along most of the latitudinal gradient of the Red Sea.

On the near-shore and fringing reef communities, stronger evidence of a latitudinal gradient is present (Sheppard and Sheppard 1991). The strongest community changes in both offshore and nearshore assemblages occur near the Farasan Islands. Unfortunately, the current political situation is a complicating factor restricting access and work in this region (the Farasan Islands complex bridges the border between Saudi Arabia and Yemen). Given the high productivity and turbidity of the Farasan Islands region, it seems likely that this area could contain one of the most unique reef assemblages, but this remains one of the most difficult places to sample.

Fishes are covered in more depth in Chap. 8. This is perhaps the best-studied group and provides several good examples of the challenges of addressing biogeography in an understudied region. These include sampling representative areas to determine with reasonable confidence that something is or is not present (critical to assess endemism), distinguishing widespread species from cryptic species complexes (Priest et al. 2016), anticipating regional variations in habitat usage, and mis-identifications when relying on literature resources alone (e.g., note the number of corrections in Golani and Bogorodsky (2010)).

Even where (or especially where) the actual coral species or reef communities may not show large changes in species composition, the composition of their symbionts may provide some insight to the environmental conditions the host animals are experiencing (Hume et al. 2016; Sawall et al. 2014, 2015; Ziegler et al. 2017) (discussed in later chapters). The role of holobionts in local adaptations for widespread species is not fully understood, but the current notion is that all animals and plants evolved with microbial partners that contribute to the physiology and adaptation of their host organisms, particularly in extreme environments (Bang et al. 2018; McFall-Ngai et al. 2013). From the Red Sea, there are unique opportunities to examine these patterns, especially with regard to the response of corals (e.g., coral bleaching and coral disease) and other reef-associated organisms to climate change, which will be discussed in detail in later chapters (see also Furby et al. 2013; Monroe et al. 2018; Roik et al. 2015a).

Mangroves, mostly monospeficic stands of *Avicennia marina*, occupy an estimated 120 km^2 in the Red Sea (Almahasheer et al. 2016a) along the narrow belt of intertidal zone in the Red Sea. In contrast with global declines in mangrove habitat, the area covered by mangroves on the Red Sea coast expanded by 12% over the 41-year period from 1972 to 2013. Mangroves shift in height from tree heights of about 15 m in the southern Red Sea to about 2 m near their northern limit at 28°N (Almahasheer et al. 2016a; Hickey et al. 2017). Mangroves are highly nutrient-limited in the northern half of the Red Sea, particularly with respect to iron, which, combined with cool winter temperatures, may result in the stunted nature of the trees (Almahasheer et al. 2016b).

Seagrass communities in the Red Sea include numerous mixed communities (dominated by *Thalassia hemprichii*, *Halophila ovalis*, and *Cymodocea rotundata*) prevalent in reef lagoons in the southern half of the Red Sea (Price et al. 1988), where they are heavily grazed by green turtles as well as thalassinidean and alpheid shrimps. Meadows of *Thalassodendrom ciliatum*, *Halophila stipulacea*, and *Enhalus acorodies* often form monospecific stands or patches (the former two more abundant in the north (Price et al. 1988)). Other dominant macrophytes include *Sargassum* and *Turbinaria* brown algae, which are prevalent on some inshore reef flats, and dense *Halimeda* populations in reef lagoons. Macroalgal abundance is highest in the southern Red Sea, where nutrient concentrations are higher. However, a recent study screened 7 species of seagrasses and 10 species of macroalgae measured at 21 locations, spanning 10° of latitude along the Saudi Arabian coast, and found that almost 90% of macrophyte species had iron concentrations indicative of iron deficiency and more than 40% had critically low iron concentrations, suggesting that iron is a limiting factor of primary production throughout the Red Sea (Anton et al. 2018). However, no latitudinal pattern was detected in any of the performance parameters studied, indicating that, unlike the case for planktonic primary producers, the south to north oligotrophic gradient of the Red Sea is not reflected in iron concentration, chlorophyll-a concentration, or productivity of Red Sea macrophytes (Anton et al. 2018).

References

Acker J, Leptoukh G, Shen S, Zhu T, Kempler S (2008) Remotely-sensed chlorophyll a observations of the northern Red Sea indicate seasonal variability and influence of coastal reefs. J Mar Syst 69:191–204

Al-Aidaroos AM, Karati KK, El-Sherbiny MM, Devassy RP, Kürten B (2017) Latitudinal environmental gradients and diel variability influence abundance and community structure of Chaetognatha in Red Sea coral reefs. Syst Biodivers 15:35–48

Almahasheer H, Aljowair A, Duarte CM, Irigoien X (2016a) Decadal stability of Red Sea mangroves. Estuar Coast Shelf Sci 169:164–172

Almahasheer H, Duarte CM, Irigoien X (2016b) Nutrient limitation in Central Red Sea mangroves. Front Mar Sci 3:271

Almazroui M, Nazrul Islam M, Athar H, Jones PD, Rahman MA (2012) Recent climate change in the Arabian Peninsula: annual rainfall and temperature analysis of Saudi Arabia for 1978–2009. Int J Climatol 32:953–966

Anisimov A, Tao W, Stenchikov G, Kalenderski S, Jish Prakash P, Yang ZL, Shi M (2017) Quantifying local-scale dust emission from the Arabian Red Sea coastal plain. Atmos Chem Phys 17:993–1015

Anton A, Hendriks IE, Marbà N, Krause-Jensen D, Garcias-Bonet N, Duarte CM (2018) Iron deficiency in seagrasses and macroalgae in the Red Sea is unrelated to latitude and physiological performance. Front Mar Sci 5:74

Antonius A (1988) Distribution and dynamics of coral diseases in the Eastern Red Sea. Proceedings of the 6th International Coral Reef Symposium 2:293–298

Aranda M, Li Y, Liew YJ, Baumgarten S, Simakov O, Wilson MC, Piel J, Ashoor H, Bougouffa S, Bajic VB, Ryu T, Ravasi T, Bayer T, Micklem G, Kim H, Bhak J, LaJeunesse TC, Voolstra CR (2016) Genomes of coral dinoflagellate symbionts highlight evolutionary adaptations conducive to a symbiotic lifestyle. Sci Rep 6:39734

Backer H, Schoell M (1972) New deeps with brines and metalliferous sediments in the Red Sea. Nat Phys Sci 240:153–158

Bang C, Dagan T, Deines P, Dubilier N, Duschl WJ, Fraune S, Hentschel U, Hirt H, Hulter N, Lachnit T, Picazo D, Pita L, Pogoreutz C, Radecker N, Saad MM, Schmitz RA, Schulenburg H, Voolstra CR, Weiland-Brauer N, Ziegler M, Bosch TCG (2018) Metaorganisms in extreme environments: do microbes play a role in organismal adaptation? Zoology 127:1–19

Banks JR, Brindley HE, Stenchikov G, Schepanski K (2017) Satellite retrievals of dust aerosol over the Red Sea and the Persian Gulf (2005–2015). Atmos Chem Phys 17:3987–4003

Batang ZB, Papathanassiou E, Al-Suwailem A, Smith C, Salomidi M, Petihakis G, Alikunhi NM, Smith L, Mallon F, Yapici T, Fayad N (2012) First discovery of a cold seep on the continental margin of the Central Red Sea. J Mar Syst 94:247–253

Beal LM, Ffield A, Gordon AL (2000) Spreading of Red Sea overflow waters in the Indian Ocean. J Geophys Res Oceans 105:8549–8564

Berumen ML, Braun CD, Cochran JEM, Skomal GB, Thorrold SR (2014) Movement patterns of juvenile whale sharks tagged at an aggregation site in the Red Sea. PLoS One 9:e103536

Borin S, Brusetti L, Mapelli F, D'Auria G, Brusa T, Marzorati M, Rizzi A, Yakimov M, Marty D, De Lange GJ, Van der Wielen P, Bolhuis H, McGenity TJ, Polymenakou PN, Malinverno E, Giuliano L, Corselli C, Daffonchio D (2009) Sulfur cycling and methanogenesis primarily drive microbial colonization of the highly sulfidic Urania deep hypersaline basin. Proc Natl Acad Sci U S A 106:9151–9156

Bougouffa S, Yang JK, Lee OO, Wang Y, Batang Z, Al-Suwailem A, Qian PY (2013) Distinctive microbial community structure in highly stratified deep-sea brine water columns. Appl Environ Microbiol 79:3425–3437

Bower AS, Farrar JT (2015) Air–Sea interaction and horizontal circulation in the Red Sea. In: Rasul NMA, Stewart ICF (eds) The Red Sea: the formation, morphology, oceanography and environment of a young ocean basin. Springer, Berlin/Heidelberg, pp 329–342

Cantin NE, Cohen AL, Karnauskas KB, Tarrant AM, McCorkle DC (2010) Ocean warming slows coral growth in the central Red Sea. Science 329:322–325

Chaidez V, Dreano D, Agusti S, Duarte CM, Hoteit I (2017) Decadal trends in Red Sea maximum surface temperature. Sci Rep 7:8144

Churchill JH, Bower AS, McCorkle DC, Abualnaja Y (2014) The transport of nutrient-rich Indian Ocean water through the Red Sea and into coastal reef systems. J Mar Res 72:165–181

Clifford M, Horton C, Schmitz J, Kantha LH (1997) An oceanographic nowcast/forecast system for the Red Sea. J Geophys Res Oceans 102:25101–25122

Daffonchio D, Borin S, Brusa T, Brusetti L, van der Wielen PW, Bolhuis H, Yakimov MM, D'Auria G, Giuliano L, Marty D, Tamburini C, McGenity TJ, Hallsworth JE, Sass AM, Timmis KN, Tselepides A, de Lange GJ, Hubner A, Thomson J, Varnavas SP, Gasparoni F, Gerber HW, Malinverno E, Corselli C, Garcin J, McKew B, Golyshin PN, Lampadariou N, Polymenakou P, Calore D, Cenedese S, Zanon F, Hoog S, Party BS (2006) Stratified prokaryote network in the oxic-anoxic transition of a deep-sea halocline. Nature 440:203–207

Devassy RP, El-Sherbiny MM, Al-Sofyani AM, Al-Aidaroos AM (2017) Spatial variation in the phytoplankton standing stock and diversity in relation to the prevailing environmental conditions along the Saudi Arabian coast of the northern Red Sea. Mar Biodivers 47:995–1008

DiBattista JD, Berumen ML, Gaither MR, Rocha LA, Eble JA, Choat JH, Craig MT, Skillings DJ, Bowen BW, McClain C (2013) After continents divide: comparative phylogeography of reef fishes from the Red Sea and Indian Ocean. J Biogeogr 40:1170–1181

DiBattista JD, Choat JH, Gaither MR, Hobbs J-PA, Lozano-Cortés DF, Myers RF, Paulay G, Rocha LA, Toonen RJ, Westneat MW, Berumen ML (2015a) On the origin of endemic species in the Red Sea. J Biogeogr 43:13–30

DiBattista JD, Roberts MB, Bouwmeester J, Bowen BW, Coker DJ, Lozano-Cortés DF, Choat JH, Gaither MR, Hobbs J-PA, Khalil MT, Kochzius M, Myers RF, Paulay G, Robitzch VSN, Saenz-Agudelo P, Salas E, Sinclair-Taylor TH, Toonen RJ, Westneat MW, Williams ST, Berumen ML (2015b) A review of contemporary patterns of endemism for shallow water reef fauna in the Red Sea. J Biogeogr 43:423–439

Froukh T, Kochzius M (2008) Species boundaries and evolutionary lineages in the blue green damselfishes *Chromis viridis* and *Chromis atripectoralis* (Pomacentridae). J Fish Biol 72:451–457

Furby KA, Bouwmeester J, Berumen ML (2013) Susceptibility of central Red Sea corals during a major bleaching event. Coral Reefs 32:505–513

Garcia-Castellanos D, Villaseñor A (2011) Messinian salinity crisis regulated by competing tectonics and erosion at the Gibraltar arc. Nature 480:359–363

Giles EC, Saenz-Agudelo P, Hussey NE, Ravasi T, Berumen ML (2015) Exploring seascape genetics and kinship in the reef sponge *Stylissa carteri* in the Red Sea. Ecol Evol 5:2487–2502

Golani D, Bogorodsky SV (2010) The fishes of the Red Sea – reappraisal and updated checklist. Zootaxa 2463:1-135

Grotzinger SW, Karan R, Strillinger E, Bader S, Frank A, Al Rowaihi IS, Akal A, Wackerow W, Archer JA, Rueping M, Weuster-Botz D, Groll M, Eppinger J, Arold ST (2018) Identification and experimental characterization of an extremophilic brine pool alcohol dehydrogenase from single amplified genomes. ACS Chem Biol 13:161–170

Hickey SM, Phinn SR, Callow NJ, Van Niel KP, Hansen JE, Duarte CM (2017) Is climate change shifting the poleward limit of mangroves? Estuar Coasts 40:1215–1226

Hughes TP, Anderson KD, Connolly SR, Heron SF, Kerry JT, Lough JM, Baird AH, Baum JK, Berumen ML, Bridge TC, Claar DC, Eakin CM, Gilmour JP, Graham NAJ, Harrison H, Hobbs J-PA, Hoey AS, Hoogenboom M, Lowe RJ, McCulloch MT, Pandolfi JM, Pratchett M, Schoepf V, Torda G, Wilson SK (2018) Spatial and temporal patterns of mass bleaching of corals in the Anthropocene. Science 359:80–83

Hume BCC, Voolstra CR, Arif C, D'Angelo C, Burt JA, Eyal G, Loya Y, Wiedenmann J (2016) Ancestral genetic diversity associated with the rapid spread of stress-tolerant coral symbionts in response to Holocene climate change. Proc Natl Acad Sci U S A 113:4416–4421

Jin D, Kite-Powell H, Hoagland P, Solow A (2012) A bioeconomic analysis of traditional fisheries in the Red Sea. Mar Resour Econ 27:137–148

Jish Prakash P, Stenchikov G, Kalenderski S, Osipov S, Bangalath H (2015) The impact of dust storms on the Arabian Peninsula and the Red Sea. Atmos Chem Phys 15:199–222

Jish Prakash P, Stenchikov G, Tao W, Yapici T, Warsama B, Engelbrecht JP (2016) Arabian Red Sea coastal soils as potential mineral dust sources. Atmos Chem Phys 16:11991–12004

Khalil MT, Bouwmeester J, Berumen ML (2017) Spatial variation in coral reef fish and benthic communities in the central Saudi Arabian Red Sea. PeerJ 5:e3410

Kheireddine M, Ouhssain M, Claustre H, Uitz J, Gentili B, Jones BH (2017) Assessing pigment-based phytoplankton community distributions in the Red Sea. Front Mar Sci 4:132

Kürten B, Khomayis HS, Devassy R, Audritz S, Sommer U, Struck U, El-Sherbiny MM, Al-Aidaroos AM (2014) Ecohydrographic constraints on biodiversity and distribution of phytoplankton and zooplankton in coral reefs of the Red Sea, Saudi Arabia. Mar Ecol 36:1195–1214

Kürten B, Al-Aidaroos AM, Kürten S, El-Sherbiny MM, Devassy RP, Struck U, Zarokanellos N, Jones BH, Hansen T, Bruss G, Sommer U (2016) Carbon and nitrogen stable isotope ratios of pelagic zooplankton elucidate ecohydrographic features in the oligotrophic Red Sea. Prog Oceanogr 140:69–90

Li W, El-Askary H, ManiKandan K, Qurban M, Garay M, Kalashnikova O (2017) Synergistic use of remote sensing and modeling to assess an anomalously high chlorophyll-a event during summer 2015 in the South Central Red Sea. Remote Sens 9:778

Malcolm HA, Jordan A, Smith SDA (2010) Biogeographical and cross-shelf patterns of reef fish assemblages in a transition zone. Mar Biodivers 40:181–193

McFall-Ngai M, Hadfield MG, Bosch TCG, Carey HV, Domazet-Lošo T, Douglas AE, Dubilier N, Eberl G, Fukami T, Gilbert SF, Hentschel U, King N, Kjelleberg S, Knoll AH, Kremer N, Mazmanian SK, Metcalf JL, Nealson K, Pierce NE, Rawls JF, Reid A, Ruby EG, Rumpho M, Sanders JG, Tautz D, Wernegreen JJ (2013) Animals in a bacterial world, a new imperative for the life sciences. Proc Natl Acad Sci U S A 110:3229–3236

Mervis J (2009) The big gamble in the Saudi Desert. Science 326:354–357

Monroe A, Ziegler M, Roik A, Röthig T, Hardestine R, Emms M, Jensen R, Voolstra CR, Berumen ML (2018) In-situ observations of coral bleaching in the central Saudi Arabian Red Sea during the 2015/2016 global coral bleaching event. PLoS One 13:e0195814

Nanninga GB, Saenz-Agudelo P, Manica A, Berumen ML (2014) Environmental gradients predict the genetic population structure of a coral reef fish in the Red Sea. Mol Ecol 23:591–602

Ngugi DK, Stingl U (2012) Combined analyses of the ITS loci and the corresponding 16S rRNA genes reveal high micro- and macrodiversity of SAR11 populations in the Red Sea. PLoS One 7:e50274

Ngugi DK, Antunes A, Brune A, Stingl U (2012) Biogeography of pelagic bacterioplankton across an antagonistic temperature-salinity gradient in the Red Sea. Mol Ecol 21:388–405

Ngugi D, Blom J, Alam I, Rashid M, Ba-Alawi W, Zhang G, Hikmawan T, Guan Y, Antunes A, Siam R, El Dorry H, Bajic V, Stingl U (2015) Comparative genomics reveals adaptations of a halotolerant thaumarchaeon in the interfaces of brine pools in the Red Sea. ISME J 9:396–411

Osipov S, Stenchikov G (2018) Simulating the regional impact of dust on the Middle East climate and the Red Sea. J Geophys Res Oceans 123:1032–1047

Osman EO, Smith DJ, Ziegler M, Kürten B, Conrad C, El-Haddad KM, Voolstra CR, Suggett DJ (2018) Thermal refugia against coral bleaching throughout the northern Red Sea. Glob Chang Biol 24:e474–e484

Patzert WC (1974) Wind-induced reversal in Red Sea circulation. Deep-Sea Res 21:109–121

Pautot G, Guennoc P, Coutelle A, Lyberis N (1984) Discovery of a large brine deep in the northern Red Sea. Nature 310:133–136

Pearman JK, Kurten S, Sarma YV, Jones BH, Carvalho S (2016) Biodiversity patterns of plankton assemblages at the extremes of the Red Sea. FEMS Microbiol Ecol 92

Pearman JK, Ellis J, Irigoien X, Sarma YVB, Jones BH, Carvalho S (2017) Microbial planktonic communities in the Red Sea: high levels of spatial and temporal variability shaped by nutrient availability and turbulence. Sci Rep 7:6611

Post AF, Dedej Z, Gottlieb R, Li H, Thomas DN, El-Absawi M, El-Naggar A, El-Gharabawi M, Sommer U (2002) Spatial and temporal distribution of *Trichodesmium* spp. in the stratified Gulf of Aqaba, Red Sea. Mar Ecol Prog Ser 239:241–250

Price ARG, Crossland CJ, Dawson Shepherd AR, McDowall RJ, Medley PAH, Stafford Smith MG, Ormond RFG, Wrathall TJ (1988) Aspects of seagrass ecology along the eastern coast of the Red Sea. Bot Mar 31:83

Priest MA, DiBattista JD, McIlwain JL, Taylor BM, Hussey NE, Berumen ML (2016) A bridge too far: dispersal barriers and cryptic speciation in an Arabian Peninsula grouper (*Cephalopholis hemistiktos*). J Biogeogr 43:820–832

Qurban MA, Wafar M, Jyothibabu R, Manikandan KP (2017) Patterns of primary production in the Red Sea. J Mar Syst 169:87–98

Racault M-F, Raitsos DE, Berumen ML, Brewin RJW, Platt T, Sathyendranath S, Hoteit I (2015) Phytoplankton phenology indices in coral reef ecosystems: application to ocean-color observations in the Red Sea. Remote Sens Environ 160:222–234

Raitsos DE, Hoteit I, Prihartato PK, Chronis T, Triantafyllou G, Abualnaja Y (2011) Abrupt warming of the Red Sea. Geophys Res Lett 38:L14601

Raitsos DE, Pradhan Y, Brewin RJW, Stenchikov G, Hoteit I (2013) Remote sensing the phytoplankton seasonal succession of the Red Sea. PLoS One 8:e64909

Raitsos DE, Brewin RJW, Zhan P, Dreano D, Pradhan Y, Nanninga GB, Hoteit I (2017) Sensing coral reef connectivity pathways from space. Sci Rep 7:9338

Riegl BM, Bruckner AW, Rowlands GP, Purkis SJ, Renaud P (2012) Red Sea coral reef trajectories over 2 decades suggest increasing community homogenization and decline in coral size. PLoS One 7:e38396

Roberts CM, Alexander RDS, Rupert FGO (1992) Large-scale variation in assemblage structure of Red Sea butterflyfishes and angelfishes. J Biogeogr 19:239–250

Roberts MB, Jones GP, McCormick MI, Munday PL, Neale S, Thorrold S, Robitzch VSN, Berumen ML (2016) Homogeneity of coral reef communities across 8 degrees of latitude in the Saudi Arabian Red Sea. Mar Pollut Bull 105:558–565

Robitzch V, Banguera-Hinestroza E, Sawall Y, Al-Sofyani A, Voolstra CR (2015) Absence of genetic differentiation in the coral along environmental gradients of the Saudi Arabian Red Sea. Front Mar Sci 2:5

Robitzch VS, Lozano-Cortes D, Kandler NM, Salas E, Berumen ML (2016) Productivity and sea surface temperature are correlated with the pelagic larval duration of damselfishes in the Red Sea. Mar Pollut Bull 105:566–574

Roder C, Berumen ML, Bouwmeester J, Papathanassiou E, Al-Suwailem A, Voolstra CR (2013) First biological measurements of deep-sea corals from the Red Sea. Sci Rep 3:2802

Roik A, Roethig T, Ziegler M, Voolstra CR (2015a) Coral bleaching event in the Central Red Sea. In Mideast Coral Reef Society Newsletter, vol 3, p 3

Roik A, Röthig T, Roder C, Müller PJ, Voolstra CR (2015b) Captive rearing of the deep-sea coral *Eguchipsammia fistula* from the Red Sea demonstrates remarkable physiological plasticity. PeerJ 3:e734

Roik A, Röthig T, Roder C, Ziegler M, Kremb SG, Voolstra CR (2016) Year-long monitoring of Physico-chemical and biological variables provide a comparative baseline of coral reef functioning in the Central Red Sea. PLoS One 11:e0163939

Röthig T, Yum LK, Kremb SG, Roik A, Voolstra CR (2017) Microbial community composition of deep-sea corals from the Red Sea provides insight into functional adaption to a unique environment. Sci Rep 7:44714

Saenz-Agudelo P, DiBattista JD, Piatek MJ, Gaither MR, Harrison HB, Nanninga GB, Berumen ML (2015) Seascape genetics along environmental gradients in the Arabian Peninsula: insights from ddRAD sequencing of anemonefishes. Mol Ecol 24:6241–6255

Sawall Y, Al-Sofyani A, Banguera-Hinestroza E, Voolstra CR (2014) Spatio-temporal analyses of *Symbiodinium* physiology of the coral *Pocillopora verrucosa* along large-scale nutrient and temperature gradients in the Red Sea. PLoS One 9:e103179

Sawall Y, Al-Sofyani A, Hohn S, Banguera-Hinestroza E, Voolstra CR, Wahl M (2015) Extensive phenotypic plasticity of a Red Sea coral over a strong latitudinal temperature gradient suggests limited acclimatization potential to warming. Sci Rep 5:8940

Schardt C (2016) Hydrothermal fluid migration and brine pool formation in the Red Sea: the Atlantis II deep. Mineral Deposita 51:89–111

Searle RC, Ross DA (2007) A geophysical study of the Red Sea axial trough between 20.5° and 22°N. Geophys J R Astron Soc 43:555–572

Sheppard CRC, Sheppard ALS (1991) Corals and coral communities of Arabia. Fauna Saudi Arabia 12:3–170

Shibl AA, Haroon MF, Ngugi DK, Thompson LR, Stingl U (2016) Distribution of *Prochlorococcus* ecotypes in the Red Sea Basin based on analyses of rpoC1 sequences. Front Mar Sci 3:104

Silva L, Calleja ML, Huete-Stauffer TM, Ivetic S, Ansari MI, Viegas M, Morán XAG (2019) Low abundances but high growth rates of coastal heterotrophic bacteria in the Red Sea. Front Microbiol 9:3244

Sofianos SS, Johns WE (2002) An Oceanic General Circulation Model (OGCM) investigation of the Red Sea circulation, 1. Exchange between the Red Sea and the Indian Ocean. J Geophys Res Oceans 107(C11):3196

Sofianos SS, Johns WE (2003) An Oceanic General Circulation Model (OGCM) investigation of the Red Sea circulation: 2. Three-dimensional circulation in the Red Sea. J Geophys Res Oceans 108:3066

Sofianos SS, Johns WE (2007) Observations of the summer Red Sea circulation. J Geophys Res Oceans 112:C06025

Spalding MD, Fox HE, Allen GR, Davidson N, Ferdaña ZA, Finlayson M, Halpern BS, Jorge MA, Lombana A, Lourie SA, Martin KD, McManus E, Molnar J, Recchia CA, Robertson J (2007) Marine ecoregions of the world: a bioregionalization of coastal and shelf areas. Bioscience 57:573–583

Swift SA, Bower AS, Schmitt RW (2012) Vertical, horizontal, and temporal changes in temperature in the Atlantis II and Discovery hot brine pools, Red Sea. Deep-Sea Res I Oceanogr Res Pap 64:118–128

Thompson LR, Williams GJ, Haroon MF, Shibl A, Larsen P, Shorenstein J, Knight R, Stingl U (2016) Metagenomic covariation along densely sampled environmental gradients in the Red Sea. ISME J 11:138

Tragou E, Garrett C, Outerbridge R, Gilman C (1999) The heat and freshwater budgets of the Red Sea. J Phys Oceanogr 29:2504–2522

Vestheim H, Kaartvedt S (2016) A deep sea community at the Kebrit brine pool in the Red Sea. Mar Biodivers 46:59–65

Voolstra CR, Li Y, Liew YJ, Baumgarten S, Zoccola D, Flot J-F, Tambutté S, Allemand D, Aranda M (2017) Comparative analysis of the genomes of *Stylophora pistillata* and *Acropora digitifera* provides evidence for extensive differences between species of corals. Sci Rep 7:17583

Wafar M, Ashraf M, Manikandan KP, Qurban MA, Kattan Y (2016a) Propagation of Gulf of Aden Intermediate Water (GAIW) in the Red Sea during autumn and its importance to biological production. J Mar Syst 154:243–251

Wafar M, Qurban MA, Ashraf M, Manikandan KP, Flandez AV, Balala AC (2016b) Patterns of distribution of inorganic nutrients in Red Sea and their implications to primary production. J Mar Syst 156:86–98

Wilson SN (2017) Assessment of genetic connectivity between Sudan and Saudi Arabia for commercially important fish species. MSc thesis. King Abdullah University of Science and Technology, Saudi Arabia

Xu W, Ruch J, Jónsson S (2015) Birth of two volcanic islands in the southern Red Sea. Nat Commun 6:7104

Yao FC, Hoteit I, Pratt LJ, Bower AS, Zhai P, Kohl A, Gopalakrishnan G (2014) Seasonal overturning circulation in the Red Sea: 1. Model validation and summer circulation. J Geophys Res Oceans 119:2238–2262

Yum LK, Baumgarten S, Röthig T, Roder C, Roik A, Michell C, Voolstra CR (2017) Transcriptomes and expression profiling of deep-sea corals from the Red Sea provide insight into the biology of azooxanthellate corals. Sci Rep 7:6442

Zarokanellos ND, Papadopoulos VP, Sofianos SS, Jones BH (2017a) Physical and biological characteristics of the winter-summer transition in the Central Red Sea. J Geophys Res Oceans 122:6355–6370

Zarokanellos ND, Kürten B, Churchill JH, Roder C, Voolstra CR, Abualnaja Y, Jones BH (2017b) Physical mechanisms routing nutrients in the Central Red Sea. J Geophys Res Oceans 122:9032–9046

Zhai P, Bower A (2013) The response of the Red Sea to a strong wind jet near the Tokar Gap in summer. J Geophys Res Oceans 118:421–434

Zhai P, Pratt LJ, Bower A (2015) On the crossover of boundary currents in an idealized model of the Red Sea. J Phys Oceanogr 45:1410–1425

Zhan P, Subramanian AC, Yao F, Hoteit I (2014) Eddies in the Red Sea: a statistical and dynamical study. J Geophys Res Oceans 119:3909–3925

Ziegler M, Arif C, Burt JA, Dobretsov S, Roder C, LaJeunesse TC, Voolstra CR (2017) Biogeography and molecular diversity of coral symbionts in the genus *Symbiodinium* around the Arabian Peninsula. J Biogeogr 44:674–686

Environmental Setting for Reef Building in the Red Sea

James Churchill, Kristen Davis, Eyal Wurgaft, and Yonathan Shaked

Abstract

The Red Sea is a distinct marine system, which, due to its limited lateral extent, is strongly influenced by the surrounding arid and semiarid terrestrial environment. Among large marine bodies, it is unusually saline, owing to a high rate of evaporation relative to precipitation, and warm. The physical environment of the Red Sea has been subject to scientific research for more than a century, with considerable advances in understanding achieved in the past two decades. In this chapter, we review the current state of knowledge of the Red Sea's physical/chemical system. The bulk of the chapter deals with the marine environment. Attention is given to a variety of topics, including: tides and lower-frequency motions of the sea surface, circulation over a range of space and time scales, the surface wave field, and the distributions of water properties, nutrients, chlorophyll-*a* (chl-*a*) and light. We also review the current understanding of atmospheric conditions affecting the Red Sea, focusing on how atmospheric circulation patterns of various scales influence the exchange of momentum, heat, and mass at the surface of the Red Sea. A subsection is devoted to geology and reef morphology, with a focus on reef-building processes in the Red Sea. Finally, because reef building and health are tightly linked with carbonate chemistry, we review the Red Sea carbonate system, highlighting recent advances in the understanding of this system.

Keywords

Red Sea geology · Red Sea reef morphology · Atmospheric forcing · Basin-scale circulation · Mesoscale processes · Eddies · Surface waves · Carbonate system · Nutrient and light distribution

2.1 Geology and Reef Morphology

Like most reefs worldwide, modern reefs in the Red Sea were initiated during the Holocene (Braithwaite 1987). Rapid changes in sea level following the end of the last glacial maximum, ca. 18 ka, resulted in old reefs being abandoned and the initiation of new reefs at higher levels over what was previously sub-aerial substrates (Montaggioni 2000, 2005; Dullo 2005). During the Holocene, as rates of sea level rise declined, reefs developed into what we recognize today as large structures that often represent thousands of years of growth.

In the Red Sea and its northern tributaries, the Gulf of Suez and the Gulf of Aqaba, the effects of changing sea level were especially pronounced. The Red Sea is connected to the Gulf of Aden and the Indian Ocean through the narrow and shallow (137 m deep and 29 km wide) strait of Bab al Mandab. During glacial times, when sea level was more than 100 m lower than during interglacial periods, water exchange between the Red Sea and the Indian Ocean was very limited (Almogi-Labin et al. 1991; Arz et al. 2007; Biton et al. 2008). Limited water exchange coupled with little freshwater input from the surrounding arid lands and high evaporation rates, resulted in high salinity within the Red Sea during glacial periods (Morcos 1970; Felis et al. 2000). It has been estimated that Red Sea salinities during the last glacial maximum were 50 psu or greater (Siddall et al. 2003; Almogi-Labin et al. 2008; Biton et al. 2008; Legge et al. 2008), a level generally considered too high for coral growth (Kleypas et al. 1999).

In the Red Sea, temperature and salinity vary over latitude, with the northern end generally being cooler and more

J. Churchill (✉)
Woods Hole Oceanographic Institution, Woods Hole, MA, USA
e-mail: jchurchill@whoi.edu

K. Davis
University of California, Irvine, Irvine, CA, USA

E. Wurgaft
Ben-Gurion University of the Negev, Beer-Sheva, Israel

Y. Shaked
Hebrew University of Jerusalem, Jerusalem, Israel

© Springer Nature Switzerland AG 2019
C. R. Voolstra, M. L. Berumen (eds.), *Coral Reefs of the Red Sea*, Coral Reefs of the World 11, https://doi.org/10.1007/978-3-030-05802-9_2

saline (Biton et al. 2010). Coral reefs in the northern Red Sea and the Gulf of Aqaba are today among the northernmost reefs in the world. The worldwide reef distribution is thought to be limited primarily by temperature: coral reefs are not found in seas where the annual minimum temperature is lower than 18–20° (Kleypas et al. 1999; Schlager 2003). Sea surface temperature in the northern Red Sea during the last glacial maximum is estimated to have been 4 °C lower than today. It therefore seems unlikely that coral reefs were abundant in the northern Red Sea during glacial times. Whether no coral enclaves survived glacial periods within the Red Sea is debatable, but it seems that colonization of the present coastline by coral reefs most likely involved a south to north migration, rather than strictly a local reef rise coupled with the rising sea level (Kiflawi et al. 2006). Whatever the recolonization pattern, fossil reefs in areas of tectonic uplift along the Red Sea coast indicate repeated establishment of reefs during interglacial times (Al-Rifaiy and Cherif 1988; Gvirtzman et al. 1992; Gvirtzman 1994; El-Asmar 1997).

The Red Sea may be viewed as an elongated, nearly landlocked, basin that is in the process of becoming an ocean (Bonatti 1985). Magmatism and sea floor spreading is occurring along the southern and central parts of the basin, but the northern basin is still being rifted (Joffe and Garfunkel 1987; Cochran and Martinez 1988; Bosworth et al. 2005). Along large portions of the coast, this young rifting process results in an often-steep topography of exposed crystalline basement rocks, punctuated by large alluvial fans transporting coarse erosion products from the highlands to the sea (Ben-Avraham et al. 1979; Bosworth et al. 2005). Sea level changes coupled with intense tectonic activity sometimes result in the drowning, exposure or burial of reefs (Shaked et al. 2004; Makovsky et al. 2008). This forces the establishment of younger reefs, displaying initial stages of development (Shaked et al. 2005).

The coastal plain in the area of the southern Red Sea is wider with gentler slopes than the coastal plain adjacent to the northern Red Sea (Bohannon 1986). Incision and alluvial transport in the area of the southern Red Sea were enhanced with the glacial-interglacial cycles of sea level fall and rise. Dropping sea level during glacial times lowered the base level and enhanced erosion and transport, whereas postglacial sea level rise inundated the newly created alluvial fans and the crystalline hillsides. Since much of the area is arid, vegetation is scarce and the substrate available for coral settlement is mostly bare basement rock, conglomerates and poorly sorted alluvial fans of coarse material intercalated with fine sands and silts (Khalil and McClay 2009).

Coral reef morphology is greatly influenced by antecedent topography. A gradually sloping substrate will result in wider reefs, either as sea level rises and the reef retreats shoreward. When sea level is stable, the reef can easily expand seaward. The reefs formed over gently sloping topography are often separated from shore by a wide lagoon. The Red Sea with its generally steep slopes is characterized by fringing reefs, hugging the shoreline (Dullo and Montaggioni 1998). Exceptions occur where antecedent topography included hills and topographic highs lining the sea's margins. Following the post-glacial sea level rise, archipelagos and even barrier-type reefs were formed in such areas.

Corals tend to settle more easily, and coral reefs appear to develop more rapidly, on marine biogenic hard substrates (e.g., coralline algae or dead corals) (Harriott and Fisk 1987; Morse et al. 1996; Hoegh-Guldberg et al. 2007; Neo et al. 2009) than on crystalline rocks or loose substrates. Therefore, reefs may have taken longer to develop on steeply sloping crystalline margins than on gently sloping terraces of marine substrates. Reef growth along the margins is punctuated by active ephemeral river outlets and alluvial fans where loose unstable substrate inhibit coral settlement (Dullo and Montaggioni 1998). In the vicinity of such active fans, reef development seems to be slower than it is away from such features (Shaked et al. 2005). However, raised coral reefs along tectonically uplifting margins reveal repeated transgressional sequences, where sandy terrestrial sediments are overlaid by coarse beach sediments followed by fine lagoonal marine sediments and finally by coral reefs dated to previous interglacial periods (Al-Rifaiy and Cherif 1988; Gvirtzman 1994; Yehudai et al. 2017).

2.2 Atmospheric Setting

Because no point in the Red Sea is more than 150 km from land, a distance small compared to the scale of major weather systems, atmospheric conditions over the Red Sea are strongly influenced by the surrounding terrestrial environment.

For example, the atmospheric circulation over the Red Sea is appreciably affected by the coastal mountain chains running along the Red Sea margins. On a large-scale, these mountains tend to channel the air flow along the length of the Red Sea. This channeling is revealed by monthly-averaged winds, which tend to be aligned with the Red Sea's longitudinal axis (Patzert 1974; Sofianos and Johns 2001; Bower and Farrar 2015; Viswanadhapalli et al. 2017). The monthly-averaged winds also indicate a seasonal shift in the regional atmospheric regime associated with phases of the Arabian monsoon. Monthly-averaged winds of the summer period, roughly June–September, show an air flow directed to the southeast prevailing over the entire Red Sea. Monthly-averaged winds over the rest of the year reveal an air flow convergence, with winds directed to the southeast over the northern Red Sea and to the northwest over the southern Red Sea. In monthly-averaged wind fields, these opposing flows converge in the 18–20°N latitude range (Patzert 1974;

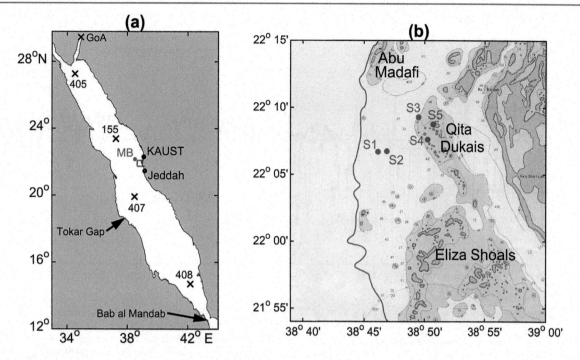

Fig. 2.1 Locations of measurements and features described in this chapter. (**a**) Large-scale map of the Red Sea showing the location of the meteorological buoy (MB) from which wind, wave and heat flux measurements were derived. Also shown are biogeochemical sampling stations (x's). Stations 405, 407 and 408 were sampled by the Geochemical Ocean Sections Study (GEOSECS) program in December 1977 (Weiss et al. 1983). Station 155 was sampled during the Mer Rouge (MEROU) program in 1982 (Papuad and Poisson 1986). Data from the northern Gulf of Aqaba (GOA) were sampled in 2011–2 as described by Wurgaft et al. (2016). (**b**) Locations of moored ADCPs (S1-S5), superimposed on a bathymetric map (United Kingdom Hydrographic Office) of the area outlined by the box in (**a**). Each ADCP was programmed to acquire data from which spectra of surface gravity waves could be computed. The red line traces the 100-m isobath, marking the eastern edge of the Red Sea basin. The buoy and ADCP were deployed as part of a collaborative field study involving Woods Hole Oceanographic Institution and King Abdullah University of Science and Technology

Sofianos and Johns 2001). However, wind data from a meteorological buoy deployed at 22° 10′ N (at "MB" in Fig. 2.1a), as part of a WHOI-KAUST cooperative investigation of the Red Sea, show that the northeastward air flow of the southern zone frequently extends well past the latitude limit suggested by the monthly-averaged winds. Although along-axis winds measured by the buoy are predominately directed to the southeast, episodes of northwestward wind frequently occur. Fifteen such events are seen in the 1-year record displayed here (Fig. 2.2).

Steering of atmospheric flow by the coastal mountains surrounding the Red Sea also produces intense winds directed across the Red Sea axis. These emerge from, or are directed into, mountain gaps. Of particular prominence are winds channeled through the Tokar Gap (Fig. 2.1a). During summer, strong eastward winds are funneled through the Gap and emerge onto the Red Sea (Jiang et al. 2009; Davis et al. 2015; Langodan et al. 2017a). In atmospheric models, these winds occur almost daily from mid-June to mid-September and are modulated diurnally, typically reaching a maximum strength, of up to 26 m s^{-1} (Davis et al. 2015), over 0700–0900 local (Saudi Arabian) time. When the Tokar Gap wind jet is most intense, large areas of the southern Red Sea are subject to strong eastward winds. As indicated by the model results of Jiang et al. (2009), the proportion of Red Sea surface area south of 20°N that experiences eastward winds of >10 m s^{-1} often exceeds 20% during July and August. During winter, air flow through the Tokar gap tends to be directed westward, resulting in funneling of air currents from the southern Red Sea into the Gap (Jiang et al. 2009).

Mountain-gap winds, emerging from the Arabian subcontinent are common phenomena over the northern Red Sea during winter (November – March) (Jiang et al. 2009; Bower and Farrar 2015). These westward wind jets occur at intervals of 10–20 days (see Fig. 2.2). They can persist for a number of days, though with diurnal modulation, and can encompass a large fraction, 10–40%, of the Red Sea area north of 20°N (Jiang et al. 2009).

Mountain-gap winds directed into the Red Sea carry many constituents of terrestrial origin into the marine environment. Prominent among these is dust. The Red Sea is situated in the Middle Eastern and North African 'dust belt' (0–40°N and 15°W–60°E). Roughly half the global dust emissions originate from this area (Prospero et al. 2002). Satellite imagery frequently show dust plumes originating at the major mountain gaps and extending across the Red Sea (Hickey and Goudie 2007; Jiang et al. 2009; Bower and Farrar 2015). Prakash et al. (2015) estimate that the Red Sea

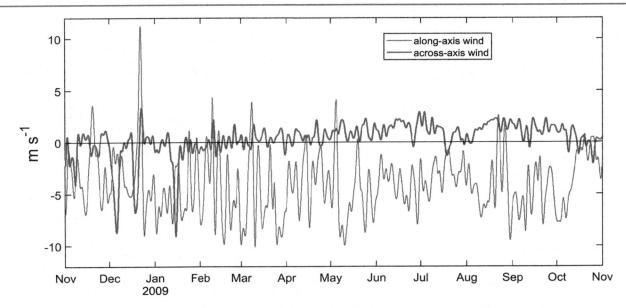

Fig. 2.2 A one-year record of low-passed filtered (66-h half-power point) along-axis (positive directed to the SSE) and across-axis (positive to the ENE) wind measured at meteorological buoy MB (Fig. 2.1a). The along-axis wind record shows the predominance of SSEward winds in the northern Red Sea, but also reveals a number of episodes of wind reversal to the NNWward direction associated with a northward extension of winter monsoon winds in the southern Red Sea. The across-axis wind record shows 'mountain-gap wind' events of strong WSWward winds (over Nov 2008 to Jan 2009) blowing across the Red Sea

experiences 5–6 significant dust storms per year, which may deposit as much as 6 Mt of solids into the Red Sea. The impact of dust deposition on the nutrient balance and productivity of the Red Sea may be significant, but has not yet been examined.

In addition to the mountain-gap winds, diurnally-modulated sea breezes, not necessarily associated with mountain-gap airflow, may be prevalent over the Red Sea during all seasons. Analysis of wind data from the WHOI-KAUST meteorological buoy (Fig. 2.1a) reveals a regularly occurring sea breeze in the central Red Sea that accounts for approximately 25% of the overall wind-stress variance (Churchill et al. 2014b). The observed sea breeze is highly polarized, with the major axis directed across-shore. The averaged (as a function of time of day) wind stress along the major axis is always directed onshore and exhibits a fourfold variation over the course of the day, reaching a maximum at 1700 local time (Churchill et al. 2014b).

While the surface wind stress can drive short-term currents and possibly basin-scale eddies (both of which are dealt with in the Marine Environment section below), model results indicate that the large-scale mean circulation of the Red Sea may be primarily the product of buoyancy loss associated with surface heat flux and evaporation (Sofianos and Johns 2002 and 2003).

It has long recognized that the rate of evaporation from the Red Sea far exceeds fresh water influx due to precipitation and runoff (Sofianos et al. 2002; and references therein). Estimates of the annual rate of evaporation from the Red Sea (computed from application of bulk formula to field data or from balancing the fresh water flux through the strait of Bab al Mandab with volume flux over the Red Sea surface) span a range 1.50–2.66 m year^{-1} (Table 1 of Sofianos et al. 2002). What may be the most tightly constrained estimate of the annual evaporation rate, based on moored measurements of currents and salinity through Bab al Mandab, is 2.06 ± 0.22 m year^{-1} (Sofianos et al. 2002). As indicated by analysis of data from the WHOI-KAUST meteorological buoy (Fig. 2.1), the evaporation rate varies seasonally about the mean, averaging roughly 3 m year^{-1} during winter and 1 m year^{-1} during summer (Bower and Farrar 2015). It also fluctuates on much shorter time scales. Bower and Farrar (2015) show that enhanced evaporation rates, exceeding 4.5 m year^{-1}, occur during the passage of cool and dry air carried past the mooring by the mountain gap wind events described above.

The net heat flux across the surface of the Red Sea also shows seasonal and shorter-term variations. The seasonal signal of net heat flux is characterized by net heat loss (transfer to the atmosphere) during winter months, roughly October–March, and net heat gain throughout the rest of the year (Ahmad et al. 1989; Tragou et al. 1999; Churchill et al. 2014b; Bower and Farrar 2015). Variations of the latent heat flux (associated with evaporation) and the incoming shortwave radiation are principally responsible for this seasonal net heat flux signal (Tragou et al. 1999; Bower and Farrar 2015). As documented by Papadopoulos et al. (2013) and Bower and Farrar (2015), the most intense events of wintertime heat loss over the northern Red Sea are associated with the westward passage of air from the Arabian subcontinent.

Papadopoulos et al. show that these events tend to correspond with a high pressure cell extending across the Mediterranean to central Asia.

The attenuation of incoming shortwave radiation due to dust may significantly contribute to the intense surface heat loss. Modeling results indicate that dust plumes over the Red Sea can extend vertically over 2–3 km and reduce the incoming shortwave, relative to clear-sky values, by up to 100 W m^{-2} (Kalenderski et al. 2013; Brindley et al. 2015; Prakash et al. 2015; Kalenderski and Stenchikov 2016). Based on results from a long-term (17 year) model simulation, Osipov and Stenchikov (2018) find that the overall impact of atmospheric dust is to cool the Red Sea surface, reduce the surface wind speed, and weaken both the water mass exchange through Bab-el-Mandeb and the overturning circulation.

Typically, the largest fluctuation of net heat flux occurs at a daily frequency, principally due to the day-night variation in incoming shortwave radiation (Figure 5 of Bower and Farrar 2015). Churchill et al. (2014b) demonstrate that the daily variation in net heat flux can lead to daily modifications in near-surface stratification (i.e. the formation of diurnal mixed layers) during periods of relatively weak surface wind stress. The warm surface layers observed by Churchill et al. form during the daylight hours of net heat gain and extend to between 5 and 20 m depths. They disappear during the nighttime hours of net heat loss, presumably due to convective vertical mixing initiated by the surface cooling.

As shown by Tragou et al. (1999), the yearly-averaged net heat flux varies spatially over the length of the Red Sea, with net heat loss north of roughly 18°N and net heat gain further south. According to calculations of Tragou et al., this spatial variation in net heat flux is the principal cause for a similar variation in net buoyancy flux, with a net buoyancy loss (gain) in the northern (southern) Red Sea.

2.3 Marine Environment

2.3.1 Basin-Scale Circulation

The geometry of the Red Sea, with an elongated basin and shallow sill at the Strait of Bab al Mandab (Fig. 2.1), combined with north-to-south gradients in surface heat flux and evaporation, result in meridional overturning and an exchange flow between the Red Sea and the Gulf of Aden. In a classic work, Phillips (1966) found similarity solutions for convectively driven flow forced by a uniform surface buoyancy flux that agree qualitatively with early observational studies of Red Sea circulation and water properties (Neumann and McGill 1962). The early summer observations by Neumann and McGill (1962) show a surface layer of warm (~30 °C) and relatively fresh (~36.5 psu) water, known as Red Sea Surface Water (RSSW), flowing into the Red Sea through the Strait of Bab al Mandab (Fig. 2.1). As it flows north, this water cools and becomes more saline due to evaporation. The result is a gradual increase in density, causing the RSSW to sink in the northern Red Sea. It returns southward in a subsurface current of colder (~26 °C), more saline (~40–40.5 psu) water known as the Red Sea Outflow Water (RSOW).

The exchange of water through the Strait of Bab al Mandab changes seasonally in response to large-scale variations in buoyancy and wind forcing (Yao et al. 2014a, b). The two-layer, inverse-estuary type circulation observed by Neumann and McGill (1962) is strongest from October to May, with an average transport of RSOW water between the Red Sea and Gulf of Aden of 0.37 Sv (Sofianos et al. 2002). During summer (June to September), a three-layer exchange flow is observed through Bab al Mandab, with a mid-level influx of relatively cool and fresh Gulf of Aden water, known as Gulf of Aden Intermediate Water (GAIW), sandwiched between outward flowing layers of RSSW above and RSOW below (Murray and Johns 1997).

The relative importance of thermohaline and wind forcing in driving basin-scale circulation within the Red Sea has long been a topic of debate (e.g., Phillips 1966; Patzert 1974; Tragou and Garrett 1997). Seasonally varying winds are thought to modify the buoyancy-driven circulation – enhancing the surface inflow through the Strait of Bab al Mandab during the winter and inducing upwelling in the Gulf of Aden in the summer, resulting in the subsurface intrusion of GAIW described above. Three-dimensional numerical simulations of Red Sea circulation by Sofianos and Johns (2002, 2003), in which wind and buoyancy forcing are applied in isolation, reveal that buoyancy forcing is dominant in driving the overall circulation patterns and exchange through the Strait of Bab al Mandab. In their simulations, stronger evaporation in the northern Red Sea drives higher buoyancy fluxes and the formation of dense RSOW. The north-south gradient in buoyancy forcing results in a downward sloping sea surface to the north.

More recent simulations conducted by Yao et al. (2014a, b) with a highly realistic numerical model reveal a strong seasonality of the overturning circulation in the Red Sea. Their results show that the convectively-driven formation of RSOW in the northern Red Sea occurs principally over October–March and largely confined to the region north of 24°N. In the model-generated climatological fields (with the impact of basin eddies averaged out), the sinking of newly-formed RSOW takes place in a narrow downwelling-band at the eastern basin margin. In the model results, the newly formed GSOW is initially transported to the southeast in ~100–300 m deep boundary currents on the eastern and western basin margins. Yao et al. (2014b) show that basin-scale eddies (described below) can appreciably alter the overturning flow structure, in a manner consistent with previous investigations (Maillard and Soliman 1986; Sofianos and Johns 2007). The modeled overturning circulation of the summer months (June–September) is dominated by the mid-level intrusion of GAIW, which is transported northward through the southern Red Sea in a eastern boundary current in the model results of Yao et al. (2014b), a result consistent with the observations of Churchill et al. (2014a).

2.3.2 Mesoscale Processes – Basin Eddies

The prevalence of eddies in the Red Sea basin has long been recognized (Quadfasel and Baudner 1993). Evidence of eddies within the basin has appeared in hydrographic survey data (Quadfasel and Baudner 1993), ADCP measurements (Sofianos and Johns 2007; Zhai and Bower 2013; Chen et al. 2014; Zarokanellos et al. 2017b), trajectories of satellite-tracked drifters (Chen et al. 2014) and numerical model results (Clifford et al. 1997; Sofianos and Johns 2003; Zhai and Bower 2013; Chen et al. 2014; Yao et al. 2014a, b). Eddies have been observed extending over diameters of order 200 km (Quadfasel and Baudner 1993), reaching depths of order 150 m (Sofianos and Johns 2007; Zhai and Bower 2013) and containing maximum velocities of ~ 1 m s^{-1} (Sofianos and Johns 2007).

A variety of statistical properties of Red Sea eddies has recently been determined from analysis satellite-altimeter-derived sea level anomaly (SLA) data acquired over 1992–2012 (Zhan et al. 2014). The results indicate that although eddies are formed during all seasons and over the full extent of the Red Sea, they are not uniformly distributed in space and time. Eddies appear with greatest frequency during the spring and summer (April–September) and in the central Red Sea (18–24°N), where the probability of a given point being within an eddy is close to 100%. Eddy diameter is also shown to vary spatially. Average eddy diameter is ~160 km in the northern and southern extremes of the Red Sea and is ~200 km in the central Red Sea. Zhan et al. (2014) note that this trend in diameter matches the variation in basin width and may indicate that eddy scale is limited by basin size. They also note that the limited zonal extent of the Red Sea may limit eddy lifetime, as eddies typically propagate zonally (westward in the northern hemisphere) which will would lead to significant coastal interaction and enhanced frictional dissipation. According to their analysis, Red Sea eddies have a mean lifetime of 45 days, with 95% the eddies expiring within 16 weeks.

A number of mechanisms for generating Red Sea eddies have been proposed. These include: baroclinic instability of flow with large vertical shear (Zhan et al. 2014), flow adjustment to seasonally varying buoyancy forcing (evaporation and heating) (Chen et al. 2014), and small-scale variations in the surface wind stress (Clifford et al. 1997; Zhai and Bower 2013). As opposed to the first two mechanisms listed above, the third may be highly localized. From analysis of satellite-derived (QuikSCAT) winds and SLA fields, Zhai and Bower (2013) link the formation of a dipole eddy pair (cyclonic and anticyclonic circulation cells) in the southern Red Sea with the summertime Tokar Gap wind jet. Their result is consistent with the model simulations of Clifford et al. (1997) in which the inclusion of forcing by orographically steered winds significantly increases the eddy prevalence in the modeled flow field. Because mountain gap winds over the Red Sea vary seasonally, they would produce a seasonally varying contribution to the overall eddy field.

Due to their prevalence and scale, which often spans the width of the Red Sea basin, eddies undoubtedly have a significant impact on coastal and basin ecosystems of the Red Sea. Acker et al. (2008) and Raitsos et al. (2013) offer evidence that basin-scale eddies transfer nutrients and/or chl-a from productive coastal reef regions to the oligotrophic waters of the Red Sea basin. Both set of investigators postulate that this phenomenon may be part of a 'mutual feedback mechanism' between coral reefs and the open basin, in which the seaward transport of nutrients stimulates blooms of phytoplankton in the deep basin, a fraction of which is transported back to the coastal reef region. The analysis of Churchill et al. (2014b) suggests that eddies may also be important in transporting mass and momentum from the deep basin to the coastal zone. They argue that the strongest currents observed in a coastal region of the central Red Sea may have been due to an along-shore pressure gradient arising from eddy-induced mass exchange between the deep basin and the coastal ocean.

A potentially important process that has yet to be extensively studied in the Red Sea is the enhancement of local productivity by mesoscale eddies. A number of investigators have found that mesoscale eddies in oceanic regions can promote productivity through vertical nutrient flux associated with changes in density structure linked with eddy generation and decay (e.g., Falkowski et al. 1991; Oschlies and Garçon 1998; McGillicuddy et al. 2007). For example, the upward doming of isopycnals that occurs within a cyclonic eddy in the northern hemisphere can deliver nutrient-rich deep water (typically found below the pycnocline) into the euphotic zone.

To our knowledge, in situ evidence of such a process is thus far limited to a study by Zarokanellos et al. (2017a). From data acquired within a cyclonic/anticyclonic eddy pair in the central Red Sea, they report a 50-m vertical isopycnal displacement associated with upward doming of isopycnals in the cyclonic eddy, and note the presence of relatively low oxygen concentrations (indicative of higher nutrient concentrations) in the surface mixed layer above the center of the cyclonic eddy.

2.3.3 Wind-Driven Flow

Wind fields over the Red Sea vary on seasonal, synoptic, and diurnal timescales as described above (Atmospheric Setting). While the basin-scale circulation patterns are dominated by seasonal buoyancy forcing, wind forcing also plays an important role by enhancing near-surface flows through the Strait of Bab al Mandab in winter and driving upwelled Gulf of Aden water into the Red Sea in summer (Basin-scale Circulation). Within the Red Sea basin, winds modify sea surface height on seasonal and synoptic (3–25 day) time scales (Sea Level Motions) and shape surface wave fields (Surface Waves). Cross-basin wind jets, or "mountain gap winds", impose a three-dimensionality on the essentially two-dimensional basin-scale thermohaline circulation, creating vorticity in near surface flows to form eddies (Mesoscale Processes – Basin Eddies) and coastal boundary layer currents (Bowers and Farrar 2015).

One aspect of wind-driven flow of particular importance to reef environments is upwelling/downwelling. The vertical movement of water associated with wind-driven upwelling has the potential to deliver deep, relatively cool and nutrient-rich water into the coastal reef regions. Statistical analysis applied by Churchill et al. (2014b) to water velocity and wind data acquired over 2008–2010 in the central Red Sea show a clear signal of wind-driven upwelling/downwelling in the Red Sea coastal zone. The signal is marked by an along-shore flow that is accelerated in the down-wind direction (positive wind stress/current correlation) and an across-shore flow that is directed to the right of the along-shore wind near the surface and to the left of the wind near the bottom.

To more fully demonstrate the effect of wind-driven upwelling/downwelling on the coastal temperature and velocity fields, we consider here (Fig. 2.3) temperature, velocity and wind stress data collected over 2008–2009 in the central Red Sea region (at site S2 in Fig. 2.1). The temperature records show periods of cooling water temperatures, of 1–3 weeks duration, superimposed on the annual temperature variation (Fig. 2.3a). The upwelling/downwelling signal observed by Churchill et al. (2014b) is apparent in the 2008–2009 current and wind stress data. Upwelling-favorable wind stresses (positive in Fig. 2.3b) are associated with offshore currents (positive "cross-wind currents"), which extend through most of the water column in the weakly stratified winter conditions and are concentrated in the upper mixed layer in summer and late spring when stratification is stronger (Fig. 2.3c). Southeast (along-shore) wind stress and near-

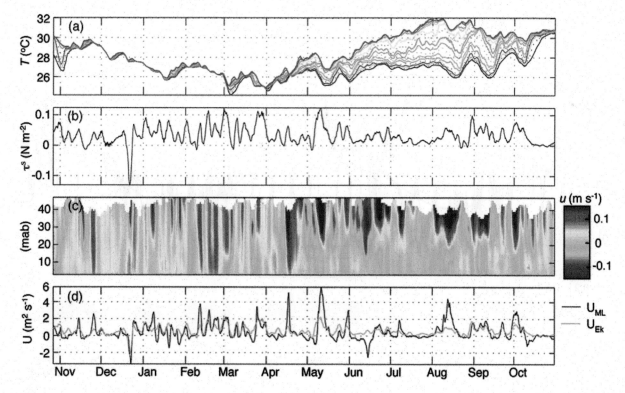

Fig. 2.3 2008–2009 time series of (**a**) water temperature at Mooring S1 (Fig. 2.1b) throughout the water column with warm colors (red/orange) near the surface and blues/black towards the bottom, (**b**) wind stress at meteorological buoy, rotated to 150°, approximately along-shore, (**c**) cross-wind currents at Mooring S2, positive towards 240°, approximately offshore, and (**d**) a comparison of upper mixed layer (U_{ML}) and Ekman (U_{Ek}) transports. In (**c**), (mab) represents meters above the bed. All time series shown were produced by filtering the original data with a low-pass filter with a 50-hr half-power point

surface cross-wind currents at S2 are significantly correlated over the entire year (r = 0.45 with 60 degrees of freedom (dof)), with a higher correlation over the time when the water is stratified (April–September; r = 0.6 with 30 dof). Churchill et al. show similar correlations.

The injection of wind energy into the upper ocean drives mixing, which can result in weak temperature gradients near the surface. The depth of the upper mixed layer, h_{ML}, calculated as the depth over which temperature is within 0.05 °C of the temperature at the top-most sensor (0.6-m depth) on the S2 mooring, varies appreciably, ranging from 3 m in summer to the full depth of the water column in winter. The cross-wind transport in the upper mixed layer, U_{ML} (estimated as the integral of the cross-wind component of the velocity from h_{ML} to the surface), compares well with Ekman transport computed from the wind stress, $U_{EK} = \dfrac{\tau^S}{\rho_0 f}$ (where τ^S, ρ_0 and f are the along-shore component of the surface wind stress, the upper layer water density and the Coriolis parameter, respectively). U_{Ek} and U_{ML} are correlated with a linear regression slope of 1.1 and r = 0.60, suggesting that cross-shelf transport is partially due to Ekman transport (Fig. 2.3d). However, as noted by Churchill et al. (2014b) the wind-driven upwelling/downwelling signal accounts for less than half of the overall variance of the sub-inertial (periods >2 days) flow observed at the central Red Sea mooring site. They observe significant departures of the observed transport from Ekman theory, indicating that other processes such as mesoscale eddies and coastal boundary currents may be important (see notable examples in mid-June and mid-August in Fig. 2.3d).

2.3.4 Sea Level Motions

Variations of the Red Sea water level span a range of order 1 m (Sultan et al. 1995a). Though relatively modest, these changes in the surface elevation may critically impact shallow ecosystems of the Red Sea. The crests of platform reefs, which are prevalent in the Red Sea and typically extend to depths of 1–2 m (DeVantier et al. 2000; Bruckner et al. 2012), may be particularly sensitive to order 1-m water level changes. As revealed by a number of investigators, hydrodynamics over shallow reef crests are sensitive to water level changes (e.g., McDonald et al. 2006; Monismith et al. 2013; Lentz et al. 2016b, 2017). In a recent work, Lentz et al. (2017) found that the drag coefficient for depth-averaged flow over a platform reef strongly depends on mean water depth, varying by an order of magnitude over a depth range of 0.2–2 m. Furthermore, order 1-m water level variations will alter the thermal environment over shallow reef tops by changing the water volume influenced by surface heat flux over the reef crest (e.g., Davis et al. 2011).

Viewed as a function of frequency, the changes in Red Sea water level may be divided into three broad categories (illustrated in Fig. 2.4). Occupying the lowest-frequency cat-

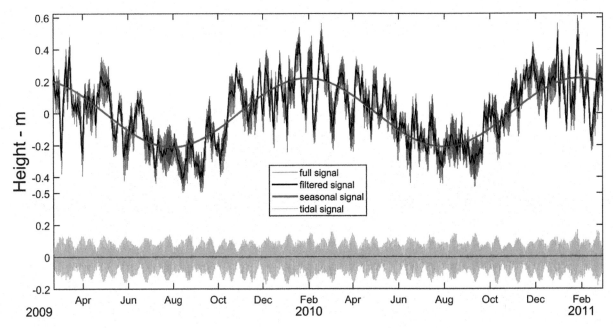

Fig. 2.4 The magenta line is a record of sea level height derived from pressure data acquired near Jeddah, SA (Fig. 2.1a). Other lines illustrate the three categories of sea level described in the text. The solid black line shows the pressure record filtered with a 66-hr half-power-point filter and encompasses the seasonal signal (red line) and the intermediate-frequency band signal (difference between the filtered and seasonal signal). The green line is the higher frequency signal (difference between the filtered and unfiltered signal), which is principally due to tidal motions

egory are motions varying over seasonal periods (>0.5 year), whereas the highest-frequency category is comprised principally of diurnal and semidiurnal tidal motions (periods <1.1 days). Sea level changes in the 'intermediate-frequency' category span a range of order 0.7 m and are contained in a period band of roughly 3–25 days. Current knowledge of sea level motions in each of these categories is reviewed below.

2.3.4.1 Seasonal Sea Level Variations

A number of researchers have reported on a seasonal signal of water level in the Red Sea marked by higher surface elevations in the winter than in the summer (Morcos 1970; Sultan et al. 1995b, 1996; Abdelrahman 1997; Sofianos and Johns 2001; Manasrah et al. 2009). Most observations of this phenomenon are from the central Red Sea (16.5–21.5°N) and show a seasonal sea level signal with a range of 0.3–0.4 m (illustrated in Fig. 2.4). Analysis of satellite-altimeter-derived sea surface height (SSH) data, indicate that the seasonal sea level range is roughly constant over the central and northern Red Sea, but declines in the southerly direction over the southern Red Sea to a magnitude of roughly 0.2 m near Bab al Mandab (Sofianos and Johns 2001). This trend is consistent with the analysis of Pazart (1974), who examined sea level records from a number of locations spanning the length of the Red Sea.

Numerous studies have considered the mechanisms driving the seasonal sea level variations in the Red Sea. Those factors most likely to contribute to the seasonal signal, based on dynamical considerations (i.e., momentum balance), are atmospheric pressure (i.e., the inverse barometer effect), surface wind stress and steric effects (variation in water density). As noted by Sultan et al. (1995a), atmospheric pressure is not likely to be of importance in driving the seasonal sea level signal because observed seasonal atmospheric pressure variations would produce a sea level response with a trend opposite to that observed (relatively high sea levels in summer). The analysis of Sofianos and Johns (2001), based on a simple 1-dimensional momentum balance (ignoring the Coriolis effect and bottom friction) with forcing by monthly wind stresses and variations in climatological water properties, indicates that the along-axis variation in wind stress is the principal driver of the seasonal sea level signal over all but the extreme southern portion of the Red Sea (south of 14°N) where the steric contribution dominates. The findings of Wahr et al. (2014), derived from combining SSH data, climatological sea water temperatures and GRACE (Gravity Recovery and Climate Experiment) mass data, are consistent with the dominance of wind forcing over steric effects in controlling the seasonal sea level variation over most of the Red Sea. However as opposed to Sofianos and Johns, Wahr et al. find that the steric influence on seasonal sea level is minimal over the southern Red Sea.

2.3.4.2 Intermediate Band Sea Level Variations

As compared with seasonal and tidal sea level motions (below), sea level variations in the intermediate frequency band (3–25 day periods) have received very little scientific attention. Reported analysis of intermediate band motions is largely confined to the examination of sea level records from Jeddah and Port Sudan (on opposite sides of the central Red Sea; Fig. 2.5) by Sultan et al. (1995a). Their analysis indicates statistically significant correlations between sea level variations at Jeddah and the along-shore wind stress, and between Port Sudan sea level variations and both the along- and across-shore wind stress components.

Despite the lack of scientific interest they have received thus far, sea level motions in the intermediate frequency band may be of importance for many environments, particularly over the crest of shallow platform reefs. As revealed by the analysis of pressure records taken near Jeddah, intermediate band sea level fluctuations include relatively large (order 0.6 m) water level changes occurring over periods of a few days (i.e., the changes seen in March 2009 in Fig. 2.4).

2.3.4.3 Tides

Considerable scientific attention was directed at Red Sea tides during the early twentieth century. Reviewed by Morcos (1970) and Defant (1961), much of this work entailed comparing theoretical calculations of tidal propagation throughout the Red Sea with tidal analysis of sea level data. A conclusion drawn by a number of investigators is that the tides over the Red Sea are predominantly co-oscillations with tides of the Gulf of Aden, with locally-forced, 'independent' tidal motions accounting for order 25% of the tidal amplitude. Cotidal charts, formulated based on the early twentieth century tidal analysis (Fig. 2.5), show an amphidromic point at roughly 20 °N, and tidal ranges increasing from less than 0.25 m over the central Red Sea to more than 0.5 m at the northern and southern extremes. As demonstrated by Vercelli (1925), Red Sea tides are predominately semidiurnal, with $[(K_1 + O_1)/(M_2 + S_2)] < 0.25$. An exception is within the nodal zone of the semidiurnal tide in the central Red Sea, where $[(K_1 + O_1)/(M_2 + S_2)]$ exceeds 0.5.

More recent tidal analyses of water level data from the central Red Sea are in agreement with tidal properties derived from the earlier work and clearly show the variation of tidal range with distance from the amphidromic point near 20 °N. The tidal signal measured at Port Sudan, close to the amphidromic point (Fig. 2.5), ranges over roughly 4 and 12 cm during the neap and spring tidal cycle, respectively (Eltaib 2010). This contrasts with the neap/spring tidal ranges measured at Jeddah (~15/35 cm; Sultan et al. 1995a) and Gizan (~20/120 cm; Eltaib 2010).

Fig. 2.5 Cotidal chart of the Red Sea (adapted from Morcos 1970). (**a**) The average tidal range in m. (**b**) Cotidal lines indicating the times of high water in hours after the transit of the moon at Greenwich

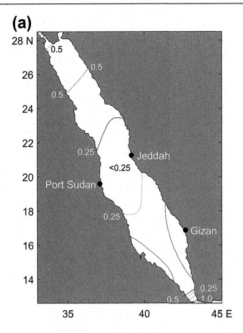
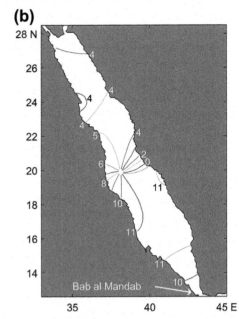

Tidal ranges at the southern extreme of the Red Sea, in the Strait of Bab al Mandab, are large and vary significantly over the length of the strait. Jarosz et al. (2005) report that the tidal range declines considerably going northward over the 150-km extent of the strait, from more than 1.5 m at Perim Narrows to less than 1 m at Hanish Sill. The character of the tidal signal also changes. Tidal energy at Perim Narrows is nearly evenly split between diurnal and semidiurnal bands, while more than 90% of the tidal energy at Hanish Sill is contained in the semidiurnal band. Earlier work by Vercelli (1925) indicates the presence of an M_2 tidal node in the central portion of the strait.

Reported analyses of tidal currents in the Red Sea are rare. Examination of ADCP velocity records by Churchill et al. (2014b) reveal particularly weak tidal velocities in the coastal zone of the central Red Sea (at ~22 °N, see Fig. 2.1b). Consistent with the tidal analysis of water level data, these currents are predominately semidiurnal. Their magnitude seldom exceeds 8 cm s^{-1}.

The weak tidal flows in the central Red Sea reported by Churchill et al. contrast with strong tidal currents observed in the Strait of Bab al Mandab by Jarosz et al. (2005). These tidal flows reach magnitudes of ~1 m s^{-1} near the southern entrance of the strait, at Perim, and exceed 0.5 m s^{-1} further north at Hanish Sill. They are nearly rectilinear and oriented along the strait. Comparable in strength to the mean exchange flows, these tidal currents produce occasional reversals in the mean inflow and outflow through the strait. The analyses of Jarosz et al. (2005) further indicate that the tidal current in the strait includes a significant baroclinic component, which appears primarily in the diurnal band and is most energetic in winter.

Results of a numerical tidal model encompassing the Red Sea and the northern Gulf of Aden, reported by Madah et al. (2015), reproduce many of the tidal features described above, including the dominance of the M_2 tide over most of the Red Sea, locations of tidal nodes, and the presence of strong tidal currents in the Strait of Bab al Mandab. The model results also show interesting, and potentially important, tidal features not yet confirmed by observations. These include strong tidal flows (peaking at order 0.3 m s^{-1}) oriented across isobaths in the coastal zone of the southern Red Sea (between 14 and 16°N; Figure 9 of Madah et al. 2015).

2.3.5 Surface Waves

The action of surface gravity waves has been shown to strongly impact the coastal environment. For example, numerous studies have indicated that the generation of bottom stress is often significantly enhanced by the interaction of near-bottom orbital currents due to surface waves with the more slowly varying flow (Cacchione and Drake 1982; Grant et al. 1984; Grant and Madsen 1986; Lyne et al. 1990; Madsen et al. 1993; Churchill et al. 1994; Chang et al. 2001). This enhancement can be appreciable at bottom depths as great as 100 m (Grant et al. 1984; Churchill et al. 1994). In addition, the breaking of surface waves at the edge of a shallow reef is a principal mechanism driving currents over the reef top (e.g., Symonds et al. 1995; Callaghan et al. 2006; Monismith 2007; Hench et al. 2008; Lowe et al. 2009; Vetter et al. 2010; Lentz et al. 2016b). This is due to a setup of sea-level elevation in the wave breaking zone, which in turn forces a across-reef current towards the protected (lee-side) of the reef (Monismith 2007; Hearn 2010; Lentz et al. 2016b).

Much of the current knowledge of the Red Sea wave climate has come from wave models that have been assessed by comparison with the wave height series measured at the central Red Sea meteorological buoy (MB in Fig. 2.2) and wave properties derived from satellite-scatterometer data (Ralston et al. 2013; Langodan et al. 2014, 2016, 2017b; Aboobacker et al. 2017; Shanas et al. 2017a, b). These modeling studies have revealed some important large-scale features of the wind-driven wave field over the Red Sea. In the model results, much of the seasonal variability in the wave properties (height, direction and dominant period) in the Red Sea basin is linked to the seasonal variation in winds along the Red Sea axis (discussed in Atmospheric Setting). In the southern Red Sea, modeled waves are highest during November–April, when driven by the seasonal monsoon winds from the southeast. The monthly-averaged significant wave height (H_S = mean height of the highest third of the waves) during this period tends to be greatest in the 13–15.5 °N latitude band, reaching values close to 2 m (Ralston et al. 2013). In the northern and central Red Sea, modeled waves tend to propagate southeastward, generated by the persistent winds from the northwest. These waves are largest (with monthly mean H_S of 1.5–2 m) when 'strong northeasterly winds shifted the atmospheric convergence zone to the south' (Ralston et al. 2013). The dominant periods of the model-generated waves are predominately in the 4–8 s range.

Numerous model results (Ralston et al. 2013; Langodan et al. 2016, 2017b; Aboobacker et al. 2017; Shanas et al. 2017b) indicate that the largest waves within the Red Sea are driven, not by the winds along the sea's axis, but by the cross-axial winds emerging from the Tokar Gap. As noted above (Atmospheric Setting), the Tokar Gap wind jet is a summer-time phenomenon associated with nocturnal discharge of cool air through the Tokar Gap. Ralston et al. (2013) find that waves generated by the Tokar Gap wind vary diurnally, with the largest waves (monthly-mean H_S ~ 4 m for July) occurring in the morning hours and the smallest waves (July mean H_S < 0.5 m) seen at night. The model results also indicate that the Red Sea wave field is impacted by winds flowing through mountain gaps along the northeastern Red Sea coast. The spatial scale over which these winds influence the modeled wave field is smaller than that of the Tokar Gap wind (~40 km vs. 100 km laterally, and ~100 km vs. 250 km along the jet).

While the modeling results reviewed above offer valuable information on the large-scale patterns of surface wave properties in the Red Sea, they have limited applicability to the Red Sea coastal areas, where wave propagation and attenuation are likely to be strongly influenced by small-scale and complex bathymetric features. A recent analysis of surface wave properties, derived using data from bottom-mounted ADCPs (Acoustic Doppler Current Profilers) and a meteorological buoy (Fig. 2.1), indicate that waves propagating from the deep basin into the coastal zone may be significantly attenuated as they move over the complex bottom topography characteristic of Red Sea coastal areas (Lentz et al. 2016a). The attenuation of wave height propagating shoreward over the coastal area examined by Lentz et al. is illustrated here by time series of H_S (Fig. 2.6) measured at an offshore meteorological buoy and three coastal sites: one (S2 in Fig. 2.1) on the outer portion of a 40–90-m deep 'plateau' situated between the Qita Dukais reef system and the

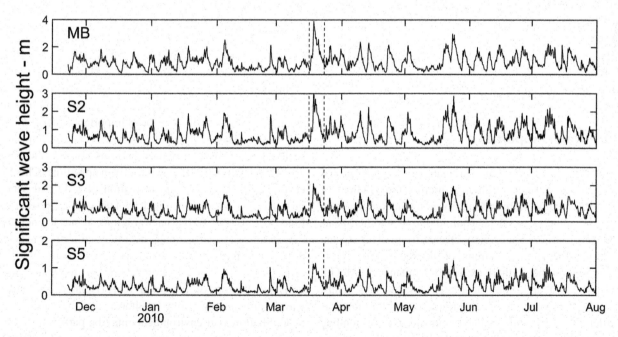

Fig. 2.6 Sample time series of significant wave height measured at locations shown in Fig. 2.1b. The vertical dashed lines bracket the event with the highest significant wave heights of the period shown

edge of the Red Sea basin and the other two (S3 and S5) within the reef system. These series are marked by frequent events in which peaks in H_S measured at the offshore buoy are matched by smaller peaks in the H_S series at the onshore sites that progressively decline in amplitude going shoreward into the plateau/reef system. This trend is nicely illustrated by the H_S measured during an event of particularly large waves in March 2010 (bracketed by dashed lines in Fig. 2.6). Over this event, maximum H_S declines from 3.9 m at the offshore buoy to values of 3.0, 2.1 and 1.2 m, respectively, at sites on the mid-plateau (S2), the outer edge of the Qita Dukais reef system (S3) and in the interior of the reef system (S5). A conclusion that may be drawn from the onshore decline in wave height, consistently seen in the H_s series, is that waves generated over the Red Sea basin are attenuated (to a greater extent than they are enhanced by wind-forcing) as they propagate onshore into the coastal/reef region. Nevertheless, reefs situated well onshore of the basin's edge can be exposed to significant wave energy. The H_S measured at site S5, which is roughly 12 km from the edge of the basin and at the edge of a platform reef, exceeds 0.5 m 26% of the time.

From analysis of pressure data on the platform reef adjacent to S5, Lentz et al. (2016a) show that wave transformation across the reef depends on incident wave height and reef depth. Incident waves with heights of >40% of reef depth break at the edge of the reef and decay gradually while propagating across the reef interior. Smaller incident waves, with heights <20% of the reef depth, propagate across the reef face without breaking but still gradually decay while moving across the reef interior. As demonstrated by Lentz et al. (2016b), wave breaking has a dominant impact in driving flows over the reef crest. From velocity and pressure data acquired over a platform reef in the central Red Sea, they find that breaking of waves on the seaward reef edge sets up a 2–10 cm elevation of sea level, which drives a 5–20 cm s^{-1} current across the reef.

2.3.6 Water Properties

As noted above (Basin-scale Circulation), evaporation and surface heat exchange result in a north-south gradient of near-surface temperature and salinity along the Red Sea axis. In hydrographic survey data acquired during summer, near-surface temperature and salinity change from ~31.5 °C and 37.6 psu at the southern extreme of the Red Sea to ~27 °C and 40.5 psu at the northern extreme (Neumann and McGill 1962; Maillard and Soliman 1986; Sofianos and Johns, 2007). During winter, the near-surface temperature range is of order 24–28 °C, with the maximum temperatures tending to occur in the central Red Sea (near 19°N) (Quadfasel and Baudner 1993; Raitsos et al. 2013; Ali et al. 2018).

In most (or perhaps all) areas of the Red Sea, near-surface temperatures exhibit a seasonal variation (Raitsos et al. 2013; Churchill et al. 2014b; Ali et al. 2018). Recent work by Ali et al. (2018) reveals that the seasonal variations in near-surface temperature and salinity in the southern Red Sea (off of Port Sudan) are not the sole product of surface heat and mass fluxes, but also due to the alongshore advection of water with spatially varying temperature and salinity.

In the summertime hydrographic data, the seasonal thermocline typically begins in the upper 50 m and extends to order 150 m. The downward decline in temperature over seasonal thermocline is of order 8 °C, whereas the downward increase in salinity over the thermocline depth range is of order 0.5 psu. The observed wintertime temperature structure (Quadfasel and Baudner 1993) is characterized by a deep surface mixed layer, that can extend to order 150 m in the northern Red Sea, and a temperature decline of order 4 °C over the seasonal thermocline. In all seasons, the temperature and salinity show little variation beneath the seasonal thermocline, ranging over ~21.6–22 °C and 40.4–40.6 psu between ~200–2000 m depths (Neumann and McGill 1962; Maillard and Soliman 1986; Sofianos and Johns 2007).

Hydrographic data from the summer and autumn show the subsurface intrusion of GAIW. Driven northward through the Strait of Bab al Mandab by upwelling-favorable monsoon winds over the western Gulf of Aden (Patzert 1974), GAIW enters the southern Red Sea during summer (June–September) as a cold (17.4 °C at 75 m vs 31.4 °C in the top 10 m), fresh (35.8 vs 37.4 psu) and nutrient-rich (e.g., NO_3, 23.5 vs 0.9 µmol l^{-1}) intrusion typically contained in the 30–120-m depth range (Poisson et al. 1984; Maillard and Soliman 1986; Souvermezoglou et al. 1989). As documented by Sofianos and Johns (2007) and Churchill et al. (2014a), the temperature and salinity signal of GAIW extends a considerable distance northward of Bab al Mandab.

In addition to the large-scale patterns described above, the temperature/salinity fields of the Red Sea contain numerous small-scale features that may be of importance to reef growth and health. These include features associated with basin-scale eddies, coastal boundary currents (Bower and Farrar 2015; Eladawy et al. 2017), wind-driven upwelling (Churchill et al. 2014b; Fig. 2.3) and formation of a diurnal surface mixed layer. As demonstrated by Davis et al. (2011) and Pineda et al. (2013), diurnal variations in surface heat flux may produce a 'microclimate' over the shallow (<2-m depth) tops of platform reefs that are prevalent in the Red Sea coastal zone. Davis et al. show that daily temperature variations over the tops of 'protected' reefs (isolated from surface waves propagating from offshore) can be as high as 5 °C. Shallow reef environments may be particularly sensitive to long-term water temperature shifts associated with climate change. Through examination of satellite radiometer-derived sea surface temperature data, Raitsos et al. (2011) document an abrupt increase in Red Sea surface temperatures, by ~0.7 °C, in the mid-1990's, with the warmer surface temperatures persisting to at least 2008. They attribute this shift to increases in air temperature associated with global climate trends.

2.3.7 Oxygen and Nutrients

The Red Sea basin may be characterized as oligotrophic (Stambler 2005) owing in part to the limited supply of nutrients delivered to the Red Sea through terrestrial runoff. It is well established that principal source new water-borne nutrients to the Red Sea is the summertime intrusion of GAIW (Khimitsa and Bibik 1979; Souvermezoglou et al. 1989). As demonstrated by Souvermezoglou et al. (1989), the seasonality of the GAIW intrusion results in a seasonal variation of the Red Sea nutrient budget, marked by a net nutrient gain in the summer and loss in the winter.

Analysis of data from a September 2011 hydrography cruise of the central and northern Red Sea by Churchill et al. (2014a) indicates that GAIW is distributed broadly through the Red Sea. Churchill et al. identify four modes of GAIW transport: (*1*) transit of nutrient-rich (NO_3 up to 19 µmol l^{-1}) GAIW through the southern Red Sea (to 19°N) in a subsurface current flowing along the eastern basin margin with a speed of ~25 cm s^{-1}, (*2*) movement of GAIW across the Red Sea basin in the circulation of basin-scale eddies, (*3*) northward flow of GAIW over the central and northern Red Sea (identifiable to 24 °N), and (*4*) incursion of GAIW into coastal reef systems. In view of the fourth mode, Churchill et al. note that GAIW could be an important source of new nutrients to coral reef ecosystems, particularly in the southern Red Sea where nutrient concentrations in GAIW have been observed at the highest concentration (Poisson et al. 1984; Sofianos and Johns 2007). Churchill et al. also observe that the high nutrient concentrations within GAIW extend into the euphotic zone and appear to fuel enhanced productivity over depths of 35–67 m. In the absence of GAIW, near-surface Red Sea water typically contains low nutrient concentrations (i.e., $NO_3 < 1$ µmol l^{-1}) in the upper 70–120 m (Morcos 1970; Poisson et al. 1984; Sofianos and Johns 2007; Churchill et al. 2014a; Triantafyllou et al. 2014).

Analysis of more recent data, from a November 2013 cruise, by Zarokanellos et al. (2017a) also reveal the presence of GAIW in the central Red Sea. Consistent with the notion that GAIW fuels productivity, the GAIW water parcels observed by Zarokanellos et al. contain elevated concentrations of chl-*a* and colored dissolved organic matter relative to surrounding water.

Beneath the surface 120 m, vertical nutrient profiles from the Red Sea are marked by a concentration maximum ($NO_3 > 15.5$ µmol l^{-1}) in the 300–650-m depth range (Neumann and McGill 1962; Morcos 1970; Poisson et al. 1984; Weikert 1987). The magnitude of the maximum concentration tends to increase going from northern Red Sea (maximum $NO_3 = 15.5$–18.0 µmol l^{-1}) to the southern Red Sea (maximum $NO_3 = 18.9$–22.1 µmol l^{-1}) (Weikert 1987). This trend is likely due to the accumulation of dissolved nutrients by remineralization of sinking particles in the layer encompassing the nutrient maximum as it flows southward from its formation region in the northern Red Sea to Bab al Mandab.

Because there is a tight inverse relationship between concentrations of dissolved nutrients and dissolved oxygen (DO), as indicated by a positive correlation between nutrient concentration and apparent oxygen utilization (Naqvi et al. 1986; Churchill et al. 2014a), the distribution of nutrients described above is closely associated with the distribution of DO in the Red Sea. For example, DO distributions along the Red Sea axis show a vertical minimum in the 300–650-m depth range of the nutrient maximum as well as a north-to-south increase in DO in this depth range (Neumann and McGill 1962; Poisson et al. 1984; Sofianos and Johns 2007).

A noteworthy feature of the nutrient mix within the Red Sea is the large departure of nutrient-to-nutrient ratios from open ocean (Redfield) values. Analysis of nutrient data by Naqvi et al. (1986) gives carbon:nitrogen:phosphorus (C:N:P) ratios of 188:21:1, considerably different from the Redfield ratios of 106:16:1. Grasshoff (1969) reports a similar excess of N over P (relative to open ocean proportions). Navqi et al. note that this N:P 'imbalance' cannot be attributed to the properties of the influx through Bab al Mandab, as the Gulf of Aden source water of this influx has N:P ratios close to the Redfield value. Using available estimates of flows and nutrient concentrations though Bab al Mandab, Navqi et al. computed a net 0.74×10^{12} g year^{-1} outflow of N from the Red Sea, similar to a later estimate by Bethoux (1988). Navqi et al. posit that the excess N exported from the Red Sea could be the result of fixation of nitrogen (entering the Red Sea from the atmosphere) by the cyanobacteria *Trichodesmium*, which is prevalent in the Red Sea. However, later calculations by Bethoux (1988) indicate that nitrogen fixation by open-water blooms of *Trichodesmium* likely accounts for only a small fraction of the imbalance of the Red Sea N budget. Bethoux finds that fixation by coral reef communities is a more probable mechanism for generating the excess N. Nitrogen fixation by Red Sea coral communities has been observed by El-Shenawy and El-Samra (1996) and Grover et al. (2014). The observations of El-Shenawy and El-Samra indicate that nitrogen fixation over reefs in the northern Red Sea and Gulf of Suez tends to occur at a greater rate during the daylight (vs. nighttime) hours and during summer (vs. spring and winter).

2.3.8 Light and Chlorophyll Distribution

The productivity and distribution of reef-building corals is highly dependent on the ambient light environment. The quantity of light to which corals are exposed is a function of the surface irradiance and the degree to which this is attenuated through the water column. Measurements of Red Sea light levels come predominately from the northern Red Sea and the Gulf of Aqaba. Winters et al. (2009) show that levels of surface irradiance in this region are exceptional high, exceeding surface irradiance levels over reef environments off of Mexico, Australia and Hawaii by order 40%. The seasonal variation of surface irradiance over the Gulf of Aqaba is significant, with maximum daily global irradiance ranging from ~550 W m^{-2} over November-January to ~950 W m^{-2} over April–July (Stambler 2006; Winters et al. 2009).

The exceptionally clear water of the Red Sea allows for deep penetration of incident light. Measurements analyzed by Stambler (2005, 2006) show that the euphotic zone depth (beyond which the downwelling shortwave radiation is <1% of the surface irradiance) in the northern Red Sea and Gulf of Aqaba ranges over 74–115 m. Raitsos et al. (2013) report a slightly narrower euphotic zone depth range of 77–96 m based on measurements from the central Red Sea (near 22 °N). The degree of light penetration in the northern Red Sea appears to vary with season. Stambler (2006) reports that the exponential attenuation coefficient for PAR [K_d(PAR)] measured in the Gulf of Aqaba ranges from a summertime minimum of 0.04 m^{-1} to a springtime maximum of 0.064 m^{-1}. The penetration of light into Red Sea also varies significantly as a function of wavelength, with minimum attenuation experience by blue light (Stambler 2006; Mass et al. 2010; Roder et al. 2013). Based on light measurements in the Gulf of Aqaba, Mass et al. (2010) report attenuation coefficients of 0.056 m^{-1} and 0.449 m^{-1} for blue (490-nm wavelength) and red (665 nm), respectively. In their data set, no light with wavelength > 600 nm appears at 40 m depth. They present evidence that coral colonies of *Stylophora pistillata* are capable of photoadaptation to the differing light conditions with depth. Specifically, they show the photosynthetic performance for colonies acquired from depths of 3 and 40 m is maximal when illuminated with full-PAR and filtered blue light, respectively.

As documented by Fricke and Knauer (1986) and Fricke et al. (1987), zooxanthellate corals may occur in the dimly-lit twilight zone beneath the euphotic zone. The observations of Fricke and Knauer (1986) reveal corals in the 100–200 m depth range where they are exposed to 0.1–1.7% of surface irradiance.

The vertical distribution of chl-*a* is principally controlled by combination of PAR and available nutrients, and is typically marked by a 'deep maximum' at the base of the main pycnocline. Fluorometer measurements taken from the central and northern Red Sea show the deep chl-*a* maximum occurring over a depth range of ~40–140 m (Churchill et al. 2014b; Zarokanellos et al. 2017a, b). During the autumn, the spatial variation of the maximum chl-*a* concentration seen in this depth range appears to be related to the distribution of the seasonal GAIW intrusion, with maximum chl-*a* concentrations tending to decrease going from south to north (presumably due to the dilution and uptake of nutrients borne by the GAIW intrusion as it is advected northward) and from east to west (reflecting the tendency of the GAIW intrusion to flow along the eastern margin of the Red Sea basin) (Churchill et al. 2014b; Zarokanellos et al. 2017a).

As revealed by analysis of satellite spectroradiometer measurements (Acker et al. 2008; Raitsos et al. 2013; Abdulsalam and Majambo 2014), the near-surface chl-*a* field in the Red Sea (above the deep chl-*a* maximum) exhibits a distinct seasonal and spatial signal. Over the entire Red Sea, near-surface chl-*a* concentrations tend to be highest in winter (Oct.-Mar.) and lowest in summer (May-Aug.). Raitsos et al. (2013) attribute the low near-surface chl-*a* concentrations of summer to the strong summertime stratification blocking the upward transfer of deep nutrients. They postulate that the high near-surface chl-a concentrations of winter are the result of vertical mixing of nutrients in the north and the productivity associated with GAIW-borne nutrients in the south. In all seasons, the near-surface chl-*a* concentrations tend to be highest in the southern Red Sea (south of 17.5°N), exceeding concentrations of the northern Red Sea (north of 22°N) by an order of magnitude. However, as shown by Acker et al. (2008), the surface chl-*a* field observed in the northern Red Sea during the winter/spring bloom period can be strongly heterogeneous with relatively high chl-*a* concentrations (up to 5 mg m^{-3}) appearing in small-scale filaments.

2.4 The Carbonate System

With the notable exception of riverine input, the mechanisms that govern the inorganic carbonate chemistry in the Red Sea are similar to those that control typical coastal and continental shelf environments. These mechanisms include air-sea gas exchange, primary productivity and respiration, and formation and dissolution of $CaCO_3$ minerals. However, the distinct climatological, geological and hydrographic settings of the Red Sea produce some unique carbonate system characteristics specific to this basin.

The shallow (approximately 140-m deep) sill of the Strait of Bab al Mandab (Fig. 2.1) limits the passage of deep Indian Ocean water into the Red Sea. As a result, the water entering the basin is predominately surface seawater, which contains relatively low levels of dissolved inorganic carbon (DIC) and nutrients. The exception is the summer-time intrusion of GAIW, which is the primary source of water-borne nutrients to the Red Sea. The limited supply of nutrients renders the Red Sea oligotrophic with primary production rates of 20–60 mmol C m^{-2} d^{-1} (Klinker et al. 1976; Lazar et al. 2008; Qurban et al. 2014). The high evaporation rates in the Red Sea (~ 2 m year^{-1}) concentrate dissolved ions in the surface water (Steiner et al. 2014). As a result, the measured DIC range in the Red Sea of 2000–2200 µmol kg^{-1} (Weiss et al. 1983; Papaud and Poisson 1986; Krumgalz et al. 1990; Wurgaft et al. 2016) is considerably higher than the salinity normalized DIC (nDIC = DIC*35/salinity) range, which spans 1790 to 1960 µmol kg^{-1}. Consequently, the measured DIC falls within the range of DIC in the world oceans (~2000–2300 µmol kg^{-1}; Emerson and Hedges 2008), whereas the salinity normalized DIC is considerably lower.

The vertical distribution of DIC in the Red Sea resembles that of pelagic environments. It is characterized by low DIC levels in the upper parts of the water column (Fig. 2.7), resulting from photosynthetic CO_2 uptake and air-sea gas exchange, and higher DIC concentration below the photic zone, due to bacterial re-mineralization of organic material. A DIC maximum is typically seen at ~ 300 m. An exception is the northern Gulf of Aqaba (GOA, Fig. 2.1) DIC distribution during winter (Fig. 2.8). The water column in this part of the Red Sea is subject to deep vertical mixing, occasionally mixing to the bottom (Genin et al. 1995; Lazar et al. 2008). However, even in the stratified center of the Red Sea, the difference between the shallow and deep-water nDIC is smaller than 100 µmol kg^{-1} (Fig. 2.8), whereas in the adjacent Gulf of Aden this difference exceeds 200 µmol kg^{-1} (Talley 2013). The relatively low DIC concentrations in the deep Red Sea are principally the result of two factors. The first is the short residence time (~ 40 year) of Red Sea deep water (Cember 1988), which limits the accumulation time of respiration products below the thermocline. The second is the low primary productivity rates of the oligotrophic Red Sea. This results in low rates of downward export of organic material from the photic zone, which reduces the accumulation potential of DIC in Red Sea deep water.

The distribution of total alkalinity (TA) in the Red Sea is affected by evaporation, $CaCO_3$ precipitation and organic matter remineralization. The TA of the Red Sea surface water varies between 2300–2600 µmol kg^{-1} (Weiss et al. 1983; Steiner et al. 2014), and increases from south to north due to evaporation (Weiss et al. 1983). However, as shown by Steiner et al. (2014), this increase is not conservative due to biological $CaCO_3$ sequestration that removes alkalinity from the seawater. As a result, the salinity normalized alkalinity (nTA = TA*35/salinity) in the Red Sea is approximately 2100 µmol kg^{-1}, considerably lower than typical TA in the surface ocean (~2300 µmol kg^{-1}, Millero et al. 1998). Moreover, nTA shows a south to north decrease, opposite to the trend of TA. Based on the deviation of TA from conservative behavior and the accompanying increase in the Sr/Ca ratios, Steiner et al. (2014) estimate that the rate of $CaCO_3$ production in the Red Sea is 7×10^{10} kg year^{-1}, and that 80% of this production is attributed to pelagic calcareous plankton with the remaining 20% attributed to coral-reef growth.

The vertical distribution of TA in the Red Sea exhibits a unique pattern. In contrast to the typical continental shelf and pelagic environments, where TA increases with depth

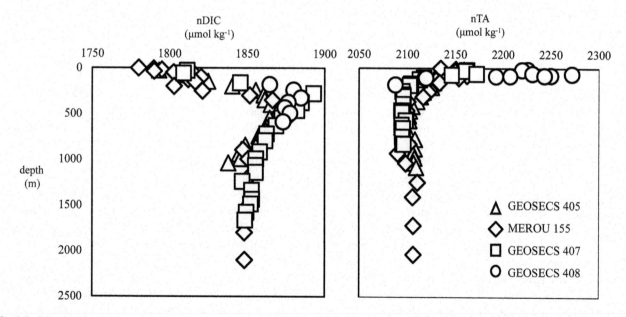

Fig. 2.7 Dissolved inorganic carbon (nDIC) and alkalinity (nTA) data from the Red Sea basin (see Fig. 2.1a for the locations). All data shown are normalized to a salinity of 35 (see text). nDIC shows a typical oceanic profile, with higher nDIC concentrations below the photic zone. The difference in nDIC between the surface and the deep water, however, is considerably smaller than the corresponding difference in the ocean (see text). The vertical distribution of nTA, however, is opposite the the typical oceanic distribution, with lower values in the deep water

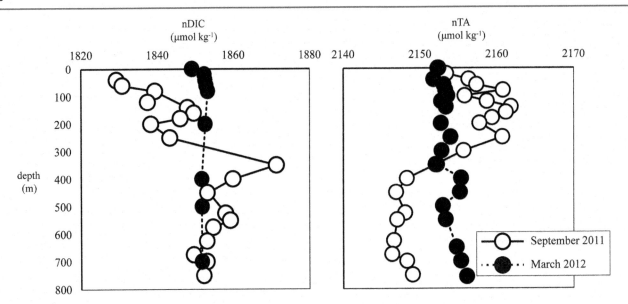

Fig. 2.8 Dissolved inorganic carbon (nDIC) and alkalinity (nTA) data from the northern Gulf of Aqaba (GOA, Fig. 2.1a). All data shown are normalized to a salinity of 35 (see text). The September 2011 profiles are representative of thermally stratified water column (usually between April and October), whereas the March 2012 profiles represent mixed water column. Note that the water column does not mix to the bottom of the northern GOA every winter. However, the mixed layer depth usually exceeds 200 m

due to $CaCO_3$ dissolution (Broecker and Takahashi 1978; Broecker et al. 1982), nTA in the Red Sea decreases with depth (Fig. 2.8). A substantial difference between the deep Red Sea and the deep ocean is that unlike the deep ocean, where the water temperatures are ~ 4 °C, the temperature of the deep Red Sea water is ~20 °C. This high temperature has an important effect on the stability of $CaCO_3$ minerals (calcite and aragonite). Because the solubility of $CaCO_3$ minerals decreases at high temperatures (Mucci 1983), the warm temperature of the Red Sea deep water, combined with its relatively high TA:DIC ratios, maintains a high degree of $CaCO_3$ saturation. The degree of saturation is commonly expressed as $\Omega = [Ca^{2+}][CO_3^{2-}]/K_{SP}$, where $[Ca^{2+}]$ and $[CO_3^{2-}]$ are the measured concentrations of Ca^{2+} and CO_3^{2-}, and K_{SP} is the solubility product for either calcite or aragonite. In the Red Sea, Ω is >1 over the entire water column, indicating supersaturated conditions with respect to $CaCO_3$ minerals, whereas in most parts of the ocean Ω values are <1 below a certain depth (the 'lysocline'). Ω values in the Red Sea range over 4.6–6.7 and 3.1–4.4, for calcite and aragonite, respectively. The lowest deep-water Ω values in the Red Sea are 3.2 and 2.2, for calcite and aragonite, respectively (calculations for this chapter were conducted using the CO_2-SYS program, Pierrot et al. 2006). As a result, $CaCO_3$ minerals do not dissolve in the water column of the Red Sea. While this explains the lack of TA increase with depth, it does not account for the observed decrease with depth. The alkalinity "deficiency" in the deep water can be partially explained by the fact that the Red Sea deep water forms in the northern Red Sea (Plahn et al. 2002) where the nTA is lower due to $CaCO_3$ sequestration (Steiner et al. 2014). However, this deficiency is apparent even in the northern Gulf of Aqaba (Fig. 2.8). The accumulation of NO_3^- and PO_4^{3-} by respiration can decrease nTA (Brewer and Goldman 1976). However, Wurgaft et al. (2016) show that the deep water TA deficiency is significant even if the nTA is corrected to account for the increase in NO_3^- and PO_4^{3-}. Moreover, Wurgaft et al. note that nTA in the deep Red Sea decreases from north to south, and that this horizontal distribution is significant when nTA is corrected for the addition of nutrients. Wurgaft et al. hypothesize that the north-south decrease in deep-water nTA stems from heterogeneous $CaCO_3$ precipitation on sinking particles, mainly dust and biogenic $CaCO_3$ from the photic zone. However, this hypothesis has yet to be thoroughly examined.

Notably, the behavior of the carbonate system parameters over the Red Sea coral-reefs can be very different from that of the adjacent open-water. For example, the carbonate system parameters over coral-reefs are subject to large diurnal cycle, whereas the diurnal variations in the open-sea are negligible. Silverman et al. (2007) show that biological $CaCO_3$ production induces a 20 μmol kg^{-1} difference between the TA in the open-water and the TA in the back lagoon of the coral reef in the northern Gulf of Aqaba. The amplitude of the diurnal TA cycle is similar, mainly because the main mechanism that increases the TA over the coral-reef is water exchange with the open-sea. Notwithstanding, Silverman et al. (2007) find that unlike the open-sea, where $CaCO_3$ minerals are stable, $CaCO_3$ dissolution constitutes an important mechanism in the diurnal nTA cycle over the coral reef.

An important issue that has not been addressed in the scientific literature is the effect that long-term changes in the Red Sea carbonate system may have on coastal reef ecosystems. In particular, the process of 'ocean acidification' (Orr et al. 2005), the global decrease in the sea water pH and CO_3^{2-} levels (Dore et al. 2009; Bates 2007), may have severe implications on the flourishing coral-reefs of the Red Sea. Reduction in pH and CO_3^{2-} levels has been shown to retard $CaCO_3$ precipitation in many calcifying organisms, such as planktonic foraminifera (De Moel et al. 2009; Moy et al. 2009), coralline algae (Kuffner et al 2008) and corals (Cooper et al. 2008; Silverman et al. 2012). It is believed that the slowing of precipitation in corals is due to the sensitivity of the enzymes involved in the $CaCO_3$ precipitation pathway to a decrease in pH or CO_3^{2-} (Erez et al. 2011). The future impact of ocean acidification on the global coralline ecosystem is estimated to be harsh (e.g., Gattuso et al. 1998; Silverman et al. 2009), with a potential threat of a global collapse of this ecosystem (Erez et al. 2011). Given the lack of recent published data on the carbonate system of the Red Sea, the full extent of acidification within the Red Sea and its impact on Red Sea corals are unknown. A process that may locally mitigate acidification is the dissolution of $CaCO_3$ minerals in sediments. Whereas this process is deemed to be an insignificant buffer mechanism on a global scale (Andersson et al. 2003), it may be important in certain areas of the Red Sea due to the abundance of soluble, Mg-rich carbonates in the Red Sea sediments (Luz et al. 1984) and intense vertical mixing (that would increase the interaction between seawater and sediments), as occurs during winter in the northern Red Sea. However, given the dearth of recent carbonate measurements from the Red Sea, the extent to which this process may locally stem acidification is uncertain.

Carbonate chemistry may also be impacted at a local scale by the seasonal intrusion of GAIW. As noted above, GAIW has been observed flowing into the coastal reef system of the southern Red Sea. However, because there a few published measurements of the carbonate properties of GAIW (Morcos 1970; Grasshoff 1969), the impact of GAIW Red Sea carbonate chemistry is unknown.

2.5 Summary and Conclusions

It is evident from the above literature review that ongoing research is giving an increasingly complex view of the Rea Sea system. As a case in point, the impacts of mountain gap winds on the Red Sea were not considered in the scientific literature until roughly 10 years ago. Since that time, modeling studies have demonstrated that winds emerging from mountain gaps onto the Red Sea can have magnitudes exceeding 20 m s^{-1}, cover a significant fraction of the Red Sea surface area, and generate surface waves with significant wave heights of order 4 m. Furthermore, it has been shown that the passage of mountain gap winds over the northern Red Sea during winter results in intensive heat loss to the atmosphere. As such, these winds may be of particular importance in the surface buoyancy loss that leads to the formation of Red Sea Overflow Water. Mountain gap winds have also been identified as a principal (but not the sole) mechanism for generating mesoscale eddies.

Recent studies reviewed above have revealed the dominance of mesoscale eddies and coastal currents in the surface height and near-surface circulation fields. As shown by the analysis of recent cruise data, these circulation features appear to be important in transporting GAIW, the principal source of water-borne nutrients to the Red Sea, over much of the length and breadth of the Red Sea. Furthermore, recent modeling studies have indicated that mesoscale eddies may appreciably influence Red Sea Overflow Water formation and the overturning flow structure of the Red Sea.

As reviewed above, understanding of the light field of the Red Sea environment has been furthered by recent investigations. Modeling has shown that the dust storms that frequent the Red Sea region can significantly attenuate the incoming shortwave radiation, by up to 100 w m^{-2}. Observations in the northern and central Red Sea have quantified the deep penetration of incident light (with euphotic zone depths exceeding 100 m) and shown that coral colonies are capable of adapting to the changes in the distribution of light wavelength with increasing depth.

Other recent advances in the understanding of the Red Sea system reviewed above include (but are not limited to): demonstrating that the seasonal signal of Red Sea surface level is linked to the seasonal variation of the along-axis wind stress, documenting the seasonal and spatial variations of the near-surface chlorophyll distribution, quantifying and modeling the flows and temperature variations over platform reef tops, and estimating the rate of $CaCO_3$ production in the Red Sea and its partition between production of pelagic calcareous plankton and coral-reef growth.

However, despite recent advances, it is clear from the above review that understanding of many critical processes in the Red Sea system is limited. While a full list of such processes would overburden this chapter, a few examples are worth noting. Little is known as to how mesoscale eddies may affect shelf-basin water exchange or promote primary productivity through upward deflection of density/nutrient surfaces. There is still considerable uncertainty regarding the processes that produce the 'excess' nitrogen (relative to phosphorus) observed in the Red Sea. The role of dust deposition in altering the chemical properties (i.e., nutrient concentration and $CaCO_3$ saturation) of the Red Sea has yet to be investigated. Perhaps the most critical process for which further research is warranted is the manner in which the Red Sea carbonate system is responding, and will respond, to long-term changes in atmospheric conditions, most notably air temperature and CO_2 concentration.

References

Abdelrahman SM (1997) Seasonal fluctuations of mean sea level at Gizan, Red Sea, 2009. J Coast Res 13:1166–1172

Abdulsalam A, Majambo G (2014) Remote sensing of phytoplankton distribution in the Red Sea and Gulf of Aden. Acta Oceanol Sin 33(9):93–99. https://doi.org/10.1007/s13131-014-0527-1

Aboobacker VM, Shanas PR, Alsaafani MA, Albarakati AMA (2017) Wave energy resource assessment for Red Sea. Renew Energy 114A:46–58

Acker J, Leptoukh G, Shen S, Zhu T, Kempler S (2008) Remotely-sensed chlorophyll a observations of the northern Red Sea indicate seasonal variability and influence of coastal reefs. J Mar Syst 69:191–204

Ahmad F, Sultan SAR, Moammar MO (1989) Monthly variations of net heat flux at the air-sea interface in coastal waters near Jeddah, Red Sea. Atmos Ocean 27(2):406–413. https://doi.org/10.1080/07055900.1989.9649343

Ali EB, Churchill JH, Barthel K, Skjelvan I, Omar AM, de Lange TE, Eltaib E (2018) Seasonal variations of hydrographic parameters off the Sudanese coast of the Red Sea, 2009–2015. Reg Stud Mar Sci 18:1–10. https://doi.org/10.1016/j.rsma.2017.12.004

Almogi-Labin A, Hemleben C, Meischner D, Erlekeuser H (1991) Paleoenvironmental events during the last 13000 years in the Central Red Sea as recorded by pteropoda. Paleoceanography 6:83–98

Almogi-Labin A, Edelman-Furstenberg Y, Hemleben C (2008) Variations in the biodiversity of thecosomatous pteropods during the Late Quarternary as a response to environmental change in the Gulf of Aden – Red Sea – Gulf of Aqaba ecosystem. In: Por FD (ed) Aqaba-Eilat, the improbable Gulf. The Hebrew University Magnes Press, Jerusalem, pp 31–49

Al-Rifaiy IA, Cherif OH (1988) The fossil coral reefs of Al-Aqaba, Jordan. Facies 18:219–230

Andersson AJ, Mackenzie FT, Ver LM (2003) Solution of shallow-water carbonates: an insignificant buffer against rising atmospheric CO_2. Geology 31:513–516

Arz HW, Lamy F, Ganopolski A, Nowaczyk N, Pätzold J (2007) Dominant Northern Hemisphere climate control over millennial-scale glacial sea-level variability. Quat Sci Rev 26:312–321

Bates NR (2007) Interannual variability of the oceanic CO_2 sink in the subtropical gyre of the North Atlantic Ocean over the last 2 decades. J Geophys Res Oceans 112:C9013–C9039

Ben-Avraham Z, Almagor G, Garfunkel Z (1979) Sediments and structure of the gulf of Elat (Aqaba) - Northern Red Sea. Sediment Geol 23:239–267

Bethoux JP (1988) Red Sea geochemical budgets and exchanges with the Indian Ocean. Mar Chem 24:83–92

Biton E, Gildor H, Peltier WR (2008) Red Sea during the last glacial maximum: implications for sea level reconstruction. Paleoceanography 23:PA1214

Biton E, Trommer G, Siccha M, Kucera M, van der Meer MTJ, Schouten S (2010) Sensitivity of Red Sea circulation to monsoonal variability during the Holocene: an integrated data and modeling study. Paleoceanography 25:PA4209

Bohannon RG (1986) Tectonic configuration of the Western Arabian continental margin, southern Red Sea. Tectonics 5:477–499

Bonatti E (1985) Punctiform initiation of seafloor spreading in the Red Sea during transition from a continental to an oceanic rift. Nature 316:33–37

Bosworth W, Huchon P, McClay K (2005) The Red Sea and Gulf of Aden basins. J Afr Earth Sci 43:334–378

Bower AS, Farrar JT (2015) Air–Sea interaction and horizontal circulation in the Red Sea. In: Rasul NMA, Stewart ICF (eds) The Red Sea, Springer earth system sciences. Springer, Berlin/Heidelberg, pp 349–342. https://doi.org/10.1007/978-3-662-45201-1_19

Braithwaite CJR (1987) Geology and paleogeography of the Red Sea region. In: Edwards AJ, Head SM (eds) Red Sea. Pergamon Press, pp 22–44

Brewer PG, Goldman JC (1976) Alkalinity changes generated by phytoplankton growth. Limnol Oceanogr 21:108–117

Brindley H, Osipov S, Bantges R, Smirnov A, Banks J, Levy R, Jish Prakash P, Stenchikov G (2015) An assessment of the quality of aerosol retrievals over the Red Sea and evaluation of the climatological cloud-free dust direct radiative effect in the region. J Geophys Res Atmos 120:10,862–10,878. https://doi.org/10.1002/2015JD023282

Broecker WS, Takahashi T (1978) Relationship between Lysocline depth and Insitu carbonate ion concentration. Deep-Sea Res 25:65–95

Broecker WS, Peng T-H, Beng Z (1982) Tracers in the Sea. Lamont-Doherty Geological Observatory, Columbia University, Palisades, pp 58–93

Bruckner A, Rowlands G, Riegl B, Purkis S, Williams A, Renaud P (2012) Khaled bin Sultan Living Oceans Foundation Atlas of Saudi Arabian Red Sea marine habitats. Panoramic Press, Phoenix, p 262

Cacchione DA, Drake DE (1982) Measurements of storm-generated bottom stresses on the continental shelf. J Geophys Res 87:1952–1960

Callaghan DP, Nielsen P, Cartwright N, Gourlay MR, Baldock TE (2006) Atoll lagoon flushing forced by waves. Coast Eng 53:691–704. https://doi.org/10.1016/j.coastaleng.2006.02.006

Cember RP (1988) On the sources, formation, and circulation of Red-Sea deep-water. J Geophys Res Oceans 93:8175–8191

Chang GC, Dickey TD, Williams AJ III (2001) Sediment resuspension over a continental shelf during Hurricanes Edouard and Hortense. J Geophys Res 106:9517–9531

Chen C et al (2014) Process modeling studies of physical mechanisms of the formation of an anticyclonic eddy in the Central Red Sea. J Geophys Res Oceans 119:1445–1464. https://doi.org/10.1002/2013JC009351

Churchill JH, Wirick CW, Flagg CN, Pietrafesa LJ (1994) Sediment resuspension over the continental shelf east of the Delmarva Peninsula. Deep-Sea Res Part II 41(2/3):341–363

Churchill JH, Bower AS, McCorkle DC, Abualnaja Y (2014a) The transport of nutrient-rich Indian Ocean water through the Red Sea and into coastal reef systems. J Mar Res 72:165–181. https://doi.org/10.1357/002224014814901994

Churchill JH, Lentz SJ, Farrar JT, Abualnaja Y (2014b) Properties of Red Sea coastal currents. Cont Shelf Res 78:51–61. https://doi.org/10.1016/j.csr.2014.01.025

Clifford M, Horton C, Schmitz J, Kantha LH (1997) An oceanographic nowcast/forecast system for the Red Sea. J Geophys Res 102(25):101–25,122

Cochran JR, Martinez F (1988) Evidence from the northern Red Sea on the transition from continental to oceanic rifting. Tectonophysics 153:25–53

Cooper TF, De'ath G, Fabricius KE, Lough JM (2008) Declining coral calcification in massive Porites in two nearshore regions of the northern Great Barrier Reef. Glob Chang Biol 14:529–538

Davis KA, Lentz SJ, Pineda J, Farrar JT, Starczak VR, Churchill JH (2011) Observations of the thermal environment on Red Sea platform reefs: a heat budget analysis. Coral Reefs 30:25–36. https://doi.org/10.1007/s00338-011-0740-8

Davis SR, Pratt LJ, Jiang H (2015) The Tokar gap jet: regional circulation, diurnal variability, and moisture transport based on numerical simulations. J Clim 28:5885–5907

De Moel H, Ganssen G, Peeters F, Jung S, Kroon D, Brummer G, Zeebe R (2009) Planktonic foraminiferal shell thinning in the Arabian Sea due to anthropogenic ocean acidification. Biogeosciences 6:1917–1925

Defant A (1961) Physical oceanography, volume 2. Pergamon Press, Oxford, p 598

DeVantier L, Tourak E, Al-Shaikh K, De'ath G (2000) Coral communities of the central-northern Saudi Arabian Red Sea. Fauna of Arabia 18:23–66

Dore JE, Lukas R, Sadler DW, Church MJ, Karl DM (2009) Physical and biogeochemical modulation of ocean acidification in the central North Pacific. Proc Natl Acad Sci U S A 106:12235–12240

Dullo WC (2005) Coral growth and reef growth: a brief review. Facies 51:37–52

Dullo WC, Montaggioni L (1998) Modern Red Sea coral reefs: a review of their morphologies and zonation. In: Purser BH, Bosence DWJ (eds) Sedimentation and tectonics in rift basins Red Sea: Gulf of Aden. Springer, Dordrecht, pp 583–594

El-Asmar HM (1997) Quaternary isotope stratigraphy and paleoclimate of coral reef terraces, Gulf of Aqaba, South Sinai, Egypt. Quat Sci Rev 16:911–924

El-Shenawy MA, El-Samra ML (1996) Nitrogen fixation in the northern Red Sea. J. K.A.U Mar. Sci., 7, Special issue, Symposuim on Red Sea marine environment, pp 33–39

Eladawy A, Nadaoka K, Negm A, Abdel-Fattah S, Hanafy M, Shaltout M (2017) Characterization of the northern Red Sea's oceanic features with remote sensing data and outputs from a global circulation model. Oceanologia 59:213–237. https://doi.org/10.1016/j.oceano.2017.01.002

Eltaib EBA (2010) Tides analysis in the Red Sea in Port Sudan and Gizan. M.S. thesis, niversity of Bergen, Geophysical Institute, 61 pp

Emerson S, Hedges J (2008) Chemical oceanography and the marine carbon cycle. Cambridge University Press

Erez J, Reynaud S, Silverman J, Schneider K, Allemand D (2011) Coral calcification under ocean acidification and global change. In: Dubinsky Z, Stambler N (eds) Coral reefs: an ecosystem in transition. Springer, pp 151–176

Falkowski PG, Ziemann D, Kolber Z, Bienfang PK (1991) Role of eddy pumping in enhancing primary production in the ocean. Nature 352:55–58

Felis T, Pätzold J, Loya Y, Fine M, Nawar AH, Wefer G (2000) A coral oxygen isotope record from the northern Red Sea documenting NAO, ENSO, and North Pacific teleconnections on Middle East climate variability since the year 1750. Paleoceanography 15:679–694

Fricke HW, Knauer B (1986) Diversity and spatial pattern of coral communities in the Red Sea upper twilight zone (a quantitative assessment study by submersible). Oecologia 71:29–37

Fricke HW, Vareschi E, Schlichter D (1987) Photoecology of the coral *Leptoseris fragilis* in the Red Sea twilight zone (an experimental study by submersible). Oecologia 73:371–381

Gattuso JP, Frankignoulle M, Bourge I, Romaine S, Buddemeier RW (1998) Effect of calcium carbonate saturation of seawater on coral calcification. Glob Planet Chang 18:37–46

Genin A, Lazar B, Brenner S (1995) Vertical mixing and coral death in the Red Sea following the eruption of Mount Pinatubo. Nature 377:507–510

Grant WD, Madsen OS (1986) The continental-shelf bottom boundary-layer. Annu Rev Fluid Mech 18:265–305

Grant WD, Williams AJ, Glenn SM (1984) Bottom stress estimates and their prediction on the Northern California continental shelf during CODE-1. The importance of wave-current interaction. J Phys Oceanogr 14:506–527

Grasshoff K (1969) Zur Chemie des Roten Meeres und des Inneren Golfs von Aden nach Beobachtungen von FS "Meteor" während der Indischen Ozean Expedition 1964/65. Meteor Forschungsergebn, Reihe A, 6

Grover R, Ferrier-Pagès C, Maguer J-F, Ezzat L, Fine M (2014) Nitrogen fixation in the mucus of Red Sea corals. J Exp Biol 217:3962–3963. https://doi.org/10.1242/jeb.111591

Gvirtzman G (1994) Fluctuations of sea level during the past 400,000 years: the record of Sinai, Egypt (northern Red Sea). Coral Reefs 13:203–214

Gvirtzman G, Kronfeld J, Buchbinder B (1992) Dated coral reefs of southern Sinai (Red Sea) and their implication to late Quaternary Sea levels. Mar Geol 108:29–37

Harriott VJ, Fisk DA (1987) A comparison of settlement plate types for experiments on the recruitment of scleractinian corals. Mar Ecol Prog Ser 37:201–208

Hearn CJ (2010) Hydrodynamics of coral reefs. In: Hopley D (ed) Encyclopedia of modern coral reefs. Springer, Berlin, pp 563–573

Hench JL, Leichter JJ, Monismith SG (2008) Episodic circulation and exchange in a wave-driven coral reef and lagoon system. Limnol Oceanogr 53:2681–2694. https://doi.org/10.4319/lo.2008.53.6.2681

Hickey B, Goudie AS (2007) The use of TOMS and MODIS to identify dust storm source areas: the Tokar delta (Sudan) and the Seistan basin (South West Asia). In: Goudie AS, Kalvoda J (eds) Geomorphological variations. P3K, Prague, pp 37–57

Hoegh-Guldberg O, Mumby PJ, Hooten AJ, Steneck RS, Greenfield P, Gomez E, Harvell CD, Sale PF, Edwards AJ, Caldeira K, Knowlton N, Eakin CM, Iglesias-Prieto R, Muthiga N, Bradbury RH, Dubi A, Hatziolos ME (2007) Coral reefs under rapid climate change and ocean acidification. Science 318:1737–1742

Jarosz E, Murray SP, Inoue M (2005) Observations on the characteristics of tides in the Bab el Mandab Strait. J Geophys Res 110:C03015. https://doi.org/10.1029/2004JC002299.

Jiang H, Farrar JT, Beardsley RC, Chen R, Chen C (2009) Zonal surface wind jets across the Red Sea due to mountain gap forcing along both sides of the Red Sea. Geophys Res Lett 36:L19605

Joffe S, Garfunkel Z (1987) Plate kinematics of the circum Red Sea- a re-evaluation. Tectonophysics 141:5–22

Kalenderski S, Stenchikov G, Zhao C (2013) Modeling a typical wintertime dust event over the Arabian Peninsula and the Red Sea. Atmos Chem Phys 13:1999–2014

Kalenderski S, Stenchikov G (2016) High-resolution regional modeling of summertime transport and impact of African dust over the Red Sea and Arabian Peninsula. J Geophys Res Atmos 121:6435–6458. https://doi.org/10.1002/2015JD024480

Khalil SM, McClay KR (2009) Structural control on syn-rift sedimentation, northwestern Red Sea margin, Egypt. Mar Petrol Geol 26:1018–1034

Khimitsa VA, Bibik VA (1979) Seasonal exchange in dissolved oxygen and phosphate between the Red Sea and the Gulf of Aden. Oceanol 9:544–546

Kiflawi M, Belmaker J, Brokovich E, Einbinder S, Holzman R (2006) The determinants of species richness of a relatively young coral-reef ichthyofauna. J Biogeogr 33:1289–1294

Kleypas JA, McManus JW, Meñez LAB (1999) Environmental limits to coral reef development: where do we draw the line? Am Zool 39:146–159

Klinker J, Reiss Z, Kropach C, Levanon I, Harpaz H, Halicz E, Asaf G (1976) Observations on circulation pattern in Gulf of Elat (Aqaba), Red-Sea. Isr J Earth Sci 25:85–103

Krumgalz B, Erez J, Chen C (1990) Anthropogenic CO_2 penetration in the northern Red Sea and in the Gulf of Elat (Aqaba). Oceanol Acta 13:283–290

Kuffner IB, Andersson AJ, Jokiel PL, Rodgers KS, Mackenzie FT (2008) Decreased abundance of crustose coralline algae due to ocean acidification. Nat Geosci 1:114–117

Langodan S, Cavaleri L, Viswanadhapalli Y, Hoteit I (2014) The Red Sea: a natural laboratory for wind and wave modeling. J Phys Oceanogr 44:3139–3159

Langodan S, Viswanadhapalli Y, Dasari HP, Knio O, Hoteit I (2016) A high resolution assessment of wind and wave energy potentials in the Red Sea. Appl Energy 181:244–255

Langodan S, Cavaleri L, Viswanadhapalli Y, Pomaro A, Bertotti L, Hoteit I (2017a) The climatology of the Red Sea – part 1: the wind. Int J Climatol 37:4509–4517. https://doi.org/10.1002/joc.5103

Langodan S, Cavaleri L, Pomaro A, Viswanadhapalli Y, Bertotti L, Hoteit I (2017b) The climatology of the Red Sea - part 2: the waves. Int J Climatol 37:4518–4528. https://doi.org/10.1002/joc.5101

Lazar B, Erez J, Silverman J, Rivlin T, Rivlin A, Dray M, Meeder E, Iluz D (2008) Recent environmental changes in the chemical-biological oceanography of the Gulf of Aqaba (Eilat). In: Por FD (ed) Aqaba-Eilat, the improbable gulf. Environment, biodiversity and preservation. Magnes Press, Jerusalem

Legge H-L, Mutterlose J, Arz HW, Pätzold J (2008) Nannoplankton successions in the northern Red Sea during the last glaciation (60–14.5 ka BP): reactions to climate change. Earth Planet Sci Lett 270:271–279

Lentz SJ, Churchill JH, Davis KA, Farrar JT (2016a) Surface gravity wave transformation across a platform coral reef in the Red Sea. J Geophys Res Oceans 121:693–705. https://doi.org/10.1002/2015JC011142

Lentz SJ, Churchill JH, Davis KA, Farrar JT, Pineda J, Starczak V (2016b) The characteristics and dynamics of wave-driven flow across a platform coral reef in the Red Sea. J Geophys Res Oceans 121. https://doi.org/10.1002/2015JC011141

Lentz SJ, Davis KA, Churchill JH, DeCarlo TM (2017) Coral reef drag coefficients – water depth dependence. J Phys Oceanogr 47:1061–1075

Lowe RJ, Falter JL, Monismith SG, Atkinson MJ (2009) Wave-driven circulation of a coastal reef–lagoon system. J Phys Oceanogr 39(4):873–893. https://doi.org/10.1175/2008JPO3958.1

Luz B, Heller-Kallai L, Almogi-Labin A (1984) Carbonate mineralogy of Late Pleistocene sediments from the northern Red Sea. Isr J Earth-Sci 33:157–166

Lyne VD, Butman B, Grant WD (1990) Sediment movement along the U.S. east coast continental shelf—I. Estimates of bottom stress using the Grant-Madsen model and near-bottom wave and current measurements. Cont Shelf Res 10:397–428

Madah F, Mayerle R, Bruss G, Bento J (2015) Characteristics of tides in the Red Sea region, a numerical model study. Open J Mar Sci 5:193–209

Madsen OS, Wright LD, Boon JD, Chisholm TA (1993) Wind stress, bed roughness and sediment suspension on the inner shelf during an extreme storm event. Cont Shelf Res 13:1303–1324

Maillard C, Soliman G (1986) Hydrography of the Red Sea and exchanges with the Indian Ocean in summer. Oceanol Acta 9:249–269

Makovsky Y, Wunch A, Ariely R, Shaked Y, Rivlin A, Shemesh A, Avraham ZB, Agnon A (2008) Quaternary transform kinematics constrained by sequence stratigraphy and submerged coastline features: the Gulf of Aqaba. Earth Planet Sci Lett 271:109–122

Manasrah R, Hasanean HM, Al-Rousan S (2009) Spatial and seasonal variations of sea level in the Red Sea, 1958-2001. Ocean Sci J 44:145–159

Mass T, Kline DI, Roopin M, Veal CJ, Cohen S, Iluz D, Levy O (2010) The spectral quality of light is a key driver of photosynthesis and photoadaption in *Stylophora pistillata* colonies from different depths in the Red Sea. J Exp Biol 213:4084–4091. https://doi.org/10.1242/jeb.039891

McDonald CB, Koseff JR, Monismith SG (2006) Effects of the depth to coral height ratio on drag coefficients for unidirectional flow over coral. Limnol Oceanogr 51(3):1294–1301

McGillicuddy DJ et al (2007) Eddy/wind interactions stimulate extraordinary mid-ocean plankton blooms. Science 316:1021–1026

Millero FJ, Lee K, Roche M (1998) Distribution of alkalinity in the surface waters of the major oceans. Mar Chem 60:111–130

Monismith SG (2007) Hydrodynamics of coral reefs. Annu Rev Fluid Mech 39:37–55. https://doi.org/10.1146/annurev.fluid.38.050304.092125

Monismith SG, Herdman LMM, Ahmerkamp S, Hench JL (2013) Wave transformation and wave-driven flow across a steep coral reef. J Phys Oceanogr 43:1356–1379

Montaggioni, L (2000) Postglacial reef growth. Comptes endus De L Academie Des Sciences Serie II Fascicule A-Sciences De La Terre Et Des Planetes, 331, 319–330

Montaggioni L (2005) History of indo-Pacific coral reef systems since the last glaciation: development patterns and controlling factors. Earth Sci Rev 71:1–75

Morcos SA (1970) Physical and chemical oceanography of the Red Sea. Oceanogr Mar Biol Annu Rev 8:73–202

Morse ANC, Iwao K, Baba M, Shimoike K, Hayashibara T, Omori M (1996) An ancient chemosensory mechanism brings new life to coral reefs. Biol Bull 191:149–154

Moy AD, Howard WR, Bray SG, Trull TW (2009) Reduced calcification in modern Southern Ocean planktonic foraminifera. Nat Geosci 2:276–280

Mucci A (1983) The solubility of calcite and aragonite in seawater at various salinities, temperatures and one atmosphere total pressure. Am J Sci 283:780–799

Murray SP, Johns W (1997) Direct observations of seasonal exchange through the Bab el Mandeb Strait. Geophys Res Lett 24:2557–2560

Naqvi SWA, Hansen HP, Kureishy TW (1986) Nutrient uptake and regeneration ratios in the Red Sea with reference to the nutrient budgets. Oceanol Acta 9:271–275

Neo ML, Todd PA, Teo SL-M, Chou LM (2009) Can artificial substrates enriched with crustose coralline algae enhance larval settlement and recruitment in the fluted giant clam (Tridacna squamosa)? Hydrobiologia 625:83–90

Neumann AC, McGill DA (1962) Circulation of the Red Sea in early summer. Deep-Sea Res 8:223–285

Orr JC, Fabry VJ, Aumont O, Bopp L, Doney SC, Feely RA, Gnanadesikan A, Gruber N, Ishida A, Joos F, Key RM, Lindsay K, Maier-Reimer E, Matear R, Monfray P, Mouchet A, Najjar RG, Plattner GK, Rodgers KB, Sabine CL, Sarmiento JL, Schlitzer R, Slater RD, Totterdell IJ, Weirig MF, Yamanaka Y, Yool A (2005) Anthropogenic Ocean acidification over the twenty-first century and its impact on calcifying organisms. Nature 437:681–686. https://doi.org/10.1038/nature04095

Oschlies A, Garçon V (1998) Eddy-induced enhancement of primary production in a model of the North Atlantic Ocean. Nature 394:266–269

Osipov S, Stenchikov G (2018) Simulating the regional impact of dust on the Middle East climate and the Red Sea. J Geophys Res Oceans 123:1023–1047. https://doi.org/. https://doi.org/10.1002/2017JC013335

Papadopoulos VP, Abualnaja Y, Josey SA, Bower A, Raitsos DE, Kontoyiannis H, Hoteit I (2013) Atmospheric forcing of the winter air-sea heat fluxes over the Northern Red Sea. J Clim 26:1685–1701

Papaud A, Poisson A (1986) Distribution of dissolved CO_2 in the Red-Sea and correlations with other geochemical tracers. J Mar Res 44:385–402

Patzert WC (1974) Wind-induced reversal in Red Sea circulation. Deep-Sea Res 21:109–121

Phillips OM (1966) On turbulent convection currents and the circulation of the Red Sea. Deep Sea Res 13:1149–1160

Pierrot D, Lewis E, Wallace D (2006) CO2sys DOS program developed for CO2 system calculations. ORNL/CDIAC-105. Carbon dioxide information analysis center, oak Ridge National Laboratory. US Department of Energy, Oak Ridge

Pineda J, Starczak V, Tarrant A, Blythe J, Davis K, Farrar T, Beruman M, da Silva JCB (2013) Two spatial scales in a bleaching event: corals from the mildest and the most extreme thermal environments escape mortality. Limnol Oceanogr 58:1531–1545. https://doi.org/10.4319/lo.2013.58.5.153

Plahn O, Baschek B, Badewien TH, Walter M, Rhein M (2002) Importance of the Gulf of Aqaba for the formation of bottom water in the Red Sea. J Geophys Res Oceans 107:22-1–22-18

Poisson A, Morcos S, Souvermezoglou E, Papaud A, Ivanoff A (1984) Some aspects of biogeochemical cycles in the Red Sea with special references to new observations made in summer 1982. Deep-Sea Res 31:707–718

Prakash PJ, Stenchikov G, Kalenderski S, Osipov S, Bangalath H (2015) The impact of dust storms on the Arabian Peninsula and the Red Sea. Atmos Chem Phys 15:199–222

Prospero JM, Ginoux P, Torres O, Nicholson SE, Gill TE (2002) Environmental characterization of global sources of atmospheric soil dust identified with the Nimbus 7 Total Ozone Mapping Spectrometer (TOMS) absorbing aerosol product. Rev Geophys 40:2-1–2-31

Quadfasel D, Baunder H (1993) Gyre-scale circulation cells in the Red Sea. Oceanol Acta 16:221–229

Qurban MA, Balala AC, Kumar S, Bhavya PS, Wafar M (2014) Primary production in the Northern Red Sea. J Mar Syst 132:75–82

Raitsos DE, Hoteit I, Prihartato PK, Chronis T, Triantafyllou G, Abualnaja Y (2011) Abrupt warming of the Red Sea. Geophys Res Lett 38:L14601

Raitsos DE, Pradhan Y, Brewin RJW, Stenchikov G, Hoteit I (2013) Remote sensing the phytoplankton seasonal succession of the Red Sea. PLoS One 8:e64909. https://doi.org/10.1371/journal.pone.0064909

Ralston DK, Jiang H, Farrar JT (2013) Waves in the Red Sea: response to monsoonal and mountain gap winds. Cont Shelf Res 65:1–13

Roder C, Berumen ML, Bouwmeester J, Papathanassiou E, Al-Suwailem A, Voolstra CR (2013) First biological measurements of deep-sea corals from the Red Sea. Sci Rep. https://doi.org/10.1038/srep02802

Schlager W (2003) Benthic carbonate factories of the Phanerozoic. Int J Earth Sci 92:445–464

Shaked Y, Agnon A, Lazar B, Marco S, Avner U, Stein M (2004) Large earthquakes kill coral reefs at the north-west gulf of Aqaba. Terra Nova 16:133–138

Shaked Y, Lazar B, Marco S, Stein M, Tchernov D, Agnon A (2005) Evolution of fringing reefs: space and time constraints from the Guilf of Aqaba. Coral Reefs 24:165–172

Shanas PR, Aboobacker VM, Albarakati AMA, Zubier KM (2017a) Superimposed wind-waves in the Red Sea. Ocean Eng 138:9–22. https://doi.org/10.1016/j.oceaneng.2017.04.020

Shanas PR, Aboobacker VM, Albarakati AMA, Zubier KM (2017b) Climate driven variability of wind-waves in the Red Sea. Ocean Model 119:105–117. https://doi.org/10.1016/j.ocemod.2017.10.001

Siddall M, Rohling EJ, Almogi-Labin A, Hemleben C, Meischner D, Schmelzer I, Smeed DA (2003) Sea-level fluctuations during the last glacial cycle. Nature 423:853–858

Silverman J, Lazar B, Erez J (2007) Community metabolism of a coral reef exposed to naturally varying dissolved inorganic nutrient loads. Biogeochemistry 84:67–82

Silverman J, Lazar B, Cao L, Caldeira K, Erez J (2009) Coral reefs may start dissolving when atmospheric CO_2 doubles. Geophys Res Lett 36:L05606–L05611

Silverman J, Lazar B, Erez J (2012) Carbon turnover rates in the one tree island reef: a 40-year perspective. Journal of geophysical research. Biogeosciences 117:G03023–G03039

Sofianos S, Johns W (2001) Wind induced sea level variability in the Red Sea. Geophys Res Lett 28:3175–3178

Sofianos SS, Johns WE (2002) An oceanic general circulation model (OGCM) investigation of the Red Sea circulation, 1. Exchange between the Red Sea and the Indian Ocean. J Geophys Res 107:3196. https://doi.org/10.1029/2001JC001184

Sofianos SS, Johns WE (2003) An oceanic general circulation model (OGCM) investigation of the Red Sea circulation: 2. Three dimensional circulation in the Red Sea. J Geophys Res 108(C3):3066. https://doi.org/10.1029/2001JC001185

Sofianos SS, Johns WE (2007) Observations of the summer Red Sea circulation. J Geophys Res 112:C06025. https://doi.org/10.1029/2006JC003886.

Sofianos SS, Johns WE, Murray SP (2002) Heat and freshwater budgets in the Red Sea from direct observations at Bab el Mandeb. Deep Sea Res Part II 49:1323–1340

Souvermezoglou E, Metzl N, Poisson A (1989) Red Sea budgets of salinity, nutrients, and carbon calculated in the strait of Bab-el-Mandab during the summer and winter season. J Mar Res 47:441–456

Stambler N (2005) Bio-optical properties of the northern Red Sea and the Gulf of Eilat Aqaba – Winter 1999. J Sea Res 54:186–203

Stambler N (2006) Light and picophytoplankton in the Gulf of Eilat (Aqaba). J Geophys Res 111:C11009

Steiner Z, Erez J, Shemesh A, Yam R, Katz A, Lazar B (2014) Basin-scale estimates of pelagic and coral reef calcification in the Red Sea and Western Indian Ocean. Proc Natl Acad Sci U S A 111:16303–16308

Sultan SAR, Ahmad F, Elghribi NM (1995a) Sea level variability in the Central Red Sea. Oceanol Acta 18(6):607–615

Sultan SAR, Ahmad F, El-Hassan A (1995b) Seasonal variations of the sea level in the central part of the Red Sea. Estuar Coast Shelf Sci 40:1–8

Sultan SAR, Ahmad F, Nassar D (1996) Relative contribution of external sources of mean sea-level variations at Port Sudan, Red Sea. Estuar Coast Shelf Sci 42:19–30

Symonds G, Black KP, Young IR (1995) Wave-driven flow over shallow reefs. J Geophys Res 100(C2):2639–2648. https://doi.org/10.1029/94JC02736

Talley LD (2013) Hydrographic Atlas of the World Ocean Circulation Experiment (WOCE): volume 4: Indian Ocean. International WOCE Project Office

Tragou E, Garrett C (1997) The shallow thermohaline circulation of the Red Sea. Deep-Sea Res Part I 44:1355–1376

Tragou E, Garrett C, Outerbridge R, Gilman G (1999) The heat and freshwater budgets of the Red Sea. J Phys Oceanogr 29:2504–2522

Triantafyllou G, Yao F, Petihakis G, Tsiaras KP, Raitsos DE, Hoteit I (2014) Exploring the Red Sea seasonal ecosystem functioning using a three-dimensional biophysical model. J Geophys Res Oceans 119:1791–1811. https://doi.org/10.1002/2013JC009641

Vercelli F (1925) Richerche di oceanografia fisica eseguite della R.N. AMMIRAGILIO MAGNAGHI (1923 – 24), part I, Correnti e maree. Ann Idrog 11:1–188

Vetter O, Becker JM, Merrifield MA, Pequignet A-C, Aucan J, Boc SJ, Pollock CE (2010) Wave setup over a Pacific Island fringing reef. J Geophys Res 115:C12066. https://doi.org/10.1029/2010JC006455

Viswanadhapalli Y, Dasari HP, Langodan S, Challa VS, Hoteit I (2017) Climatic features of the Red Sea from a regional assimilative model. Int J Climatol 37:2563–2581. https://doi.org/10.1002/joc.4865

Wahr J, Smeed DA, Leuliette E, Swenson S (2014) Seasonal variability of the Red Sea, from satellite gravity, radar altimetry, and in situ observations. J Geophys Res Oceans 119:5091–5104. https://doi.org/10.1002/2014JC010161

Weikert H (1987) Plankton and the pelagic environment. In: Edwards AJ, Head SM (eds) Key environments: Red Sea. Pergamon, Oxford, pp 90–111

Weiss R, Broecker W, Craig H, Spencer D (1983) Hydrographic data 1977–1978 GEOSECS Indian Ocean expedition, vol 5. National Science Foundation, Washington, DC

Winters G, Beer S, Zvi BB, Brickner I, Loya Y (2009) Spatial and temporal photoacclimation of *Stylophora pistillata*: zooxanthella size, pigmentation, location and clade. Mar Ecol Prog Ser 384:107–119

Wurgaft E, Steiner Z, Luz B, Lazar B (2016) Evidence for inorganic precipitation of $CaCO_3$ on suspended solids in the open water of the Red Sea. Mar Chem 186:145–155

Yao F, Hoteit I, Pratt LJ, Bower AS, Zhai P, Kohl A, Gopalakrishnan G (2014a) Seasonal overturning circulation in the Red Sea: 1. Model validation and summer circulation. J Geophys Res Oceans 119. https://doi.org/10.1002/2013JC009004

Yao F, Hoteit I, Pratt LJ, Bower AS, Kohl A, Gopalakrishnan G, Rivas D (2014b) Seasonal overturning circulation in the Red Sea: 2. Winter circulation. J Geophys Res Oceans 119. https://doi.org/10.1002/2013JC009331

Yehudai M, Lazar B, Bar N, Kiro Y, Agnon A, Shaked Y, Stein M (2017) U–Th dating of calcite corals from the Gulf of Aqaba. Geochim Cosmochim Acta 198:285–298. https://doi.org/10.1016/j.gca.2016.11.005

Zarokanellos ND, Kurten B, Churchill JH, Roder C, Voolstra CR, Abualnaja Y, Jones BH (2017a) Physical mechanisms routing nutrients in the central Red Sea. J Geophys Res Oceans 122:9032–9046. https://doi.org/10.1002/2017JC013017

Zarokanellos ND, Papadopoulos VP, Sofianos SS, Jones BH (2017b) Physical and biological characteristics of the winter-summer transition in the central Red Sea. J Geophys Res Oceans 122:6355–6370. https://doi.org/10.1002/2017JC012882

Zhai P, Bower AS (2013) The response of the Red Sea to a strong wind jet near the Tokar gap in summer. J Geophys Res Oceans 118:422–434. https://doi.org/10.1029/2012JC008444

Zhan P, Subramanian AC, Yao F, Hoteit I (2014) Eddies in the Red Sea: a statistical and dynamical study. J Geophys Res Oceans 119:3909–3925. https://doi.org/10.1002/2013JC009563

Ecophysiology of Reef-Building Corals in the Red Sea

Maren Ziegler, Anna Roik, Till Röthig, Christian Wild, Nils Rädecker, Jessica Bouwmeester, and Christian R. Voolstra

Abstract

The Red Sea is one of the warmest and most saline seas on the planet. Yet, scleractinian corals have managed to flourish under these distinct conditions supporting one of the largest networks of coral reef ecosystems worldwide. Here, we summarize current knowledge on the ecophysiology of reef-building corals gained from 60 years of research in the Red Sea starting from insights in the 1960s to the most recent studies of the past few years. We provide a brief overview over seasonal dynamics and environmental gradients in the Red Sea that are used to study ecophysiological processes of corals under changing environmental and extreme conditions (i.e., temperature, salinity, nutrient, and light availability). We then focus on how this environmental variability shapes the central processes of coral physiology in the Red Sea covering the topics of photosynthesis, calcification, nutrient cycling, and reproduction. We continue by reporting the first physiological measurements of Red Sea deep-sea corals. Last, we discuss how, through the integration of traditional methods with recent developments in the omics field and model systems, we are now beginning to understand the complexity of processes that contribute to the ecological success of corals under these variable conditions. This synthesis may serve as a basis for future studies that aim to contribute to a better understanding of the impacts of environmental change on coral reefs in the Red Sea and the rest of the world.

Keywords

Metabolism · Photosynthesis · Calcification · Nutrient cycling · Coral reproduction · Phenotypic plasticity · Thermotolerance · Deep-sea corals · Holobiont · Hologenome

3.1 Introduction

Scleractinian, reef-building corals form the structural basis of coral reef ecosystems and a large diversity of organisms rely on the habitat they provide (Reaka-Kudla 1997; Roberts et al.

2002). Corals are metaorganisms or so-called holobionts composed of the coral animal that hosts a diversity of microbial organisms (Rohwer et al. 2002). It is this association with microbial organisms that underlies the success of corals in oligotrophic tropical and subtropical oceans. Separate chapters are dedicated to the most prominent of these associates, specifically the dinoflagellates from the family Symbiodiniaceae (LaJeunesse et al. 2018) (Chap. 5) and the bacterial communities (Chap. 4). Here, we focus on the physiology of the coral holobiont and on how physiological interactions with the symbiotic associates leads to the success of reef corals along the environmental gradients of the Red Sea.

This continuous success of corals along large environmental gradients can be in part attributed to the flexibility in their symbioses with endosymbiont algae, illustrating a remarkable degree of phenotypic plasticity (Auld et al. 2010; DeWitt et al. 1998; Todd 2008). For example, in shallow high-light environments, photosynthetic energy from Symbiodiniaceae drives holobiont productivity and enhances calcification (Muscatine and Porter 1977). In deeper waters with reduced light availability, the Red Sea coral *Stylophora pistillata* may compensate reduced photosynthetic energy production through heterotrophic carbon acquisition (McCloskey and Muscatine 1984). Within the holobiont, Symbiodiniaceae's photosynthetic and the coral's heterotrophic carbon acquisition are linked to the metabolism of the bacterial community in a complex system of nutrient cycling, which is the key to highly productive coral ecosystems in oligotrophic waters (Rädecker et al. 2015).

In this chapter, we summarize current knowledge on coral holobiont physiology in the Red Sea to provide directions for future research areas of relevance in this region. First, we emphasize the Red Sea's long history as a region of coral research and highlight key findings derived from early work on Red Sea corals. After that, we provide a brief overview over the environmental context that is driving ecophysiology of corals in the Red Sea. Then, we focus on different central processes of coral physiology covering the topics of photosynthesis, calcification, nutrient cycling, and reproduction. We continue by describing the physiology of Red Sea deep-sea corals. Last, based on the recent development of omics technologies and the availability of omics data, we provide a perspective for understanding coral holobiont function via the analysis of hologenomes and holotranscriptomes and highlight the value of model system-based research to elucidate function and structural principles of inter-organismal relationships.

3.2 The Red Sea as a Historic Area of Research on Coral Reef Ecosystems and Coral Physiology

There is a wealth of historical data on various aspects of coral reef ecosystems in the Red Sea. Probably one of the first studies of Red Sea biology was the Danish 'Arabia Felix' expedition (1762–1763) collecting plants and animals from Suez to Yemen (Hansen 1964). Between 1799 and 1872 few explorers worked in the region, including the French naturalist Marie J. C. Savigny and the German scientists Christian G. Ehrenberg, Wilhelm F. Hemprich, Eduard Rüppel, and Carl B. Klunzinger who produced a comprehensive coral collection (Head 1987). In the late nineteenth century (1895–1898) the Austrian/Hungarian vessel S.M.S. Pola dredged at 15 stations (212–978 m) for deep-sea corals in the Red Sea (Marenzeller 1907). In 1904, the British zoologist Cyril Crossland started the next Red Sea expedition, during which he produced an important coral collection (Head 1987). The Egyptian University (now Cairo University) established a marine research station in Hurghada, Egypt as early as 1928. The marine station is now called "National Institute of Oceanography and Fisheries, Red Sea branch". A lot of research originating from this station was published in the *Egyptian Journal of Aquatic Research*. However, the research topics focused on fisheries and physical oceanography rather than on coral physiology. In the last decades, other research activities started taking place in Egypt (e.g., since 1990 the Hurghada Environmental Protection and Conservation Association [HEPCA] or since 2003 the Red Sea Environmental Center [RSEC]) that focus on the ecology of mammals and sharks as well as coral reef monitoring.

In the mid-twentieth century, two underwater pioneers and scientists worked in the Red Sea and inspired following generations: The Austrian Hans Hass with his contributions to film ("*Under the Red Sea*", 1951), technical advancement in diving, and behavioural research, and his French counterpart Jacques Cousteau. In 1963, one of Jacques Cousteau's visions became reality. In the western central Red Sea, "Continental Shelf Station 2" (Conshelf 2) became operational and allowed six "oceanauts", some of them researchers, to live and work at a depth of 10 m down at Sha'ab Rumi off the coast of Sudan (Fig. 3.1). This startling activity produced the awarded documentary "*World Without Sun*" (1964) about the attempt to create an environment in which men could live and work on the sea floor.

Further research activities in the Red Sea were triggered in the late 1960s and the early 1970s, mainly by the establishment of two marine research stations at the western and eastern coast of the northern Gulf of Aqaba: The Interuniversity Institute for Marine Sciences (IUI) south of Eilat, Israel, was founded in 1968 and the Marine Science Station (MSS) south of Aqaba, Jordan, in 1975. Early work on coral physiology in the Red Sea thus concentrated in the Gulf of Aqaba, which has an area of ~ 3,600 km^2 and accounts for less than 1% of the entire Red Sea area (~ 438,000 km^2). Despite the comparably small area, important coral physiology-related work has been conducted in this region. Early studies from the Gulf of Aqaba comprise basic aspects of reproduction (Rinkevich and Loya 1979; Shlesinger and Loya 1985), photosynthesis and respiration (McCloskey and

Fig. 3.1 Remainders of Jacques Cousteau's Continental Shelf Station 2 can still be encountered at Sha'ab Rumi in Sudan. The main structure depicted served as a garage for underwater vehicles. Photo credit: Tane Sinclair-Taylor

Muscatine 1984), and zooxanthellae release (Hoegh-Guldberg et al. 1987). Key findings of these studies included the discovery of breeding synchrony, light-dependent metabolic plasticity, and low bleaching susceptibility. Interestingly, most of these studies investigated only one species of hard coral, *Stylophora pistillata*, for which a genome sequence became recently available (Voolstra et al. 2017).

Moving forward to the 1980s, the effects of local stressors started to play a role in Red Sea research activities with the large majority of studies still being conducted in the northern Gulf of Aqaba. At the Jordanian coast, studies identified the negative effects of local land-derived sewage and phosphate pollution (Walker and Ormond 1982). At the Israeli coast, studies identified the negative effects of inorganic nutrients on the photosynthetic efficiency of hard corals (Dubinsky et al. 1990). This triggered an intense scientific discourse about the effect of fish-farm nutrient enrichment in the northern Gulf of Aqaba on the physiology (mainly reproduction) of local corals, as reflected by six consecutive publications between 2003 and 2005 (Bongiorni et al. 2003; Loya and Kramarsky-Winter 2003; Loya et al. 2004, 2005; Rinkevich et al. 2003; Rinkevich 2005). As a result, several key factors for the rapid and severe degradation of the reefs in the Gulf of Aqaba were identified, including fish farms effluents, mass tourism, coastal development, and land-derived pollution.

While research in the Gulf of Aqaba was flourishing, other parts of the Red Sea remained less studied. In 1975, King Abdulaziz University (KAU) in Jeddah on the central eastern coast of the Red Sea opened their Department of Marine Science (now Faculty of Marine Science) and their work is being published in the *Journal of King Abdulaziz University: Marine Sciences*. At the end of the 1980s, occasional studies in the central Red Sea also targeted effects of local stressors, such as metal (Hanna and Muir 1990) and oil pollution (Al Sofyani 1994). Later, the chemical composition of tissue and skeleton of central Red Sea corals was characterized (Al-Lihaibi et al. 1998; Basaham and Al-Sofyani 2007). In the 1990s, an assessment of the ecosystems along the entire Saudi Arabian coast provided a plethora of new data for the Red Sea including central and southern parts (Price et al. 1998). The official inauguration of the Red Sea Research Center at King Abdullah University of Science and Technology (KAUST) at the Red Sea coast of Saudi Arabia in 2011 has extended the possibilities for Red Sea research, as it provides direct access to the coral reefs of the central Red Sea that were particularly understudied before.

Although coral research in the central Red Sea has gained a lot of momentum in the last years, the southern parts remain poorly explored. One of the earliest expeditions to the south was "The Israel South Red Sea Expedition" (Oren 1962), which mainly provided a foundation on the coastal ecology in the south, but also contributed the first physiological data for calcification in scleractinian corals and other reef calcifiers. Only much later in 2010, new data on the coral communities, physiology, and symbiosis for the most southern parts of the Red Sea were made available, as a part of the "Jeddah Red Sea (latitudinal) Transect", a collaboration of KAU University Jeddah (Saudi Arabia) and the Helmholtz-Centre for Ocean Research Kiel (GEOMAR, Germany) (Sawall et al. 2014, 2015).

The effects of global stressors on Red Sea coral physiology have moved into the focus of most recent studies (e.g., Cantin et al. 2010; Sawall et al. 2015; Pogoreutz et al. 2017, 2018; Osman et al. 2018). Overall, there seems to be an obvious trend that scientific publications before 2000 rather concentrated on basic physiological aspects of corals (with a focus of studies on the model coral species *S. pistillata*) and the Red Sea as an oligotrophic and seasonally influenced coral reef environment. The focus of more recent (since 2000) and latest (since 2010) research has shifted towards acclimatization and adaptation mechanisms of Red Sea corals in response to the predicted increase in global stressors, in particular, ocean warming (Fine et al. 2013; Pogoreutz et al. 2017, 2018; Osman et al. 2018).

3.3 Environmental Conditions in the Red Sea That Affect Coral Physiology

Physico-chemical conditions affect the physiological performance of the coral holobiont and determine the limits of coral reef distribution on a global scale (Kleypas et al. 1999). Light and inorganic nutrients are required for efficient photosynthesis (Muscatine and Porter 1977). Ocean currents can supply coral habitats with particulate nutrients and fuel heterotrophy (Alongi et al. 2011). Temperature also plays a central role in shaping coral habitats. Within ambient temperature ranges, higher temperatures enhance metabolic rates (Pörtner 2002) and coral calcification (Marshall and Clode 2004), but temperature extremes may exceed the thermal limits of a coral species and lead to destabilization of the coral holobiont and to coral bleaching, as recently reported from the Red Sea in 2010 and 2015 (Furby et al. 2013; Monroe et al. 2018; Osman et al. 2018; Roik et al. 2015b). Salinity also influences coral holobiont functioning and distribution (Kleypas et al. 1999). Interestingly, high salinity may support thermal tolerance in corals, possibly related to increased production of the reactive oxygen species-scavenging osmolyte floridoside (Gegner et al. 2017; Ochsenkühn et al. 2017). This mechanism could be a contributing factor in the higher thermal tolerance of corals in the more saline northern Red Sea, compared to the corals in the central Red Sea (Osman et al. 2018; Fine et al. 2013).

The Red Sea is one of the warmest, most saline, and oligotrophic seas due to its geographic location in between hot and arid landmasses and due to its isolation from the Indian Ocean (Edwards 1987; Sheppard et al. 1992). The water body maintains a high total alkalinity and aragonite saturation state compared to other tropical reef locations (Kleypas et al. 1999; Silverman et al. 2007b; Steiner et al. 2014) (Table 3.1). Overall, physico-chemical conditions differ from those that corals experience in most tropical reefs (Couce et al. 2012).

Environmental conditions in the Red Sea vary substantially between the northern and southern end and form a latitudinal gradient. Remote sensing data show a transition from a moderately warm (20–26 °C), highly saline (up to 41 PSU), and highly oligotrophic (<0.4 µg L^{-1} chlorophyll a) environment in the north (27–30 °N), to a very warm (28–32 °C), moderately saline (~37 PSU), and comparably nutrient rich environment (up to 4 µg L^{-1} chlorophyll a) in the south (13–17 °N) (Raitsos et al. 2013). Physiological performance of an abundant Red Sea coral, *Pocillopora verrucosa*, growing along this latitudinal gradient, indicates that this species relies on its large phenotypic plasticity rather than genetic adaptations to the different environmental conditions (Sawall et al. 2015). A finding that was further corroborated by population genetic analyses by Robitzch et al. (2015) that revealed absence of genetic differentiation of *P. verrucosa* from the central southern to the central northern Red Sea. Over the past 10 years a comprehensive body of *in situ* environmental data from coral habitats have been collected (Table 3.1). However, continuous monitoring has been limited to the

Table 3.1 *In situ* environmental conditions in coral habitats of the Red Sea. Where available minimum, maximum, and annual mean values are provided

Variable	North (Gulf of Aqaba)	Central	South
Temperature °C	19–31 [24][a]	24–32 [28][a]	26–34 [30][a]
Salinity PSU	40.5–40.9[b]	38.4–39.8 [39.3][c]	38.1–39.1[d]
Total alkalinity µmol kg^{-1}	2462–2484[b]	2315–2459 [2391][e]	–
Dissolved oxygen mg L^{-1}	4.8–9.9[f]	0.1–8.9 [3.5][c]	–
Total nitrogen µmol L^{-1}	3.1–3.3[d]	3.1–4.1[d]	3.5–5.1[d]
Nitrate and nitrite µmol L^{-1}	0.05–1.8 [0.4][b#]	0.1–1 [0.5][c]	–
Total phosphorus µmol L^{-1}	0.1[d]	0.0–0.2[d]	0.2–0.5[d]
Phosphate µmol L^{-1}	0.01–0.13 [0.05][g]	0–0.1 [0.05][c]	–
Chlorophyll a µg L^{-1}	0.1–0.15[d]; 0.05–0.5[g]	0.22–0.75[d]; 0–3.4 [0.4][c]	0.53–3.42[d]

References: (a) (Sawall et al. 2014) Min. – Max. and [Annual average] based on measurements in reef sites, (b) (Silverman et al. 2007b) Min. – Max. of daily averages based on year-long measurements in reef sites, (c) (Roik et al. 2016) Min. – Max. and [Annual average] based on year-long cross-shelf measurements in reef sites, (d) (Kürten et al. 2014) Ranges of measurements from September to October in reef sites, (e) (Roik et al. 2018) Min. – Max. and [Annual average] based on winter and summer cross-shelf measurements in reef sites, (f) (Silverman et al. 2007a) Min. – Max. of daily averages based on year-long measurements in reef sites, (g) (Badran 2001) Min. – Max. and [Annual average] of weekly year-long measurements at 1 and 25 m, 3 km from the reef site, (#) values based on measurements of nitrate only northern and the central part and high resolution *in situ* data are still missing from the south.

The Red Sea is a highly oligotrophic environment with strong seasonal fluctuations in nutrient concentrations and temperature (Raitsos et al. 2013). In the central and southern part, reefs extend along the shelf into the Red Sea, forming offshore barrier reef structures. Cross-shelf gradients as well as a pronounced seasonal variability expose the coral reefs to large environmental changes on spatial and temporal scales (Roik et al. 2016). Year-long *in situ* time series at 7–9 m depth showed that reefs in the central Red Sea are characterized by pronounced temperature and salinity differences between seasons (24–33 °C, 38.4–39.8 PSU). With increasing distance from shore along the cross-shelf gradient dissolved oxygen increases on average, while chlorophyll a, turbidity, and sedimentation decrease (Roik et al. 2016).

Based on environmental data summarized here, Red Sea coral reef habitats are characterized by environmental settings that at least in part reflect predictions of future ocean conditions, such as ocean warming and deoxygenation (Hoegh-Guldberg 1999; Keeling et al. 2010). Not only are reefs exposed to high summer water temperatures that exceed average maxima for coral reefs globally (Kleypas et al. 1999; Osman et al. 2018), but they are also exposed to low levels of dissolved oxygen, which may compromise respiratory processes of reef organisms (Pörtner 2010). In conclusion, investigation of Red Sea corals may reveal distinct physiologies and adaptations in response to these particular environmental conditions and some studies have begun to assess these traits (e.g., Hume et al. 2016; Roder et al. 2015; Roik et al. 2015a; Sawall et al. 2014, 2015; van der Merwe et al. 2014; Röthig et al. 2016b; Ziegler et al. 2014, 2015a, 2016; Bellworthy and Fine 2017; Krueger et al. 2017).

3.4 Red Sea Corals Maintain Efficient Photosynthesis Across Depth and Geographical Gradients

Photosynthetic energy provides the foundation of highly productive coral reef ecosystems. Photosynthates are produced by the dinoflagellate endosymbionts from the family Symbiodiniaceae (Chap. 5) and typically passed on to the coral host in the form of glycerol or amino acids (Markell and Trench 1993) and/or glucose (Burriesci et al. 2012). In the Red Sea, zooxanthellate corals can be found as deep as 145 m (Schlichter et al. 1986). This extends well beyond the common depth limit of around 100 m and can be explained by the transparency of the oligotrophic Red Sea waters, which allows light to penetrate deeper than in other ocean basins (Fig. 3.5A) (Kahng et al. 2014).

Three mechanisms maintain high photosynthetic efficiency over the large light gradients encountered along a coral's vertical depth distribution range. Firstly, light-harvesting (LH) complexes and photoprotective (PP) pigments can acclimatize. Based on observations of Red Sea corals, this optimization of photosynthetic efficiency is mainly achieved through increased incorporation of the LH pigments chlorophyll a, chlorophyll c_2, and peridinin (Nir et al. 2011; Stambler et al. 2008; Ziegler et al. 2015b). At the same time, the concentration of PP pigments such as diadinoxanthin, diatoxanthin, and ß-carotene decreases and leads to higher ratios of LH/PP pigments (Dubinsky and Stambler 2009; Falkowski and Dubinsky 1981), which allows the organism to achieve maximum photosynthetic rates at lower irradiances (Falkowski et al. 1990). The photosynthetic apparatus also seems to adjust to spectral changes of the light, caused by the wavelength-specific absorption in water (Mass et al. 2007). Modulation of the photosynthetic pigments is a dynamic process that can achieve photo-acclimatization to changes in the light regime within a month, as illustrated by two depth transplantation experiments of the corals *S. pistillata* and *P. verrucosa* in the Red Sea (Falkowski and Dubinsky 1981; Ziegler et al. 2014).

Secondly, while cellular pigment content increases, the densities of Symbiodiniaceae cells often decrease with light intensity (e.g., Mass et al. 2007; Ziegler et al. 2015a, 2015b). This can be explained by an increase of self-shading of LH units and algal cells, which leads to decreases in photosynthetic efficiency (McCloskey and Muscatine 1984). Under certain conditions, Symbiodiniaceae cells may turn from a source into a sink of energy, a process that was hypothesized to lead to seasonally recurring bleaching of deep-growing *S. pistillata* in the Gulf of Aqaba (Nir et al. 2014). Generally, decreases of algal cell densities with increasing summer temperatures and irradiance are part of the natural seasonal cycle of most corals (Fagoonee et al. 1999). These cycles are also present in Red Sea corals, such as *P. verrucosa*, both across latitudinal and seasonal scales (Sawall et al. 2014).

Thirdly, Symbiodiniaceae species or types differ in their photophysiological properties and ability to adjust to environmental conditions. Accordingly, to maintain high photosynthetic production rates, the Symbiodiniaceae community composition of a coral species may change in response to depth and also in response to other environmental factors, e.g. over seasons or cross-shelf locations (Iglesias-Prieto et al. 2004; Lampert-Karako et al. 2008; Macdonald et al. 2008; Rowan and Knowlton 1995; Sampayo et al. 2007). An example of a symbiont with a wide photophysiological tolerance can be found in *Symbiodinium microadriaticum* (ITS2 type A1), which is associated with *P. verrucosa* almost across the entire Red Sea (Sawall et al. 2015; Ziegler et al. 2014, 2015a, 2017a). In contrast, many different Symbiodiniaceae lineages with low photophysiological tolerance are associated with *Porites lutea* in the central Red Sea (Ziegler et al. 2015a). Moreover, different host-symbiont combinations can lead to complex physiological patterns. For instance, the same *Cladocopium* type (formerly Clade C) harbored by different coral species in the central Red Sea exhibited different pigment phenotypes (Ziegler et al. 2015b). This finding indicates that the host environment influences symbiont physiology, possibly through different skeletal light reflection (Enriquez et al. 2005; Kahng et al. 2012; Kaniewska et al. 2011; Wangpraseurt et al. 2012).

Enabled by their efficient symbiont photosynthesis and nutrient cycling, corals belong to the most important primary producers on Red Sea reefs. For example, on a reef in the Gulf of Aqaba with high live coral cover, hard corals contributed more to benthic primary production than all other benthic organisms combined at 10 m depth (Hoytema et al. 2016), and also across different reef zones between 0.5–20 m (Cardini et al. 2016). In the northern Red Sea, primary production peaks in spring and summer, which can be explained by the temperature dependence of photosynthesis (Mass et al. 2007). Yet, the latitudinal temperature control on photosynthetic rates in *P. verrucosa* is less clear and may be hampered by association with different Symbiodiniaceae lineages (Sawall et al. 2014, 2015). In summary, complex interactions of (opposing) environmental gradients with changes in Symbiodiniaceae community composition drive the functional plasticity of coral photophysiology in the Red Sea.

3.5 Coral Calcification Rates Peak During Spring Season in the Red Sea

Scleractinian corals are ecosystem engineers and constitute the foundation of reef ecosystems due to their ability to accrete calcium carbonate skeletons that provide the complex structural three dimensional framework, which serves as a habitat for thousands of species. They deposit calcium carbonate at the calicoblastic ectoderm, an extracellular space at the tissue–skeleton interface that is isolated from ambient seawater (Allemand et al. 2011; Tambutté et al. 2011). Corals control the carbonate chemistry in the calcifying space to favor carbonate precipitation by increasing pH and the concentration of calcium ions (Al-Horani et al. 2003; McCulloch et al. 2012). Energetically, these processes are driven via photosynthesis of Symbiodiniaceae that translocate sugars to their coral host (Burriesci et al. 2012; Kopp et al. 2015; Muscatine 1990), but also via heterotrophic feeding (Houlbrèque and Ferrier-Pagès 2009). Light regimes and nutrients (inorganic and particulate) can influence the rates at which corals accrete their skeletons (Ferrier-Pagès et al. 2000; Gattuso et al. 1999). In addition, saturation of carbonate ions in seawater and temperature are both major factors that control coral calcification (Clausen and Roth 1975; Schneider and Erez 2006). Higher calcification rates at warmer, lower latitudes compared to cooler, higher latitudes suggest that warmer temperatures accelerate calcification rates and support coral growth (Carricart-Ganivet 2004; Lough and Barnes 2000). However, temperatures that exceed a critical thermal limit cause a decline in calcification rates (Marshall and Clode 2004), because thermal stress disturbs the coral-algae symbiosis, thereby cutting off an important energy supply (Weis 2008).

Coral calcification in the Red Sea is of great interest because of the unique environmental conditions in this water basin (Kleypas et al. 1999). Most notable are the high sea surface temperature averages and maxima and a high saturation state of the carbonate ion that contribute to coral calcification (Kleypas et al. 1999, 2008; Silverman et al. 2007b). These conditions are known to be favorable for coral growth, but a global comparison of the major reef-building coral genera *Porites*, *Acropora*, and *Pocillopora* showed that annual average calcification rates from the central Red Sea (1–1.5 mg cm^{-2} day^{-1}) were not higher than in other coral reef locations (Roik et al. 2015a; Fig. 3.2).

Coral calcification maxima in the Red Sea indicate that summer temperatures in the central and southern parts of the Red Sea exceed the thermal optima of corals and rather slow down calcification rates. Globally, calcification maxima typically occur during the warmer summer months, when local seawater temperatures meet the thermal optimum of local calcifiers (Crossland 1984; Hibino and van Woesik 2000; Kuffner et al. 2013). In contrast, calcification maxima for the coral *P. verrucosa*, growing along the latitudinal gradient of the Red Sea, occurred at different times of the year – during summer in the northern and during winter in the southern Red Sea (Sawall et al. 2015). The study found that temperature optima for *P. verrucosa* in the Red Sea were at 28.5 °C, irrespective of geographic location, and hence calcification was highest during the season with temperatures around this optimum (Sawall et al. 2015). Complementary to these findings, growth of various coral species (*S. pistillata*, *Pocillopora damicornis*, *Acropora granulosa*) from the north (Sinai) was faster during the warm periods (Kotb 2001; Mass et al. 2007) and growth of the three major reef-building genera *Porites*, *Acropora*, and *Pocillopora* in the central Red Sea was highest during the cooler spring season and not in summer (Roik et al. 2015a). Net-calcification rates of entire benthic communities further reflect the patterns that were found for corals, i.e., calcification was highest in summer in the colder northern Red Sea and in spring in the warmer central Red Sea (Bernstein et al. 2016; Silverman et al. 2007b). In the central Red Sea, temperature (as well as pH variation) was negatively correlated with reef growth, while total alkalinity was positively correlated with reef growth (Roik et al. 2018).

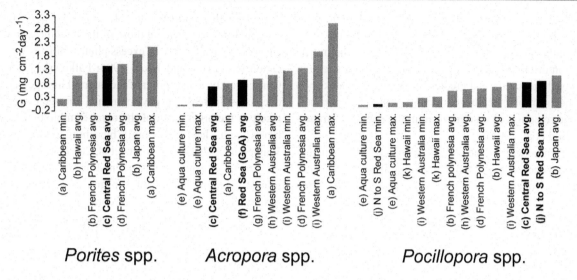

Fig. 3.2 Coral calcification rates from the central Red Sea in a global comparison. GoA = Gulf of Aqaba, N to S = north to south. References are marked with a – k; references and methods used: (*a*) (Goreau and Goreau 1959) Ca^{45}Cl$_2$-incubations, (*b*) (Comeau et al. 2014b) Buoyant weight, (*c*) (Roik et al. 2015a) Buoyant weight, (*d*) (Comeau et al. 2013) Buoyant weight, (**e**) (Schoepf et al. 2013) Buoyant weight, (*f*) (Schneider and Erez 2006) Total alkalinity depletion method, (*g*) (Comeau et al. 2014a) Buoyant weight, (*h*) (Foster et al. 2014) Buoyant weight, (*i*) (Ross et al. 2015) Buoyant weight, (*j*) (Sawall et al. 2015) Total alkalinity depletion method, (*k*) (Clausen and Roth 1975) Ca^{45}Cl$_2$-incubations

Overall, a recent study that determined carbonate budgets of Red Sea coral reefs suggests that offshore reefs show a net positive growth, whereas reefs closer to shore have a net negative growth, i.e. erode (Roik et al. 2018). Importantly, the erosive forces in Red Sea reefs seem less pronounced than elsewhere, yet overall coral reef carbonate budgets were not higher than estimates from reef systems elsewhere (Roik et al. 2018).

In summary, coral reefs in the central and southern Red Sea are exposed to water temperatures in the summer that exceed the calcification optima, and hence, coral calcification in the Red Sea will be particularly susceptible to increases in water temperatures. Interestingly, a recent comparison based on carbonate budgets suggests that overall reef growth has remained constant since 1995 (Roik et al. 2018). Yet, in the massive coral *Diploastrea heliopora*, calcification rates during the last 2 decades were lower than during the previous decades, as inferred from analyses of skeletal coral cores from the central Red Sea (Cantin et al. 2010). This decline coincided with the onset of an abrupt warming in the 1990ies that was demonstrated for the entire basin based on remotely sensed sea surface temperatures (Raitsos et al. 2011). The data available to date suggest that increasing water temperatures already take a toll on coral calcification in the central and southern parts of the Red Sea, and that they may be the ultimate controlling factor of coral growth in the region. The consequences of further warming for Red Sea coral reefs may thus become (even) more apparent in the near future.

3.6 Nutrient Cycling Sustains Coral Holobiont Productivity in the Nutrient-Poor Waters of the Red Sea

Although the Red Sea is highly oligotrophic and nutrient availability exhibits strong seasonality, productivity in coral holobionts is sustained at constant, high rates across large spatio-temporal scales (Cardini et al. 2015; Sawall et al. 2015). The key to understanding this apparent paradox is the tight nutrient-exchange symbiosis between heterotrophic corals and autotrophic Symbiodiniaceae.

The allocation of photosynthetically fixed carbon from Symbiodiniaceae to the coral host underpins the success of this symbiosis in waters where food may be scarce (Muscatine and Porter 1977). *In hospite*, Symbiodiniaceae may transfer up to 95% of the fixed carbon to the coral, and therefore these algae constitute the major energy source of the coral holobiont (Falkowski et al. 1984). The release of photosynthates appears to be induced by the coral via specific compounds in the coral tissue, so-called 'host release factors' (Cook and Davy 2001; Gates et al. 1995). The released carbon metabolites (e.g., glucose) however contain no to little nitrogen and therefore require further supplementation with nitrogen to be incorporated into the host anabolism (Burriesci et al. 2012). When nutrients such as nitrogen are limiting growth, corals store large fractions of the translocated carbon in the form of proteins, lipids, or fatty acids (Harland et al. 1993; Tolosa et al. 2011). Further, corals may release up to

50% of their excess carbon in the form of coral mucus, as recently shown in several Red Sea corals (Haas and Wild 2010; Naumann et al. 2012; Wild et al. 2010). This mucus production offers protection against environmental stressors such as sedimentation and creates a positive feedback loop within heterotrophic benthic communities thereby contributing to the retention of nutrients within the coral reef framework (Wild et al. 2004).

Photosynthetically fixed carbon alone may be insufficient to meet the energetic requirements of corals under low light conditions. Nonetheless, symbiotic corals in the Red Sea flourish even at great water depths and throughout strong seasonal variations in irradiance (Hoytema et al. 2016; Kahng et al. 2014). This can be attributed to a strong trophic plasticity in the coral-Symbiodiniaceae symbiosis, allowing corals to compensate for the reduced availability of photosynthates by heterotrophic feeding to supplement carbon acquisition (Grottoli et al. 2006; McCloskey and Muscatine 1984). The degree of plasticity depends on the heterotrophic capacity of the coral host. Whilst some coral species may be able to fulfil their energetic requirements by heterotrophic feeding alone, other species are highly dependent on photosynthetically fixed carbon (Houlbrèque and Ferrier-Pagès 2009; Ziegler et al. 2014). In the Red Sea coral *S. pistillata* for example, photosynthates alone are sufficient for host energetic requirements at a depth of 3 m and account for 75% of host carbon requirements at a depth of 35 m (McCloskey and Muscatine 1984).

Carbon cycling in corals is tightly linked to the uptake and cycling of nitrogen, a major nutrient that is limiting primary production on coral reefs (Hatcher 1990; Rädecker et al. 2015). A constant nitrogen limitation of Symbiodiniaceae is key to the persistence of the coral–alga symbiosis (Falkowski et al. 1993). This nutrient-limited state is imperative for the adjustment of cell division rates of Symbiodiniaceae to equal those of the coral host and further ensures the availability of excess photosynthates for translocation (Dubinsky and Jokiel 1994; Falkowski et al. 1993). Consequently, a disruption of the internal nitrogen limitation due to an increase in environmental nitrogen supply may result in the breakdown of the coral-Symbiodiniaceae symbiosis and lead to coral bleaching (D'Angelo and Wiedenmann 2014; Pogoreutz et al. 2017).

Nonetheless, nitrogen is essential for sustaining the nutritional requirements of primary productivity in Symbiodiniaceae. Whilst the coral holobiont exhibits a remarkable efficiency of recycling and retaining nitrogen, additional uptake is required to sustain coral net growth and productivity (Rädecker et al. 2015). In this context, heterotrophic feeding, the uptake of inorganic nitrogen from seawater, and microbial dinitrogen (N_2) fixation are the major sources of nitrogen for the coral holobiont. Even though both, the coral host and Symbiodiniaceae, have the cellular machinery to efficiently assimilate and incorporate ammonium (Lin et al. 2015; Pernice et al. 2012), the majority of inorganic nitrogen acquisition from seawater is attributed to the algal symbionts (Grover et al. 2002, 2003). Considering the highly oligotrophic conditions of the Red Sea however, the uptake of inorganic nitrogen alone may not suffice to sustain productivity, and thus has to be supplemented with nutrients from heterotrophic feeding (Johannes et al. 1970). Particularly, Cardini et al. (2015) showed the importance of N_2 fixing bacteria as an additional source of nitrogen during times of lowest nutrient availability in the summer months in the Red Sea, which may provide up to 11% of the nitrogen needed for the metabolic requirements of Symbiodiniaceae (Fig. 3.3).

Fig. 3.3 **Seasonality of nitrogen cycling in Red Sea corals**. Nitrogen (N) cycling pathways in Red Sea coral holobionts during (**a**) winter months and (**b**) summer months. Bold arrows indicate a relative increase of a process, dashed arrows indicate a relative decrease of a process. Adapted after (Cardini et al. 2015; Rädecker et al. 2015)

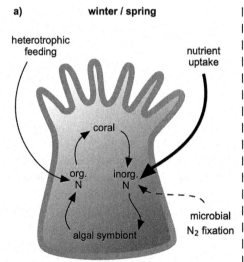
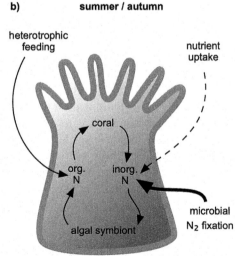

3.7 Coral Reproduction Follows Latitudinal Water Temperature Regimes in the Red Sea

Coral reproduction is a key process for the maintenance of coral assemblages and for their recovery after disturbances (Gilmour et al. 2013; Glynn et al. 2015). Reef-building corals use a diverse range of reproductive strategies, but most species are hermaphroditic broadcast spawners with external fertilization, releasing male and female gametes in the water column once a year in tightly synchronous spawning events (Baird et al. 2009; Guest et al. 2012; Harrison and Wallace 1990). The timing of synchronous spawning is vital as environmental conditions need to be suitable for successful fertilization, but also for subsequent coral larval settlement and metamorphosis (Sorek et al. 2014). Western Atlantic corals mostly spawn in summer when the average monthly sea water temperature is at its maximum (Bastidas et al. 2005; Van Woesik et al. 2006), while Indo-Pacific corals mostly spawn in spring when winter sea water temperatures increase rapidly, but before maximum temperatures are reached (Babcock et al. 1986; Harrison and Wallace 1990; Keith et al. 2016; Willis et al. 1985). Broadcast spawning generally occurs in the hours following sunset, on the nights or within a week of full moon (Kaniewska et al. 2015; Sweeney et al. 2011). Another common reproductive mode is brooding, in which sperm is released in the water column but fertilization is internal, and coral larvae are brooded within the coral polyps before being released (Baird et al. 2009; Harrison and Wallace 1990). Unlike spawning coral species, which usually display a single gametogenesis cycle per year, brooding coral species frequently exhibit multiple cycles within a single year, resulting in extended breeding periods during which motile larvae are released over several consecutive months often also following a lunar periodicity (Harrison and Wallace 1990; Villanueva et al. 2008).

In the Red Sea (excluding the Gulf of Aqaba and Gulf of Suez) at least 99 scleractinian broadcast spawning coral species release their gametes together in spring on nights of or around full moon, with up to 13 species spawning together within a single night (Bouwmeester et al. 2011b, 2015, 2016; Gladstone 1996; Hanafy et al. 2010; Sawall and Al-Sofyani 2015). A secondary spawning season has also been detected in the central Red Sea in the fall involving nine species that do not spawn in spring and for which little or no data on the timing of spawning are available in other regions of the world (Bouwmeester et al. 2015, 2016). Spawning in the fall has been observed in other locations and is even predominant in Western Australia, where recent work revealed that the season of spawning is primarily genetically determined in the *Acropora* assemblage, with environmental conditions controlling the month of spawning within the reproductive season (Gilmour et al. 2016).

Timing of spawning across coral taxa is comparable between the central and north-western Red Sea and appears to start a month earlier in the Farasan Islands, southern Red Sea (Table 3.2). For example, *Acropora* species spawn in April to May in the northern and central Red Sea (Fig. 3.4) (Bouwmeester et al. 2015; Hanafy et al. 2010), and in March to April in the Farasan Islands (Bouwmeester, unpublished data), although a month-difference could also reflect annual variations due to the full moon occurring earlier or later within a calendar month in some years (Baird et al. 2009). In the Gulf of Aqaba, northern Red Sea, temperature regimes are very different from the main Red Sea basin, with sea surface temperatures 4–5 °C colder than in the central Red Sea. As a result, corals in the Gulf of Aqaba predominantly spawn in summer, with *Acropora* species spawning in June to July (Bouwmeester and Berumen 2015; Eyal-Shaham et al. 2016; Kramarsky-Winter and Loya 1998; Shlesinger et al. 1998; Shlesinger and Loya 1985), with a two-months offset in spawning compared to the central Red Sea (Bouwmeester and Berumen 2015).

While the majority of reef-building corals in the Red Sea spawn around the full moon period, 2 species, *Platygyra lamellina* and *Pocillopora verrucosa*, spawn around the new moon period. Both species spawn in July – August in the Gulf of Aqaba (Shlesinger and Loya 1985), while *P. verrucosa* spawns in the morning of new moon in May in the central Red Sea (Bouwmeester et al. 2011a; Fadlallah 1985), confirming the two-months offset in spawning of other coral species between the two regions (Table 3.2). Reproductive patterns in the region appear to be driven by water temperatures, with the rapid increase of sea surface temperature in spring acting as a cue to synchronize broadcast spawning in the Red Sea, while in the Gulf of Aqaba, the colder regional temperatures are avoided and spawning is offset until summer.

Brooding coral species in the Red Sea are few but they are abundant in shallow reef environments. *Stylophora pistillata*, one of the most common species found on Red Sea reef flats and reef crests, releases coral larvae from December/January to June in the Gulf of Aqaba (Rinkevich and Loya 1979; Shlesinger and Loya 1985), from March to June in the north western Red Sea (Mohamed et al. 2007), and from April to June and September to November in the central Red Sea (Fadlallah and Lindo 1988). In all three regions, the breeding period lasts several months, although gametogenetic cycles are shorter in the Red Sea than in the Gulf of Aqaba, and two peaks of planulation are observed (from April to June and from September to November) in the central Red Sea, avoiding the warm months of July and August as observed for spawning species (Fadlallah and Lindo 1988) (Table 3.2).

Table 3.2 Spawning times of scleractinian corals in different regions of the Red Sea. A: Gulf of Aqaba (Eilat, Dahab); N: Northern Red Sea (Hurghada, Safaga, Marsa Alam, Al Wajh, Yanbu); C: Central Red Sea (Thuwal, Jeddah); S: Southern Red Sea (Farasan Islands)

	Region				Month												Reference
	A	N	C	S	J	F	M	A	M	J	J	A	S	O	N	D	
SPAWNING CORAL SPECIES																	
DENDROPHYLLIIDAE																	
Turbinaria stellulata			X							■							a
PORITIDAE																	
Goniopora columna			X						■	■							a
G. savignyi	X											▨					b
Porites columnaris			X						■								a
P. lobata			X						■								a
P. lutea	X		X						■		▨						a, b
P. monticulosa			X											■			a
P. nodifera			X						■	■							a
P. solida			X						■	■							a
EUPHYLLIIDAE																	
Galaxea astreata			X						■	■							c
G. fascicularis	X		X							■	■	▨					a, d
Gyrosmilia interrupta			X						■	■							c
ACROPORIDAE																	
Acropora abrotanoides		X						■									e
A. aculeus	X								▨								f
A. anthocercis			X					■									a
A. arabensis		X						■									e
A. clathrata		X						■									e
A. cytherea	X	X						■	▨								e, f
A. digitifera	X	X						■									e, f
A. eurystoma	X		X						■								a, d, f
A. downingi			X					■									a
A. gemmifera	X	X	X					■	■								a, e, f
A. haimei		X							■								e
A. hemprichii	X		X					■									a, b
A. humilis	X	X	X					■		■							a, c-g
A. hyacinthus	X	X	X					■	■								a, d, e
A. lamarcki			X					■									a
A. latistella		X						■									e
A. loripes			X					■									a
A. lutkeni	X		X					■									a, f
A. maryae		X	X					■									a, e
A. microclados	X	X	X					■	■								a, e, f
A. monticulosa	X							■									f
A. muricata			X					■									e
A. nasuta	X	X	X					■	■								c, e, f
A. ocellata		X						■									e
A. parapharaonis			X					■									a
A. pharaonis	X	X	X					■	■								a, c-e
A. plantaginea			X					■									a
A. polystoma	X		X					■									a, f
A. samoensis	X	X	X					■	■								a, e, f
A. secale			X					■									a
A. selago			X					■									a
A. spicifera		X						■									e
A. squarrosa		X	X					■									a, e
A. tenuis		X						■									e
A. valida	X	X	X					■	■								a-c, e, f

(continued)

Table 3.2 (continued)

Species															Ref
A. variolosa	X	X													a, f
A. verweyi		X													a
Alveopora allingi	X														h
A. ocellata	X														h
Astreopora myriophthalma	X	X													a, d
Montipora cocosensis		X													c
M. efflorescens		X													a
M. informis		X													c
M. erythraea	X														b
M. tuberculosa		X													a, c
M. turgescens		X													a
AGARICIIDAE															
Pachyseris inattesa		X													c
Pavona varians	X	X													a, d
POCILLOPORIDAE															
Pocillopora verrucosa	X														d
P. verrucosa		X													a, i-k
FUNGIIDAE															
Lobactis scutaria	X														l
Pleuractis granulosa	X														l
COSCINAREIDAE															
Coscinaraea monile		X													a
Craterastrea levis															c
PSAMMOCORIDAE															
Psammocora haimiana		X													c
P. profundacella		X													c
INCERTAE SEDIS															
Blastomussa loyae		X													a, c
B. merleti															c
DIPLOASTREIDAE															
Diploastrea heliopora		X													a
MERULINIDAE															
Astrea cf curta		X													a
Cyphastrea chalcidicum		X													a
C. kausti		X													a
C. microphthalma	X	X													a, b
C. serailia		X													a
Dipsastraea albida		X													a
D. favus	X														d
D. helianthoides		X													a
D. maritima		X													a
D. matthaii		X													a
D. rotundata		X													a
D. speciosa		X													a, c
D. veroni		X													a
Echinopora forskaliana		X													a
E. gemmacea		X													a, k
E. hirsutissima		X													a
Favites abdita		X													a
F. paraflexuosa		X													a
F. pentagona	X	X													a, b
F. spinosa		X													a
Goniastrea aspera		X													a
G. edwardsi		X													a
G. pectinata		X													a

(continued)

Table 3.2 (continued)

Species																	References
G. retiformis	X	X	X	X													a, d, k
G. stelligera		X															a
Hydnophora microconos		X															a
Leptoria phrygia		X															a
Merulina scheeri		X															c
Mycedium elephantotus		X															c
M. umbra		X															c
Oulophyllia bennettae		X															a
O. crispa		X															a
Platygyra acuta		X															a
P. crosslandi		X															a
P. lamellina	X	X	X	X													c, d, k
P. sinensis		X															a, c
LOBOPHYLLIIDAE																	
Acanthastrea brevis		X															a
A. echinata	X																b
Cynarina lacrymalis	X																b
Echinophyllia aspera		X															a, c
Lobophyllia corymbosa		X															a, k
L. hemprichii	X																b
Oxypora crassipinosa		X															a
Symphyllia erythraea		X															a
BROODING CORAL SPECIES																	
ACROPORIDAE																	
Alveopora daedalea	X																d
POCILLOPORIDAE																	
Pocillopora damicornis		X															k
Seriatopora caliendrum	X																d
S. hystrix		X															k
Stylophora pistillata	X	X	X	X													d, g, k, m, n

References: (a) (Bouwmeester et al. 2015), (b) (Shlesinger et al. 1998), (c) (Bouwmeester et al. 2016), (d) (Shlesinger and Loya 1985), (e) (Hanafy et al. 2010), (f) (Bouwmeester and Berumen 2015), (g) (Mohamed et al. 2007), (h) (Eyal-Shaham et al. 2016), (i) (Fadlallah 1985), (j) (Bouwmeester et al. 2011a), (k) (Sawall and Al-Sofyani 2015), (l) (Kramarsky-Winter and Loya 1998), (m) (Zakai et al. 2006), (n) (Fadlallah and Lindo 1988)

3.8 Deep-Sea Corals in the Red Sea Illustrate Remarkable Physiological Plasticity of Azooxanthellate Corals

Similar to shallow water corals, deep-sea corals are considered ecosystem engineers that form colonies and some species are even able to form reef structures (Roberts et al. 2006). Deep-sea scleractinian corals also accrete calcium carbonate skeletons; however, unlike their shallow counterparts, they are azooxanthallate and therefore exclusively depend on heterotrophic feeding to satisfy their energetic requirements.

Early in the twentieth century skeletons from at least six species of scleractinian deep-sea corals from the Red Sea were collected, including *Balanophyllia rediviva*, *Dasmosmilia valida*, *Javania insignis*, *Madracis interjecta*, *Rhizotrochus typus*, and *Trochocyathus virgatus* (Marenzeller 1907). Only recently however, live specimens of deep-sea corals from the Red Sea were collected, biological measurements were recorded, and their habitats characterized (Roder et al. 2013). Deep-sea coral habitats were characterized in the central and northern Red Sea at depths between 230 and 740 m. The corals were mostly found at sites of prominent topography, either on seamounts or slopes, attached to rocky substrates, overhangs, or on thin sediment layers covering rocks. During a cruise in late 2011, Roder et al. (2013) observed six species of scleractinian deep-sea corals: *Eguchipsammia fistula* (Fig. 3.5B), *Rhizotrochus typus* (Fig. 3.5D), *Dendrophyllia* sp., one undetermined species of Caryophyllidae, and two further undetermined species. The

Fig. 3.4 Broadcast spawning of *Acropora lamarcki* on 16 April 2011 at 22:34 h, two nights before full moon, at Al Fahal Reef, Thuwal, central Red Sea. (Photo credit: Jessica Bouwmeester)

following year, Qurban et al. (2014) observed seven deep-sea coral species: three scleractinians including *E. fistula*, *R. typus*, and *Dasmosmilia valida*, four Alcyonacea (*Acanthogorgia* sp., *Chironephthya* sp., *Pseudopterogorgia* sp., and one unidentified seafan), and one Antipatharia (*Stichopathes* sp.). Taken together, both surveys described a total of seven live scleractinian coral species in the deep Red Sea. Furthermore, considerable dead coral rubble fields were repeatedly observed (Qurban et al. 2014; Roder et al. 2013).

Deep-sea habitat conditions in the Red Sea are remarkably different from those in other regions. While corals in the depths of the Red Sea face water temperatures of >20 °C (Fig. 3.5a), the upper thermal tolerance limit for deep-sea corals from other regions is around 14 °C (Naumann et al. 2014). Further, dissolved oxygen levels were measured at 1–2 mg L^{-1} and are 3–5 times lower than in other deep-sea coral habitats (5.7–10.3 mg L^{-1}) (Davies et al. 2008). Salinity levels of about 40.6 PSU (Fig. 3.5A) are high compared to the previously highest measurements of 38.8 PSU from deep-sea coral habitats in the central Mediterranean (Freiwald et al. 2004). In addition, available nutrition is presumably sparse (Roder et al. 2013). These unusual conditions are believed to be challenging for deep-sea corals and may explain low metabolic rates and tissue reductions in the deep-sea corals of the Red Sea (Qurban et al. 2014; Roder et al. 2013; Thiel 1987). The first physiological measurements of three species of scleractinian Red Sea deep-sea corals (i.e., *E. fistula*, *Dendrophyllia* sp., and an undetermined species of Caryophyllidae) confirmed a rather distinct physiology. Respiration and calcification rates indicated a highly reduced metabolism compared to other (well-studied) deep-sea corals, e.g., *Lophelia pertusa* (Davies et al. 2008; Roder et al. 2013). The reduced tissue growth, low respiration, and low calcification rates are putative compensation mechanisms for the lack of oxygen and nutrients in combination with the prevalent high temperatures (Roder et al. 2013).

Considering the unique physico-chemical conditions in the deep Red Sea it comes as a surprise that all live scleractinian corals from this region identified to the species level are not endemics. *E. fistula* has also been described from the Indo-West Pacific and New Zealand, *R. typus* from the Indo-West Pacific, and *D. valida* from the Indian Ocean (van der Land 2008). Interestingly, aquarium-based rearing of *E. fistula* in markedly different conditions than the corals' natural Red Sea habitat (salinity, temperature, and pH were similar, but oxygen, total alkalinity, aragonite saturation state, and nutrition were much higher, and pressure was lower) led to substantial increased tissue growth accompanied by enhanced polyp proliferation (Fig. 3.5C). This suggests that this species might not be particularly adapted to the Red Sea, but rather prevails at Red Sea conditions due to a remarkably physiological plasticity (Roik et al. 2015c).

Overall, distribution patterns and physiological measurements of scleractinian corals found in the deep Red Sea indicate physiological flexibility of at least some deep-sea corals across a wide range of environmental conditions. Comparative analyses and experiments with species from the Red Sea and from other regions may further reveal the extent of connectivity, physiological plasticity, and adaptive mechanisms of deep-sea corals between regions with contrasting environmental conditions (Röthig et al. 2017; Yum et al. 2017).

3.9 'Symbiomics'–Elucidating Coral Function Using Holobiont Genomics and Model System-Based Approaches

Recent years have brought a changing imperative in life sciences sparked by the revolution of genomic tools to study the molecular setup of organisms (McFall-Ngai et al. 2013). Initially, a reductionist approach was used to understand organismal physiology, i.e., studies were conducted in an isolated, lab-based, and preferably germ-free environment. While this approach was largely successful in deciphering, e.g., the embryonic development of fruit flies or the function

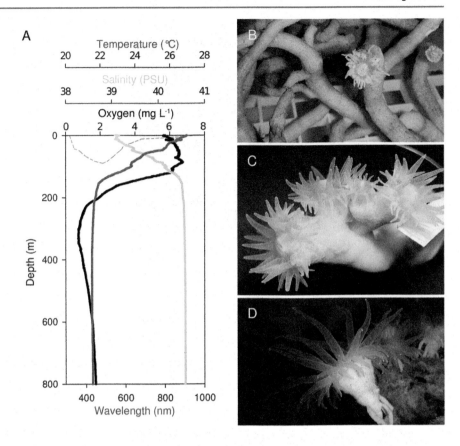

Fig. 3.5 Environmental conditions of deep-sea coral habitats in the Red Sea. (**A**) Depth profiles of temperature (dark gray), salinity (light gray), and oxygen concentrations (black) in the Red Sea. The dashed line indicates the margin of the photic zone at 1% surface radiation per wavelength (modified from Roder et al. (2013); under CCA 4.0 International Public License). (**B**) Freshly collected *Eguchipsammia fistula*, (**C**) *E. fistula* after 1 year of aquarium-based rearing, (**D**) freshly collected *Rhizotrochus typus*. Photo credits: (**B** & **D**) Paul J. Müller, (**C**) Anna Roik

of specific genes in transgenic mice, it failed to recognize the importance of host-associated bacteria (or more generally microbes) to animal and plant function (McFall-Ngai et al. 2013; Rosenberg et al. 2007; Bang et al. 2018). The development of next-generation sequencing changed our understanding of microbial diversity associated with organisms and environments. There is now a multitude of studies that support the notion that a host-specific microbiome co-evolves with the multicellular host and provides functions related to metabolism, immunity, and environmental adaptation, among others (Hooper et al. 2012; Moran and Yun 2015; Tremaroli and Backhed 2012; Ziegler et al. 2017b). Accordingly, it is currently debated whether holobionts constitute a so-called hologenome, positing that evolutionary forces are acting simultaneously on the host and its associated symbiotic microbes as the unit of selection (Rosenberg et al. 2007). The probiotic hypothesis (Reshef et al. 2006) further proposes that a dynamic relationship between hosts and their associated microbes exists that is flexible and adjusts under different prevailing environmental conditions, or in other words, that the environment selects for the most advantageous holobiont composition. What emerges as a consensus from these studies and theories is that animals and plants are holobionts that cannot be understood in isolation, but rather must be understood as a consortium of organisms that constitute their own 'ecosystem'.

While the contribution of the prokaryotic bacterial community to holobiont function has received much attention, the contribution of eukaryotic microbes to holobiont function has largely been overlooked. The study of the coral-alga symbioses is one exception with the description of the first endosymbiotic alga in the genus *Symbiodinium* dating back to 1962 (Freudenthal 1962) and the early recognition of this association of the coral animal with algal symbionts as the foundation of coral reef ecosystems. The coral-Symbiodiniaceae association provides an explicit opportunity and model to study the contribution of eukaryotes to holobiont function and the Red Sea. In a broader context, the Arabian Seas provide the opportunity to study these holobionts in rather extreme environments, which might help decipher responses of corals to global change (Hume et al. 2016). There is now a multitude of studies completed or in progress that conduct large scale surveys on eukaryotic and prokaryotic microbial diversity associated with different coral species across environmental and latitudinal gradients in the Red Sea (Roder et al. 2015; Sawall et al. 2015; Ziegler et al. 2016, 2017a). Describing and characterizing these community assembly patterns represents a first step that can be combined with ecophysiological methods to understand holobiont functioning. For instance, based on observations of Symbiodiniaceae physiology and the Symbiodiniaceae community in *Pocillopora verrucosa* and *Porites lutea* from the central Red Sea, the concept of phenotypic plasticity was extended to include specifics of the endosymbiotic coral-Symbiodiniaceae association. Ziegler et al. (2015a) could show that the coral host species was either associated with a specific *Symbiodiniaceae* type that had a broad physiological tolerance or with several different Symbiodiniaceae genera and types that each had distinct physiological properties with low tolerance that were adapted to the prevailing environmental conditions, effectively illustrating the different roles of distinct host-symbiont associations for holobiont functioning (Ziegler et al. 2015a).

Although sequencing-based diversity surveys are becoming increasingly common, complete holobiont genome-wide interrogation to elucidate gene function is still lacking. Beyond algal and bacterial associates, the coral holobiont consists of many more species of presumably functional significance, such as fungi, archaea, viruses. There are only few studies that consider the entire organismal community, although the genomic tools do allow for this (Daniels et al. 2015). Given the dramatic reduction in cost, we advocate the sequencing of hologenomes to shed light on structure and function of host symbiont relationships (Voolstra et al. 2015). The first 'complete' hologenome comes from the coral *S. pistillata*: genomes of the coral host (Bhattacharya et al. 2016; Voolstra et al. 2017), algal symbiont (Aranda et al. 2016), and the numerically abundant bacterial symbiont in the genus *Endozoicomonas* (Neave et al. 2014, 2017) are available and provide an opportunity to interrogate genomes for footprints of interorganismal relationships. The increasing availability of holobiont genomes promises to provide an unprecedented view on the structure, function, and evolution of eukaryote-eukaryote and eukaryote-prokaryote relationships and will be a key step in decoding interactions of the coral holobiont and of metaorganisms in general, which represent the functional units of multicellular life.

Another important aspect to acknowledge is that corals are not an ideal model system for various reasons: corals have very long generation times (years), are notoriously difficult and expensive to grow in culture, and they cannot be kept without their associated algal symbionts, prohibiting the study of a non-symbiotic 'control' or 'reference' state. A model system that allows for easy manipulation in a laboratory environment is needed to decipher the intricacies of the coral-algal symbiosis. To this end, the sea anemone Aiptasia (*sensu Exaiptasia pallida*) provides a tractable laboratory animal model for investigating the coral-dinoflagellate endosymbiosis (Baumgarten et al. 2015; Voolstra 2013; Weis et al. 2008). A key aspect is Aiptasia's ease of culturing and flexibility in its symbioses (e.g., Aiptasia can host the same symbionts as corals), allowing the comparative analysis of symbiotic and non-symbiotic states side-by-side in a laboratory context. Further, laboratory cultures of their algal symbionts from Florida (strain CC7), Hawaii (strain H2), and the Red Sea (strain not designated yet) allow comparative analyses to elucidate the specific adjustments of these holobionts to different environments. Moreover, Aiptasia's virome (Brüwer and Voolstra 2018) and bacterial microbiome (Röthig et al. 2016) have been characterized and both show structural similarities to those of the coral holobiont. Of note, distinct viral and bacterial communities between symbiotic and aposymbiotic anemones suggest a role of the microbiome in the host-Symbiodiniaceae symbiosis. Culturing of microbial associates and the establishment of axenic host systems may provide further insight into the contribution of microbes to holobiont functioning (Röthig et al. 2016).

The utility of Aiptasia as a model system was demonstrated by the recent assembly and analysis of the Aiptasia genome that revealed multiple features of interest to understanding the evolution and function of the cnidarian-algal endosymbiosis. This included aspects of host dependency on algal-derived nutrients and clues on the composition and function of the host-derived membrane across which such nutrients must pass, a novel and expanded cnidarian-specific family of putative pattern-recognition receptors that might be involved in animal-algal interactions (termed CniFLs for cnidarian ficolin-like proteins), and extensive lineage-specific horizontal gene transfer into both the algal symbiont and the cnidarian host organisms (Baumgarten et al. 2015). In particular, the finding of extensive integration of genes of prokaryotic origin, including genes for antimicrobial peptides, provides evidence for an intimate association of the animal-algal pair with its associated prokaryotic microbiome, and pays tribute to the evolution of this model system in a holobiont context.

We argue that functional insights gained from the tractable Aiptasia model system will allow large advancements in our understanding of the biology of reef-building corals and provide critical guidance for practical applications in coral reef conservation. Further, molecular dissection of the Aiptasia symbiosis has the potential to profoundly transform our understanding of the function and structural principles of interorganismal relationships more generally. For instance, the evolutionary relatedness of Aiptasia symbionts to human pathogens (e.g., *Plasmodium* and *Toxoplasma*) promises to clarify an array of fundamental issues in biology (e.g., cell-cell recognition, immunity, intracellular signalling, cell cycle control), even beyond the scope of coral reef ecosystems.

References

Al-Horani FA, Al-Moghrabi SM, de Beer D (2003) Microsensor study of photosynthesis and calcification in the scleractinian coral, *Galaxea fascicularis*: active internal carbon cycle. J Exp Mar Biol Ecol 288:1–15

Allemand D, Tambutté É, Zoccola D, Tambutté S (2011) Coral calcification, cells to reefs. In: Dubinsky Z, Stambler N (eds) Coral reefs: an ecosystem in transition. Springer, Dordrecht, pp 119–150

Al-Lihaibi SS, Al-Sofyani AA, Niaz GR (1998) Chemical composition of corals in Saudi Red Sea Coast. Oceanol Acta 21:495–501

Alongi DM, Trott LA, Møhl M (2011) Strong tidal currents and labile organic matter stimulate benthic decomposition and carbonate fluxes on the southern Great Barrier Reef shelf. Cont Shelf Res 31:1384–1395

Al-Sofyani A (1994) Variation in Evels and location of lipids in corals tissue of the ROPME Sea area. JKAU Mar Sci 5:121–131

Aranda M, Li Y, Liew YJ, Baumgarten S, Simakov O, Wilson M, Piel J, Ashoor H, Bougouffa S, Bajic VB, Ryu T, Ravasi T, Bayer T, Micklem G, Kim H, Bhak J, LaJeunesse TC, Voolstra CR (2016) Genomes of coral dinoflagellate symbionts highlight evolutionary adaptations conducive to a symbiotic lifestyle. Sci Rep 6:39734

Auld JR, Agrawal AA, Relyea RA (2010) Re-evaluating the costs and limits of adaptive phenotypic plasticity. Proc R Soc Lond B Biol Sci 277:503–511

Babcock RC, Bull GD, Harrison PL, Heyward AJ, Oliver JK, Wallace CC, Willis BL (1986) Synchronous spawnings of 105 scleractinian coral species on the Great Barrier Reef. Mar Biol 90:379–394

Badran MI (2001) Dissolved oxygen, chlorophyll a and nutrients: seasonal cycles in waters of the Gulf of Aquaba, Red Sea. Aquat Ecosyst Health Manage 4:139–150

Baird AH, Guest JR, Willis BL (2009) Systematic and biogeographical patterns in the reproductive biology of scleractinian corals. Annu Rev Ecol Evol Syst 40:551–571

Bang C, Dagan T, Deines P, Dubilier N, Duschl WJ, Fraune S, Hentschel U, Hirt H, Hülter N, Lachnit T, Picazo D, Pita L, Pogoreutz C, Rädecker N, Saad MM, Schmitz RA, Schulenburg H, Voolstra CR, Weiland-Bräuer N, Ziegler M, Bosch TCG (2018) Metaorganisms in extreme environments: do microbes play a role in organismal adaptation? Zoology 127:1–19

Basaham AS, Al-Sofyani AA (2007) Observations based on the variation in alkaline earth elements' (Ca, Mg, Ba and Sr) distribution in the *Porites* Skeleton of the central west coast of Saudi Arabia. JKAU Mar Sci 18:213–223

Bastidas C, Cróquer A, Zubillaga AL, Ramos R, Kortnik V, Weinberger C, Márquez LM (2005) Coral mass- and split-spawning at a coastal and an offshore Venezuelan reefs, southern Caribbean. Hydrobiologia 541:101–106

Baumgarten S, Simakov O, Esherick LY, Liew YJ, Lehnert EM, Michell CT, Li Y, Hambleton EA, Guse A, Oates ME, Gough J, Weis VM, Aranda M, Pringle JR, Voolstra CR (2015) The genome of *Aiptasia*, a sea anemone model for coral symbiosis. Proc Natl Acad Sci 112:11893–11898

Bellworthy J, Fine M (2017) Beyond peak summer temperatures, branching corals in the Gulf of Aqaba are resilient to thermal stress but sensitive to high light. Coral Reefs 36(4):1071–1082

Bernstein WN, Hughen KA, Langdon C, McCorkle DC, Lentz SJ (2016) Environmental controls on daytime net community calcification on a Red Sea reef flat. Coral Reefs 35:697–711

Bhattacharya D, Agrawal S, Aranda M, Baumgarten S, Belcaid M, Drake J, Erwin D, Foret S, Gates RD, Gruber DF, Hanna B, Lesser MP, Levy O, Liew YJ, MacManes M, Mass T, Medina M, Mehr S, Meyer E, Price DC, Putnam HM, Qiu H, Shinzato C, Shoguchi E, Stokes AJ, Tambutte S, Tchernov D, Voolstra CR, Wagner N, Walker CW, Weber APM, Weiss V, Zelzion E, Zoccola D, Falkowski PG (2016) Comparative genomics explains the evolutionary success of reef-forming corals. eLife 5:e13288

Bongiorni L, Shafir S, Angel D, Rinkevich B (2003) Survival, growth and gonad development of two hermatypic corals subjected to in situ fish-farm nutrient enrichment. Mar Ecol Prog Ser 253:137–144

Bouwmeester J, Berumen ML (2015) High reproductive synchrony of *Acropora* (Anthozoa: Scleractinia) in the Gulf of Aqaba, Red Sea. F1000Research 4:2

Bouwmeester J, Berumen ML, Baird AH (2011a) Daytime broadcast spawning of *Pocillopora verrucosa* on coral reefs of the Central Red Sea. Galaxea J Coral Reef Stud 13:23–24

Bouwmeester J, Khalil MT, De La Torre P, Berumen ML (2011b) Synchronous spawning of *Acropora* in the Red Sea. Coral Reefs 30:1011–1011

Bouwmeester J, Baird AH, Chen CJ, Guest JR, Vicentuan KC, Voolstra CR, Berumen ML (2015) Multi-species spawning synchrony within scleractinian coral assemblages in the Red Sea. Coral Reefs 34:65–77. erratum 34:79–79

Bouwmeester J, Gatins R, Giles EC, Sinclair-Taylor TH, Berumen ML (2016) Spawning of coral reef invertebrates and a second spawning season for scleractinian corals in the central Red Sea. Invertebr Biol 135:273–284

Brüwer JD, Voolstra CR (2018) First insight into the viral community of the cnidarian model metaorganism Aiptasia using RNA-Seq data. PeerJ 6:e4449

Burriesci MS, Raab TK, Pringle JR (2012) Evidence that glucose is the major transferred metabolite in dinoflagellate – cnidarian symbiosis. J Exp Biol 215:3467–3477

Cantin NE, Cohen AL, Karnauskas KB, Tarrant AM, McCorkle DC (2010) Ocean warming slows coral growth in the central Red Sea. Science 329:322–325

Cardini U, Bednarz VN, Naumann MS, van Hoytema N, Rix L, Foster RA, Al-Rshaidat MMD, Wild C (2015) Functional significance of dinitrogen fixation in sustaining coral productivity under oligotrophic conditions. Proc R Soc Lond Ser B Biol Sci 282:20152257

Cardini U, Bednarz VN, van Hoytema N, Rovere A, Naumann MS, Al-Rshaidat MMD, Wild C (2016) Budget of primary production and dinitrogen fixation in a highly seasonal Red Sea Coral Reef. Ecosystems 19(5):771–785

Cook CB, Davy SK (2001) Are free amino acids responsible for the 'host factor' effects on symbiotic zooxanthellae in extracts of host tissue? Hydrobiologia 461:71–78

Carricart-Ganivet JP (2004) Sea surface temperature and the growth of the West Atlantic reef-building coral *Montastraea annularis*. J Exp Mar Biol Ecol 302:249–260

Clausen CD, Roth AA (1975) Effect of temperature and temperature adaptation on calcification rate in the hermatypic coral *Pocillopora damicornis*. Mar Biol 33:93–100

Comeau S, Edmunds PJ, Spindel NB, Carpenter RC (2013) The responses of eight coral reef calcifiers to increasing partial pressure of CO_2 do not exhibit a tipping point. Limnol Oceanogr 58:388–398

Comeau S, Carpenter RC, Edmunds PJ (2014a) Effects of irradiance on the response of the coral *Acropora pulchra* and the calcifying alga *Hydrolithon reinboldii* to temperature elevation and ocean acidification. J Exp Mar Biol Ecol 453:28–35

Comeau S, Carpenter RC, Nojiri Y, Putnam HM, Sakai K, Edmunds PJ (2014b) Pacific-wide contrast highlights resistance of reef calcifiers to ocean acidification. Proc R Soc Lond B Biol Sci 281:20141339

Cook CB, Davy SK (2001) Are free amino acids responsible for the 'host factor' effects on symbiotic zooxanthellae in extracts of host tissue? Hydrobiologia 461:71–78

Couce E, Ridgwell A, Hendy EJ (2012) Environmental controls on the global distribution of shallow-water coral reefs. J Biogeogr 39:1508–1523

Crossland CJ (1984) Seasonal variations in the rates of calcification and productivity in the coral *Acropora formosa* on a high-latitude reef. Mar Ecol Prog Ser 15:135–140

D'Angelo C, Wiedenmann J (2014) Impacts of nutrient enrichment on coral reefs: new perspectives and implications for coastal management and reef survival. Curr Opin Environ Sustain 7:82–93

Daniels C, Baumgarten S, Yum LK, Michell CT, Bayer T, Arif C, Roder C, Weil E, Voolstra CR (2015) Metatranscriptome analysis of the reef-building coral *Orbicella faveolata* indicates holobiont response to coral disease. Front Mar Sci 2:62

Davies AJ, Wisshak M, Orr JC, Murray Roberts J (2008) Predicting suitable habitat for the cold-water coral *Lophelia pertusa* (Scleractinia). Deep-Sea Res I Oceanogr Res Pap 55:1048–1062

DeWitt TJ, Sih A, Wilson DS (1998) Costs and limits of phenotypic plasticity. Trends Ecol Evol 13:77–81

Dubinsky Z, Jokiel PL (1994) Ratio of energy and nutrient fluxes regulates symbiosis between zooxanthellae and corals. Pac Sci 48:313–324

Dubinsky Z, Stambler N (2009) Photoacclimation processes in phytoplankton: mechanisms, consequences, and applications. Aquat Microb Ecol 56:163–176

Dubinsky Z, Stambler N, Ben-Zion M, McCloskey LR, Muscatine L, Falkowski PG (1990) The effect of external nutrient resources on the optical properties and photosynthetic efficiency of *Stylophora pistillata*. Proc R Soc Lond Ser B Biol Sci 239:231–246

Edwards AJ (1987) Climate and oceanography. In: Edwards AJ, Head SM (eds) Red Sea. Pergamon Press & International Union for Conservation of Nature and Natural Resources, Oxford, pp 45–69

Enriquez S, Mendez ER, Iglesias-Prieto R (2005) Multiple scattering on coral skeletons enhances light absorption by symbiotic algae. Limnol Oceanogr 50:1025–1032

Eyal-Shaham L, Eyal G, Tamir R, Loya Y (2016) Reproduction, abundance and survivorship of two *Alveopora* spp. in the mesophotic reefs of Eilat, Red Sea. Sci Rep 6:20964

Fadlallah Y (1985) Reproduction in the coral *Pocillopora verrucosa* on the reefs adjacent to the industrial city of Yanbu (Red Sea, Saudi Arabia). Proceedings of the Fifth International Coral Reef Congress, Tahiti, pp 313–318

Fadlallah YH, Lindo RT (1988) Contrasting cycles of reproduction in *Stylophora pistillata* from the Red Sea and the Arabian Gulf, with emphasis on temperature. 6th International Coral Reef Symposium, Australia, pp 225–230

Fagoonee I, Wilson HB, Hassell MP, Turner JR (1999) The dynamics of zooxanthellae populations: a long-term study in the field. Science 283:843–845

Falkowski PG, Dubinsky Z (1981) Light-shade adaptation of *Stylophora pistillata*, a hermatypic coral from the Gulf of Eilat. Nature 289:172–174

Falkowski PG, Dubinsky Z, Muscatine L, Porter JW (1984) Light and the bioenergetics of a symbiotic coral. Bioscience 34:705–709

Falkowski PG, Jokiel PL, Kinzie RA (1990) Irradiance and corals. In: Dubinsky Z (ed) Ecosystems of the world. Coral Reefs. Elsevier, Amsterdam, pp 89–108

Falkowski PG, Dubinsky Z, Muscatine L, McCloskey L (1993) Population control in symbiotic corals. Bioscience 43:606–611

Ferrier-Pagès C, Gattuso JP, Dallot S, Jaubert J (2000) Effect of nutrient enrichment on growth and photosynthesis of the zooxanthellate coral *Stylophora pistillata*. Coral Reefs 19:103–113

Fine M, Gildor H, Genin A, (2013) A coral reef refuge in the Red Sea. Global Change Biology 19 (12):3640–3647

Foster T, Short JA, Falter JL, Ross C, McCulloch MT (2014) Reduced calcification in Western Australian corals during anomalously high summer water temperatures. J Exp Mar Biol Ecol 461:133–143

Freiwald A, Fosså JH, Grehan A, Koslow T, Roberts JM (2004) Cold-water coral reefs. UNEP-WCMC, Cambridge, p 84

Freudenthal HD (1962) *Symbiodinium* gen. nov. and *Symbiodinium microadriaticum* sp. nov., a Zooxanthella: taxonomy, life cycle, and morphology. J Protozool 9:45–52

Furby KA, Bouwmeester J, Berumen ML (2013) Susceptibility of Central Red Sea corals during a major bleaching event. Coral Reefs 32:505–513

Gates RD, Hoegh-Guldberg O, McFall-Ngai MJ, Bil KY, Muscatine L (1995) Free amino acids exhibit anthozoan "host factor" activity: they induce the release of photosynthate from symbiotic dinoflagellates *in vitro*. Proc Natl Acad Sci 92:7430–7434

Gattuso J-P, Allemand D, Frankignoulle M (1999) Photosynthesis and calcification at cellular, organismal and community levels in coral reefs: a review on interactions and control by carbonate chemistry. Am Zool 39:160–183

Gegner HM, Ziegler M, Rädecker N, Buitrago-López C, Aranda M, Voolstra CR (2017) High salinity conveys thermotolerance in the coral model Aiptasia. Biol Open 6:1943–1948

Gilmour JP, Smith LD, Heyward AJ, Baird AH, Pratchett MS (2013) Recovery of an isolated coral reef system following severe disturbance. Science 340:69–71

Gilmour JP, Underwood JN, Howells EJ, Gates E, Heyward AJ (2016) Biannual spawning and temporal reproductive isolation in *Acropora* corals. PLoS One 11:e0150916

Gladstone W (1996) Unique annual aggregation of longnose parrotfish (*Hipposcarus harid*) at Farasan Island (Saudi Arabia, Red Sea). Copeia 1996:483–485

Glynn PW, Riegl B, Purkis S, Kerr JM, Smith TB (2015) Coral reef recovery in the Galápagos Islands: the northernmost islands (Darwin and Wenman). Coral Reefs 34:421–436

Goreau TF, Goreau NI (1959) The physiology of skeleton formation in corals. II. Calcium deposition by hermatypic corals under various conditions in the reef. Biol Bull 117:239–250

Grottoli AG, Rodrigues LJ, Palardy JE (2006) Heterotrophic plasticity and resilience in bleached corals. Nature 440:1186–1189

Grover R, Maguer JF, Reynaud-Vaganay S, Ferrier-Pages C (2002) Uptake of ammonium by the scleractinian coral *Stylophora pistillata*: effect of feeding, light, and ammonium concentrations. Limnol Oceanogr 47:782–790

Grover R, Maguer J-F, Allemand D, Ferrier-Pages C (2003) Nitrate uptake in the scleractinian coral *Stylophora pistillata*. Limnol Oceanogr 48:2266–2274

Guest JR, Baird AH, Goh BPL, Chou LM (2012) Sexual systems in scleractinian corals: an unusual pattern in the reef-building species *Diploastrea heliopora*. Coral Reefs 31:705–713

Haas AF, Wild C (2010) Composition analysis of organic matter released by cosmopolitan coral reef-associated green algae. Aquat Biol 10:131–138

Hanafy MH, Aamer MA, Habib M, Rouphael AB, Baird AH (2010) Synchronous reproduction of corals in the Red Sea. Coral Reefs 29:119–124

Hanna RG, Muir GL (1990) Red Sea corals as biomonitors of trace metal pollution. Environ Monit Assess 14:211–222

Hansen T (1964) Arabia Felix. The Danish expedition of 1761–1767. Collins, London

Harland AD, Navarro JC, Spencer Davies P, Fixter LM (1993) Lipids of some Caribbean and Red Sea corals: total lipid, wax esters, triglycerides and fatty acids. Mar Biol 117:113–117

Harrison P, Wallace C (1990) Reproduction, dispersal and recruitment of scleractinian corals. In: Dubinsky Z (ed) Ecosystems of the world. Elsevier, Amsterdam, pp 133–207

Hatcher BG (1990) Coral reef primary productivity: a hierarchy of pattern and process. Trends Ecol Evol 5:149–155

Head SM (1987) Introduction. In: Edwards AJ, Head SM (eds) Red Sea. Pergamon Press & International Union for Conservation of Nature and Natural Resources, Oxford, pp 1–21

Hibino K, van Woesik R (2000) Spatial differences and seasonal changes of net carbonate accumulation on some coral reefs of the Ryukyu Islands, Japan. J Exp Mar Biol Ecol 252:1–14

Hoegh-Guldberg O (1999) Climate change, coral bleaching and the future of the world's coral reefs. Mar Freshw Res 50:839–866

Hoegh-Guldberg O, McCloskey LR, Muscatine L (1987) Expulsion of zooxanthellae by symbiotic cnidarians from the Red Sea. Coral Reefs 5:201–204

Hooper LV, Littman DR, Macpherson AJ (2012) Interactions between the microbiota and the immune system. Science 336:1268–1273

Houlbrèque F, Ferrier-Pagès C (2009) Heterotrophy in tropical scleractinian corals. Biol Rev 84:1–17

Hoytema N, Bednarz VN, Cardini U, Naumann MS, Al-Horani FA, Wild C (2016) The influence of seasonality on benthic primary production in a Red Sea coral reef. Mar Biol 163:1–14

Hume BCC, Voolstra CR, Arif C, D'Angelo C, Burt JA, Eyal G, Loya Y, Wiedenmann J (2016) Ancestral genetic diversity associated with the rapid spread of stress-tolerant coral symbionts in response to Holocene climate change. Proc Natl Acad Sci 113:4416–4421

Iglesias-Prieto R, Beltran VH, LaJeunesse TC, Reyes-Bonilla H, Thome PE (2004) Different algal symbionts explain the vertical distribution of dominant reef corals in the eastern Pacific. Proc R Soc Lond Ser B Biol Sci 271:1757–1763

Johannes RE, Coles SL, Kuenzel NT (1970) The role of zooplankton in the nutrition of some scleractinian corals. Limnol Oceanogr 15:579–586

Kahng SE, Hochberg EJ, Apprill A, Wagner D, Luck DG, Perez D, Bidigare RR (2012) Efficient light harvesting in deep-water zooxanthellate corals. Mar Ecol Prog Ser 455:65–77

Kahng SE, Copus JM, Wagner D (2014) Recent advances in the ecology of mesophotic coral ecosystems (MCEs). Curr Opin Environ Sustain 7:72–81

Kaniewska P, Magnusson SH, Anthony KRN, Reef R, Kühl M, Hoegh-Guldberg O (2011) Importance of macro-versus microstructure in modulating light levels inside coral colonies. J Phycol 47:846–860

Kaniewska P, Alon S, Karako-Lampert S, Hoegh-Guldberg O, Levy O (2015) Signaling cascades and the importance of moonlight in coral broadcast mass spawning. eLife 4:e09991

Keeling RF, Körtzinger A, Gruber N (2010) Ocean deoxygenation in a warming world. Annu Rev Mar Sci 2:199–229

Keith SA, Maynard JA, Edwards AJ, Guest JR, Bauman AG, van Hooidonk R, Heron SF, Berumen ML, Bouwmeester J, Piromvaragorn S, Rahbek C, Baird AH (2016) Coral mass spawning predicted by rapid seasonal rise in ocean temperature. Proc R Soc Lond Ser B Biol Sci 283(1830):20160011

Kleypas JA, McManus JW, Menez LAB (1999) Environmental limits to coral reef development: where do we draw the line? Am Zool 39:146–159

Kleypas JA, Danabasoglu G, Lough JM (2008) Potential role of the ocean thermostat in determining regional differences in coral reef bleaching events. Geophys Res Lett 35:L03613

Kopp C, Domart-Coulon I, Escrig S, Humbel BM, Hignette M, Meibom A (2015) Subcellular investigation of photosynthesis-driven carbon assimilation in the symbiotic reef coral *Pocillopora damicornis*. MBio 6:e02299–e02214

Kotb MMA (2001) Growth rates of three reef-building coral species in the Northern Red Sea, Egypt. Egypt J Aquat Biol Fish 5:165–185

Kramarsky-Winter E, Loya Y (1998) Reproductive strategies of two fungiid corals from the northern Red Sea: environmental constraints? Mar Ecol Prog Ser 174:175–182

Krueger T, Horwitz N, Bodin J, Giovani M-E, Escrig S, Meibom A, Fine M (2017) Common reef-building coral in the northern Red Sea resistant to elevated temperature and acidification. R Soc Open Sci 4(5):170038

Kuffner IB, Hickey TD, Morrison JM (2013) Calcification rates of the massive coral *Siderastrea siderea* and crustose coralline algae along the Florida keys (USA) outer-reef tract. Coral Reefs 32:987–997

Kürten B, Al-Aidaroos AM, Struck U, Khomayis HS, Gharbawi WY, Sommer U (2014) Influence of environmental gradients on C and N stable isotope ratios in coral reef biota of the Red Sea, Saudi Arabia. J Sea Res 85:379–394

LaJeunesse TC, Parkinson JE, Gabrielson PW, Jeong HJ, Reimer JD, Voolstra CR, Santos SR (2018) Systematic revision of Symbiodiniaceae highlights the antiquity and diversity of coral endosymbionts. Curr Biol 28:2570–2580. https://doi.org/10.1016/j.cub.2018.07.008

Lampert-Karako S, Stambler N, Katcoff DJ, Achituv Y, Dubinsky Z, Simon-Blecher N (2008) Effects of depth and eutrophication on the zooxanthella clades of *Stylophora pistillata* from the Gulf of Eilat (Red Sea). Aquat Conserv Mar Freshwat Ecosyst 18:1039–1045

Lin S, Cheng S, Song B, Zhong X, Lin X, Li W, Li L, Zhang Y, Zhang H, Ji Z, Cai M, Zhuang Y, Shi X, Lin L, Wang L, Wang Z, Liu X, Yu S, Zeng P, Hao H, Zou Q, Chen C, Li Y, Wang Y, Xu C, Meng S, Xu X, Wang J, Yang H, Campbell DA, Sturm NR, Dagenais-Bellefeuille S, Morse D (2015) The *Symbiodinium kawagutii* genome illuminates dinoflagellate gene expression and coral symbiosis. Science 350:691–694

Lough JM, Barnes DJ (2000) Environmental controls on growth of the massive coral *Porites*. J Exp Mar Biol Ecol 245:225–243

Loya Y, Kramarsky-Winter E (2003) *In situ* eutrophication caused by fish farms in the northern Gulf of Eilat (Aqaba) is beneficial for its coral reefs: a critique. Mar Ecol Prog Ser 261:299–303

Loya Y, Lubinevsky H, Rosenfeld M, Kramarsky-Winter E (2004) Nutrient enrichment caused by *in situ* fish farms at Eilat, Red Sea is detrimental to coral reproduction. Mar Pollut Bull 49:344–353

Loya Y, Rosenfeld M, Kramarsky-Winter E (2005) Nutrient enrichment and coral reproduction: empty vessels make the most sound (response to a critique by B. Rinkevich). Mar Pollut Bull 50:114–118

Macdonald AH, Sampayo E, Ridgway T, Schleyer M (2008) Latitudinal symbiont zonation in *Stylophora pistillata* from Southeast Africa. Mar Biol 154:209–217

Marenzeller EV (1907) Expedition SM Schiff "Pola" in das Rote Meer, nördliche und südliche Hälfte 1895

Markell DA, Trench RK (1993) Macromolecules exuded by symbiotic dinoflagellates in culture: amino acids and sugar composition. J Phycol 29:64–68

Marshall AT, Clode P (2004) Calcification rate and the effect of temperature in a zooxanthellate and an azooxanthellate scleractinian reef coral. Coral Reefs 23:218–224

Mass T, Einbinder S, Brokovich E, Shashar N, Vago R, Erez J, Dubinsky Z (2007) Photoacclimation of *Stylophora pistillata* to light extremes: metabolism and calcification. Mar Ecol Prog Ser 334:93–102

van der Merwe R, Röthig T, Voolstra CR, Ochsenkühn MA, Lattemann S, Amy GL (2014) High salinity tolerance of the Red Sea coral *Fungia granulosa* under desalination concentrate discharge conditions: an in situ photophysiology experiment. Front Mar Sci 1:58

McCloskey LR, Muscatine L (1984) Production and respiration in the Red Sea coral *Stylophora pistillata* as a function of depth. Proc R Soc Lond Ser B Biol Sci 222:215–230

McCulloch M, Falter J, Trotter J, Montagna P (2012) Coral resilience to ocean acidification and global warming through pH up-regulation. Nat Clim Chang 2:623–627

McFall-Ngai M, Hadfield MG, Bosch TCG, Carey HV, Domazet-Lošo T, Douglas AE, Dubilier N, Eberl G, Fukami T, Gilbert SF, Hentschel U, King N, Kjelleberg S, Knoll AH, Kremer N, Mazmanian SK, Metcalf JL, Nealson K, Pierce NE, Rawls JF, Reid A, Ruby EG, Rumpho M, Sanders JG, Tautz D, Wernegreen JJ (2013) Animals in a bacterial world, a new imperative for the life sciences. Proc Natl Acad Sci 110:3229–3236

Mohamed T, Kotb M, Ghobashy A, Deek M (2007) Reproduction and growth rate of two scleractinian coral species in the northern Red Sea, Egypt. Egypt J Aquat Res 33:70–86

Monroe A, Ziegler M, Roik A, Röthig T, Hardestine R, Emms M, Jensen T, Voolstra CR, Berumen M (2018) *In situ* observations of coral bleaching in the central Saudi Arabian Red Sea during the 2015/2016 global coral bleaching event. PLoS One 13(4):e0195814

Moran NA, Yun Y (2015) Experimental replacement of an obligate insect symbiont. Proc Natl Acad Sci 112:2093–2096

Muscatine L (1990) The role of symbiotic algae in carbon and energy flux in reef corals. In: Dubinsky Z (ed) Ecosystems of the world. Coral reefs. Elsevier, Amsterdam, pp 75–87

Muscatine L, Porter JW (1977) Reef corals: mutualistic symbioses adapted to nutrient-poor environments. Bioscience 27:454–460

Naumann MS, Richter C, Mott C, El-Zibdah M, Manasrah R, Wild C (2012) Budget of coral-derived organic carbon in a fringing coral reef of the Gulf of Aqaba, Red Sea. J Mar Syst 105-108:20–29

Naumann MS, Orejas C, Ferrier-Pagès C (2014) Species-specific physiological response by the cold-water corals *Lophelia pertusa* and

Madrepora oculata to variations within their natural temperature range. Deep-Sea Res II Top Stud Oceanogr 99:36–41

Neave MJ, Michell CT, Apprill A, Voolstra CR (2014) Whole-genome sequences of three symbiotic *Endozoicomonas* strains. Genome Announc 2:e00802–e00814

Neave MJ, Michell CT, Apprill A, Voolstra CR (2017) Endozoicomonas genomes reveal functional adaptation and plasticity in bacterial strains symbiotically associated with diverse marine hosts. Sci Rep 7:40579

Nir O, Gruber DF, Einbinder S, Kark S, Tchernov D (2011) Changes in scleractinian coral *Seriatopora hystrix* morphology and its endocellular *Symbiodinium* characteristics along a bathymetric gradient from shallow to mesophotic reef. Coral Reefs 30:1089–1100

Nir O, Gruber DF, Shemesh E, Glasser E, Tchernov D (2014) Seasonal Mesophotic coral bleaching of *Stylophora pistillata* in the northern Red Sea. PLoS One 9:e84968

Ochsenkühn MA, Röthig T, D'Angelo C, Wiedenmann J, Voolstra CR (2017) The role of floridoside in osmoadaptation of coral-associated algal endosymbionts to high-salinity conditions. Sci Adv 3:e1602047

Oren OH (1962) The Israel South Red Sea expedition. Nature 194:1134–1137

Osman EO, Smith DJ, Ziegler M, Kürten B, Conrad C, El-Haddad KM, Voolstra CR, Suggett DJ (2018) Thermal refugia against coral bleaching throughout the northern Red Sea. Glob Chang Biol 24:1354–1013

Pernice M, Meibom A, Van Den Heuvel A, Kopp C, Domart-Coulon I, Hoegh-Guldberg O, Dove S (2012) A single-cell view of ammonium assimilation in coral-dinoflagellate symbiosis. ISME J 6:1314–1324

Pogoreutz C, Rädecker N, Cárdenas A, Gärdes A, Voolstra CR, Wild C (2017) Sugar enrichment provides evidence for a role of nitrogen fixation in coral bleaching. Glob Chang Biol 23:3838–3848

Pogoreutz C, Rädecker N, Cárdenas A, Gärdes A, Wild C, Voolstra CR (2018) Dominance of *Endozoicomonas* bacteria throughout coral bleaching and mortality suggests structural inflexibility of the *Pocillopora verrucosa* microbiome. Ecol Evol 8:2240–2252

Pörtner HO (2002) Climate variations and the physiological basis of temperature dependent biogeography: systemic to molecular hierarchy of thermal tolerance in animals. Comp Biochem Physiol A Mol Integr Physiol 132:739–761

Pörtner H-O (2010) Oxygen- and capacity-limitation of thermal tolerance: a matrix for integrating climate-related stressor effects in marine ecosystems. J Exp Biol 213:881–893

Price ARG, Jobbins G, Shepherd ARD, Ormond RFG (1998) An integrated environmental assessment of the Red Sea coast of Saudi Arabia. Environ Conserv 25:65–76

Qurban MA, Krishnakumar PK, Joydas TV, Manikandan KP, Ashraf TTM, Quadri SI, Wafar M, Qasem A, Cairns SD (2014) In-situ observation of deep water corals in the northern Red Sea waters of Saudi Arabia. Deep-Sea Res I Oceanogr Res Pap 89:35–43

Rädecker N, Pogoreutz C, Voolstra CR, Wiedenmann J, Wild C (2015) Nitrogen cycling in corals: the key to understanding holobiont functioning? Trends Microbiol 23:490–497

Raitsos DE, Hoteit I, Prihartato PK, Chronis T, Triantafyllou G, Abualnaja Y (2011) Abrupt warming of the Red Sea. Geophys Res Lett 38:L14601

Raitsos DE, Pradhan Y, Brewin RJ, Stenchikov G, Hoteit I (2013) Remote sensing the phytoplankton seasonal succession of the Red Sea. PLoS One 8:e64909

Reaka-Kudla ML (1997) Global biodiversity of coral reefs: a comparison with rainforests. In: Reaka-Kudla ML, Wilson DE (eds) Biodiversity II: understanding and protecting our biological resources. Joseph Henry Press, Washington, DC, pp 83–108

Reshef L, Koren O, Loya Y, Zilber-Rosenberg I, Rosenberg E (2006) The coral probiotic hypothesis. Environ Microbiol 8:2068–2073

Rinkevich B (2005) Nutrient enrichment and coral reproduction: between truth and repose (a critique of Loya et al.). Mar Pollut Bull 50:111–113. author reply 114–118

Rinkevich B, Loya Y (1979) The reproduction of the Red Sea coral *Stylophora pistillata*. II. Synchronization in breeding and seasonality of planulae shedding. Mar Ecol Prog Ser 1:145–152

Rinkevich B, Angel D, Shafir S, Bongiorni L (2003) 'Fair is foul and foul is fair': response to a critique. Mar Ecol Prog Ser 261:305–309

Roberts CM, McClean CJ, Veron JE, Hawkins JP, Allen GR, McAllister DE, Mittermeier CG, Schueler FW, Spalding M, Wells F, Vynne C, Werner TB (2002) Marine biodiversity hotspots and conservation priorities for tropical reefs. Science 295:1280–1284

Roberts JM, Wheeler AJ, Freiwald A (2006) Reefs of the deep: the biology and geology of cold-water coral ecosystems. Science 312:543–547

Robitzch V, Banguera-Hinestroza E, Sawall Y, Al-Sofyani A, Voolstra CR (2015) Absence of genetic differentiation in the coral *Pocillopora verrucosa* along environmental gradients of the Saudi Arabian Red Sea. Front Mar Sci 2:5

Roder C, Berumen ML, Bouwmeester J, Papathanassiou E, Al-Suwailem A, Voolstra CR (2013) First biological measurements of deep-sea corals from the Red Sea. Sci Rep 3:2802

Roder C, Bayer T, Aranda M, Kruse M, Voolstra CR (2015) Microbiome structure of the fungid coral *Ctenactis echinata* aligns with environmental differences. Mol Ecol 24:3501–3511

Rohwer F, Seguritan V, Azam F, Knowlton N (2002) Diversity and distribution of coral-associated bacteria. Mar Ecol Prog Ser 243:1–10

Roik A, Roder C, Röthig T, Voolstra CR (2015a) Spatial and seasonal reef calcification in corals and calcareous crusts in the Central Red Sea. Coral Reefs 35(2):1–13

Roik A, Röthig T, Ziegler M, Voolstra CR (2015b) Coral bleaching event in the central Red Sea. Mideast Coral Reef Soc Newsl 3:3

Roik A, Röthig T, Roder C, Müller PJ, Voolstra CR (2015c) Captive rearing of the deep-sea coral *Eguchipsammia fistula* from the Red Sea demonstrates remarkable physiological plasticity. PeerJ 3:e734

Roik A, Röthig T, Roder C, Ziegler M, Kremb SG, Voolstra CR (2016) Year-long monitoring of physico-chemical and biological variables provide a comparative baseline of coral reef functioning in the Central Red Sea. PLoS One 11:e0163939

Roik A, Röthig T, Pogoreutz C, Saderne V, Voolstra CR (2018) Coral reef carbonate budgets and ecological drivers in the central Red Sea – a naturally high temperature and high total alkalinity environment. Biogeosciences 15:6277–6296

Rosenberg E, Koren O, Reshef L, Efrony R, Zilber-Rosenberg I (2007) The role of microorganisms in coral health, disease and evolution. Nat Rev Microbiol 5:355–362

Ross CL, Falter JL, Schoepf V, McCulloch MT (2015) Perennial growth of hermatypic corals at Rottnest Island, Western Australia (32°S). PeerJ 3:e781

Röthig T, Costa RM, Simona F, Baumgarten S, Torres AF, Radhakrishnan A, Aranda M, Voolstra CR (2016) Distinct bacterial communities associated with the coral model *Aiptasia* in aposymbiotic and symbiotic states with *Symbiodinium*. Front Mar Sci 3:234

Röthig T, Ochsenkühn MA, Roik A, van der Merwe R, Voolstra CR (2016b) Long-term salinity tolerance is accompanied by major restructuring of the coral bacterial microbiome. Mol Ecol 25(6):1308–1323

Röthig T, Yum LK, Kremb SG, Roik A, Voolstra CR (2017) Microbial community composition of deep-sea corals from the Red Sea provides insight into functional adaption to a unique environment. Sci Rep 7:44714

Rowan R, Knowlton N (1995) Intraspecific diversity and ecological zonation in coral algal symbiosis. Proc Natl Acad Sci 92:2850–2853

Sampayo EM, Franceschinis L, Hoegh-Guldberg O, Dove S (2007) Niche partitioning of closely related symbiotic dinoflagellates. Mol Ecol 16:3721–3733

Sawall Y, Al-Sofyani A (2015) Biology of Red Sea corals: metabolism, reproduction, acclimatization, and adaptation. In: Rasul MAN, Stewart CFI (eds) The Red Sea: the formation, morphology, oceanography and environment of a young ocean basin. Springer, Berlin, Heidelberg, pp 487–509

Sawall Y, Al-Sofyani A, Banguera-Hinestroza E, Voolstra CR (2014) Spatio-temporal analyses of *Symbiodinium* physiology of the coral *Pocillopora verrucosa* along large-scale nutrient and temperature gradients in the Red Sea. PLoS One 9:e103179

Sawall Y, Al-Sofyani A, Hohn S, Banguera-Hinestroza E, Voolstra CR, Wahl M (2015) Extensive phenotypic plasticity of a Red Sea coral over a strong latitudinal temperature gradient suggests limited acclimatization potential to warming. Sci Rep 5:8940

Schlichter D, Fricke HW, Weber W (1986) Light harvesting by wavelength transformation in a symbiotic coral of the Red Sea twilight zone. Mar Biol 91:403–407

Schneider K, Erez J (2006) The effect of carbonate chemistry on calcification and photosynthesis in the hermatypic coral *Acropora eurystoma*. Limnol Oceanogr 51:1284–1293

Schoepf V, Grottoli AG, Warner ME, Cai W-J, Melman TF, Hoadley KD, Pettay DT, Hu X, Li Q, Xu H, Wang Y, Matsui Y, Baumann JH (2013) Coral energy reserves and calcification in a high-CO_2 world at two temperatures. PLoS One 8:e75049

Sheppard C, Price A, Roberts C (1992) Marine ecology of the Arabian region. Academic, London

Shlesinger Y, Loya Y (1985) Coral community reproductive patterns: red sea versus the Great Barrier Reef. Science 228:1333–1335

Shlesinger Y, Goulet T, Loya Y (1998) Reproductive patterns of scleractinian corals in the northern Red Sea. Mar Biol 132:691–701

Silverman J, Lazar B, Erez J (2007a) Community metabolism of a coral reef exposed to naturally varying dissolved inorganic nutrient loads. Biogeochemistry 84:67–82

Silverman J, Lazar B, Erez J (2007b) Effect of aragonite saturation, temperature, and nutrients on the community calcification rate of a coral reef. J Geophys Res 112:C05004

Sorek M, Díaz-Almeyda EM, Medina M, Levy O (2014) Circadian clocks in symbiotic corals: the duet between *Symbiodinium* algae and their coral host. Mar Genomics 14:47–57

Stambler N, Levy O, Vaki L (2008) Photosynthesis and respiration of hermatypic zooxanthellate Red Sea corals from 5-75-m depth. Isr J Plant Sci 56:45–53

Steiner Z, Erez J, Shemesh A, Yam R, Katz A, Lazar B (2014) Basin-scale estimates of pelagic and coral reef calcification in the Red Sea and Western Indian Ocean. Proc Natl Acad Sci 111:16303–16308

Sweeney AM, Boch CA, Johnsen S, Morse DE (2011) Twilight spectral dynamics and the coral reef invertebrate spawning response. J Exp Biol 214:770–777

Tambutté S, Holcomb M, Ferrier-Pagès C, Reynaud S, Tambutté É, Zoccola D, Allemand D (2011) Coral biomineralization: from the gene to the environment. J Exp Mar Biol Ecol 408:58–78

Thiel H (1987) Benthos of the deep Red Sea. In: Edwards AJ, Head SM (eds) Red Sea. Pergamon Press & International Union for Conservation of Nature and Natural Resources, Oxford, pp 112–127

Todd PA (2008) Morphological plasticity in scleractinian corals. Biol Rev 83:315–337

Tolosa I, Treignier C, Grover R, Ferrier-Pagès C (2011) Impact of feeding and short-term temperature stress on the content and isotopic signature of fatty acids, sterols, and alcohols in the scleractinian coral *Turbinaria reniformis*. Coral Reefs 30:763–774

Tremaroli V, Backhed F (2012) Functional interactions between the gut microbiota and host metabolism. Nature 489:242–249

van der Land J (ed) (2008) UNESCO-IOC Register of Marine Organisms (URMO). Available online at http://www.marinespecies.org/urmo

Van Woesik R, Lacharmoise F, Köksal S (2006) Annual cycles of solar insolation predict spawning times of Caribbean corals. Ecol Lett 9:390–398

Villanueva RD, Yap HT, Montano MNE (2008) Timing of planulation by pocilloporid corals in the northwestern Philippines. Mar Ecol Prog Ser 370:111–119

Voolstra CR (2013) A journey into the wild of the cnidarian model system *Aiptasia* and its symbionts. Mol Ecol 22:4366–4368

Voolstra CR, Miller DJ, Ragan MA, Hoffmann A, Hoegh-Guldberg O, Bourne D, Ball E, Ying H, Foret S, Takahashi S, Weynberg KD, van Oppen MJ, Morrow K, Chan CX, Rosic N, Leggat W, Sprungala S, Imelfort M, Tyson GW, Kassahn K, Lundgren P, Beeden R, Ravasi T, Berumen M, Abel E, Fyffe T (2015) The ReFuGe 2020 consortium–using 'omics' approaches to explore the adaptability and resilience of coral holobionts to environmental change. Front Mar Sci 2:68

Voolstra CR, Li Y, Liew YJ, Baumgarten S, Zoccola D, Flot J-F, Tambutté S, Allemand D, Aranda M (2017) Comparative analysis of the genomes of *Stylophora pistillata* and *Acropora digitifera* provides evidence for extensive differences between species of corals. Sci Rep 7:17583

Walker DI, Ormond RFG (1982) Coral death from sewage and phosphate pollution at Aqaba, Red Sea. Mar Pollut Bull 13:21–25

Wangpraseurt D, Larkum AW, Ralph PJ, Kühl M (2012) Light gradients and optical microniches in coral tissues. Front Microbiol 3:316

Weis VM (2008) Cellular mechanisms of cnidarian bleaching: stress causes the collapse of symbiosis. J Exp Biol 211:3059–3066

Weis VM, Davy SK, Hoegh-Guldberg O, Rodriguez-Lanetty M, Pringle JR (2008) Cell biology in model systems as the key to understanding corals. Trends Ecol Evol 23:369–376

Wild C, Huettel M, Klueter A, Kremb SG, Rasheed MYM, Jorgensen BB (2004) Coral mucus functions as an energy carrier and particle trap in the reef ecosystem. Nature 428:66–70

Wild C, Niggl W, Naumann M, Haas A (2010) Organic matter release by Red Sea coral reef organisms—potential effects on microbial activity and in situ O_2 availability. Mar Ecol Prog Ser 411:61–71

Willis BL, Babcock RC, Harrison PL, Oliver JK, Wallace CC (1985) Patterns in the mass spawning of corals on the Great Barrier Reef from 1981 to 1984, 5th International Coral Reef Symposium, pp 343–348

Yum LK, Baumgarten S, Röthig T, Roder C, Roik A, Michell C, Voolstra CR (2017) Transcriptomes and expression profiling of deep-sea corals from the Red Sea provide insight into the biology of azooxanthellate corals. Sci Rep 7:6442

Zakai D, Dubinsky Z, Avishai A, Caaras T, Chadwick NE (2006) Lunar periodicity of planula release in the reef-building coral *Stylophora pistillata*. Mar Ecol Prog Ser 311:93–102

Ziegler M, Roder CM, Büchel C, Voolstra CR (2014) Limits to physiological plasticity of the coral *Pocillopora verrucosa* from the Central Red Sea. Coral Reefs 33:1115–1129

Ziegler M, Roder C, Büchel C, Voolstra CR (2015a) Niche acclimatization in Red Sea corals is dependent on flexibility of host-symbiont association. Mar Ecol Prog Ser 533:149–161

Ziegler M, Roder CM, Büchel C, Voolstra CR (2015b) Mesophotic coral depth acclimatization is a function of host-specific symbiont physiology. Front Mar Sci 2:4

Ziegler M, Roik A, Porter A, Zubier K, Mudarris MS, Ormond R, Voolstra CR (2016) Coral microbial community dynamics in response to anthropogenic impacts near a major city in the Central Red Sea. Mar Pollut Bull 105:629–640

Ziegler M, Arif C, Burt J, Dobretsov SV, Roder C, LaJeunesse TC, Voolstra CR (2017a) Biogeography and molecular diversity of coral symbionts in the genus *Symbiodinium* around the Arabian Peninsula. J Biogeogr 44:674–686

Ziegler M, Seneca FO, Yum LK, Palumbi SR, Voolstra CR (2017b) Bacterial community dynamics are linked to patterns of coral heat tolerance. Nat Commun 8:14213

Microbial Communities of Red Sea Coral Reefs

Matthew J. Neave, Amy Apprill, Greta Aeby, Sou Miyake, and Christian R. Voolstra

Abstract

This chapter explores the microorganisms that inhabit different components of the coral reef ecosystem in the Red Sea. Microbes play crucial roles in numerous reef processes, including primary production as well as nutrient and organic matter cycling. Microbes are also ubiquitous symbionts of eukaryotic organisms, providing the host with nutrients, chemical cycling, and defensive functions. The Red Sea is a particularly interesting study system due to its unusual physiochemical properties, such as a strong north-south temperature and salinity gradient. Here we examine the influence of these unusual characteristics on microbes in the water column and sediments, and those associated with corals, sponges, and fish. In the water column, the microbial community indeed appears to correlate with prevailing north-south environmental conditions. For example, heterotrophic picoplankton and the cyanobacteria *Synechococcus* tend to be more abundant in the warmer, less saline, southern waters. On the other hand, the microbes associated with corals, sponges, and fish seem to be conserved throughout the Red Sea and many other parts of the world. For example, several coral species in the Red Sea harbor *Endozoicomonas* bacteria, and this is also observed world-wide. Moreover, the dominance of *Epulopiscium* bacteria in surgeonfish and highly conserved microbial communities in sponges are also commonly reported in other regions. In terms of microbial-based diseases, Red Sea corals display many typical disorders, including white syndromes, skeletal eroding band, black band disease, and growth anomalies, but these are rare within Red Sea waters. Thus, despite strong environmental extremes driving free-living microbial communities in the Red Sea, the microbes in tightly regulated symbiotic environments appear to be conserved, although strain-level and genotype specialization are areas of continuing research.

Keywords

Red Sea · Bacteria · Archaea · *Endozoicomonas* · *Epulopiscium* · Coral disease · Environmental extremes

4.1 Reef Microbe Studies in the Red Sea

Microbes are important constituents of marine food webs, where they carry out essential functions, including primary production, nitrogen fixation, and nutrient recycling within the microbial loop (Azam et al. 1983; Rädecker et al. 2015). In nutrient-poor coral reef waters, scavenging, sequestering, and recycling of nutrients and organic matter by microbes are particularly important processes and facilitate the movement of nutrients through food webs (Zhang et al. 2015). For example, organisms at higher trophic levels, such as fish and sea cucumbers, rely on microbial processes for much of their carbon requirements (McMahon et al. 2015). Further, the contribution of bacteria to eukaryotic function, whether in shared ecosystems or intimate symbioses, plays a fundamental role in animal biology. It is now recognized that most multicellular organisms support abundant and diverse communities of bacteria and archaea within their microbiome (Bang et al. 2018; McFall-Ngai et al. 2013). Often a substantial fraction of these microbes reside in the gut cavity, but different microbial species frequently occupy other niches within their diverse hosts, such as the epidermis or skeletal compartments (Bayer et al. 2013; Meron et al. 2011; Miyake et al. 2016; Neave et al. 2016, 2017b). The microbiome typically encodes far more genes than is encoded in the host genome, greatly expanding the functional capacity of the host. Moreover, host adaptability and resilience is assumed to be improved because microbial genes can evolve quickly, or microbes with beneficial functions can be rapidly acquired (Theis et al. 2016; Ziegler et al. 2017).

Microbial processes and interactions may be distinct in the Red Sea compared to other environments because of its unusual characteristics, including high temperatures, high salinity (due to negligible freshwater input), high solar irradiation year-around, and a significant Aeolian input of dust from the surrounding desserts (Roik et al. 2016). Despite this unique environment for life, the Red Sea has traditionally been difficult to study due to logistical and political reasons. Most initial research was conducted in the Gulf of Aqaba, a small, shallow outlet at the northeastern part of the Red Sea (Fig. 4.1). Studies from this area provided important insights into Red Sea microorganisms, in particular for potential disease-associated microbes of reef organisms (Arotsker et al. 2009; Barash et al. 2005; Barneah et al. 2007; Ben-Haim 2003; Colorn et al. 2002; Thompson et al. 2006; Winkler et al. 2004). In comparison, the main body of the Red Sea has been relatively poorly studied, although it is likely to contain very different microbial communities to the Gulf of Aqaba due to the many physico-chemical differences. Recently, a new research center opened on the Saudi Arabian coast, leading to a number of Red Sea coral reef microbial studies covering much of the eastern seaboard (Apprill et al. 2013; Bayer et al. 2013; Cardenas et al. 2017; Furby et al. 2014; Jessen et al. 2013; Lee et al. 2011, 2012; Neave et al. 2017b; Roder et al. 2015; Röthig et al. 2016, 2017a, 2017b; Ziegler et al. 2016). These reports have started to unravel the identity of Red Sea bacteria and archaea in reef environments - in particular those associated with corals - and are beginning to examine the functional contribution of these cells and communities to Red Sea reef processes.

This chapter will begin with an examination of the microbes that inhabit reef waters and sediments in the Red Sea, and discuss the predominant drivers structuring these communities. We will then turn to microbes that live symbiotically with Red Sea corals, sponges, and fish, including discussions on potentially disease-causing microbes. The current status of Red Sea research will be summarized and recommendations for future work will be made.

4.2 Coral Reef Waters and Sediments

Pelagic and sediment-associated microorganisms are the unseen, yet abundant members of the coral reef community, contributing important components to the biogeochemical cycling and functioning of coral reefs (Ducklow 1990). Although the term microorganism can encompass a wide size fraction and diversity of organisms, this review will focus on the unicellular prokaryotes encompassing both the bacteria and archaea. In reef waters, these cells are frequently referred to as picoplankton, are generally sized 0.2–2.0 μm, and comprise both the heterotrophic cells as well as phototrophic cyanobacteria (Sieburth et al. 1978). Sediments primarily support cells of similar sizes, with phototrophs limited to the surface depths (Nealson 1997). The abundance, diversity, and functioning of both picoplankton and sediment-associated microbes are generally understudied in the Red Sea. Much of our knowledge of reef water and sediment-associated microorganisms has focused on the Gulf of Aqaba with more recent data from the eastern and central eastern Red Sea (Fig. 4.1).

4.2.1 Coral Reef Picoplankton

4.2.1.1 Abundance of Major Reef Picoplankton Groups

The composition of picoplankton communities is generally related to the biogeochemical features of ocean habitats, including temperature, salinity, nutrients, and the hydrological regime. In the Red Sea, the most extensive geographic study of reef water picoplankton abundance focused on surface waters over coral reefs in the eastern Red Sea, along the coast of Saudi Arabia (Furby et al. 2014). From the northern Red Sea (yet outside the Gulf of Aqaba) to the southeastern coast, mean seasonal sea surface temperatures over the reefs vary by about 4 °C (26 °C – 30 °C, from 2007–2010) and although salinity was not directly measured, it averages 40 PSU (Baranova 2015), and PO_4^{3-} and $NO_3^- + NO_2^-$ were less than 0.15 and 0.5 μM, respectively. The study found heterotrophic picoplankton and the cyanobacteria *Synechococcus* residing in greater abundances in the south compared to the north, and concentrations varied from 3.6×10^5 to 1.4×10^6

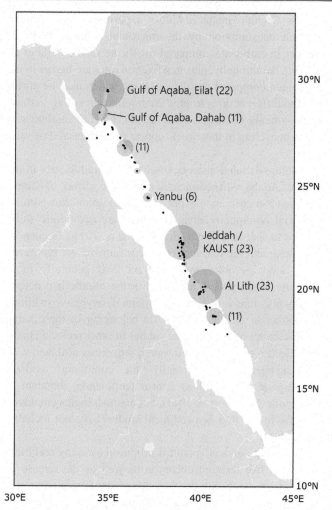

Fig. 4.1 Location of Red Sea studies examining the microbial communities associated with reefs and reef organisms (as of May 2017). Individual study sites are indicated by black dots and clusters of study sites are indicated by red dots that are proportional to cluster size (exact cluster size is given by the text)

cells ml^{-1} and ~5 × 10^4 cells ml^{-1}, respectively, a trend attributed to the temperature differences between the waters (Furby et al. 2014). Interestingly, the Cyanobacteria *Prochlorococcus* were undetectable in reefs near the coastal city of Jeddah and were otherwise present at concentrations of 3–10 × 10^4 cells ml^{-1} throughout the basin. The lack of the oligotroph *Prochlorococcus* on reefs coincided with measurable concentrations of NH$_4^+$ (Furby et al. 2014), and although *Prochlorococcus* strains can utilize NH$_4^+$ (Biller et al. 2015), cells are frequently absent from more eutrophic areas (Partensky et al. 1999).

Picoplankton abundances were also examined in reef waters in the Gulf of Aqaba, a narrow 14 km gulf at the northern eastern tip of the Red Sea, which is surrounded by desert and experiences a high evaporation rate of 1 cm d^{-1} (Klinker et al. 1976) and salinity of 41 PSU (Baranova 2015). A study by Boelen et al. (2002) found that abundances of heterotrophic picoplankton and *Synechococcus* were between 3–4 × 10^5 ml^{-1} and 2.5 × 10^4 ml^{-1}, respectively. Interestingly, *Prochlorococcus* in the Gulf of Aqaba were present at concentrations of 1×10^5 ml^{-1} and greater than *Synechococcus* (Boelen et al. 2002). Comparing these data to the Red Sea basin study by Furby et al. (2014) revealed that there is a continuous trend from the southern Red Sea to the Gulf of Aqaba of decreasing abundances of heterotrophic picoplankton and *Synechococcus*, which could reflect growth limitations due to elevated temperature, increased salinity, or lack of mixing of these waters with the Indian Ocean. Yet, *Prochlorococcus* thrives in the northern reef environment of the Gulf of Aqaba.

4.2.1.2 Diversity of Reef Picoplankton

In addition to studies examining the abundances of the major picoplankton functional groups, the diversity of reef water picoplankton in the Red Sea was recently investigated using sequencing and phylogenetic identifications of small subunit ribosomal RNA genes. The most extensive geographic study of reef water plankton focused on surface water bacteria in the eastern Red Sea at northern reefs spanning 500 km and central reefs spanning 80 km. Community diversity was found to be similar in both regions, and sequences associated with *Synechococcus* and the SAR11 and SAR116 clades of *Alphaproteobacteria* were dominant in the surface reef waters (Apprill et al. 2013). Sequences related to the NSb marine group of the *Flavobacteriaceae*, *Prochlorococcus*, S25–593 group of *Rickettsiales* and *Rhodobacteraceae* were also ubiquitous and abundant (>1% of sequences) at all sites, and *Gammaproteobacteria* belonging to the SAR86 clade and *Alteromonas* spp. were present at >1% sequence abundance at most sites (Bayer et al. 2013). Many of these bacterioplankton have also been recovered from year-long surveys in the central Red Sea (Roik et al. 2016) and are also some of the major phylogenetic groups found globally in coastal tropical ocean waters (Yeo et al. 2013) as well as in waters overlying coral reefs (Kelly et al. 2014; Nelson et al. 2011). Interestingly, the northern reefs of the Red Sea did harbor more consistent sequence recovery from the *Acidimicrobiales* family of *Acinobacteria* as well as *Rhodobacteraceae* lineages which were not as abundant on the southern reefs (Apprill et al. 2013) and may be related to the warmer and more saline conditions present there. To our knowledge there have not been any studies of planktonic archaea in reef waters of the Red Sea, but *Euryarchaeota* and especially *Halobacteriales* and lesser abundances of the *Desulfurococcales* family of *Crenarchaeota* are found in open surface waters within the basin (Qian et al. 2011). Other than differentiating photosynthetic and heterotrophic cell metabolisms, there have yet to be any functional diversity based surveys on Red Sea reef picoplankton. Several functional-based studies related to nutrient stress on open water Red Sea picoplankton (Fuller et al. 2005; Lindell 2005) will provide a helpful backdrop for future investigations of reef picoplankton.

4.2.1.3 Reef Organisms Alter the Picoplankton Community

Studies in the Gulf of Aqaba have also shown depletion of eukaryotic and prokaryotic phytoplankton cells over the coral reefs. Abundances of phytoplankton and chlorophyll *a* over the reef were found to be 15–65% lower than the surrounding open waters (Yahel et al. 1998), and this difference was attributed to grazing by the reef benthos. The study focused on a wide range of phytoplankton cell sizes (0.5–5 μm) and suggested that heterotrophic picoplankton (sized ~1 μm) may also be susceptible to grazing by the reef benthos (Yahel et al. 1998). Picoplankton grazing by reef benthos has been previously observed with corals as well as mixed benthic communities, and picoplankton are thought to be a major source of nitrogen to the reef community (McNally et al. 2016; Ribes et al. 2003). Additionally, the release of organic matter exudates by reef benthos may also provide energy and nutrients to the reef (Cardenas et al. 2017; Wild et al. 2005a). Thus, while there appear to be oceanographic-related trends in picoplankton abundances in the Red Sea reefs, the composition of benthic grazers on the reef may also greatly contribute to the observed picoplankton abundances.

4.2.2 Reef Sand-Associated Microorganisms

Coral reefs primarily contain sandy, permeable sediments that provide habitat to microorganisms, which play important roles in the re-mineralization of organic matter within the reef (Alongi et al. 2007). Reefs sands are primarily comprised of the remains of calcifying organisms including corals. However, some reefs can also contain silicate sands and other terrestrial deposits, and the two sands differ in their permeability, porosity, and grain size (Rasheed et al. 2003). The mineralogical composition thus impacts the physicochemical properties and organic matter retention, and overall habitat type for microorganisms. The majority of coral reef sediment studies have focused on the Gulf of Aqaba, where carbonate and silicate sands are both available on reefs.

4.2.2.1 Drivers of Microbial Community Variations in Reef Sands

In the Gulf of Aqaba, cell abundances in carbonate and silicate reef sands are very similar, and on the order of $1.5–3 \times 10^9$ cells cm^{-3} (Schöttner et al. 2011). In these carbonate and silicate sands, bacterial diversity was found to be similar between the sands, but taxonomic composition of the community varied between the sand types (Schöttner et al. 2011). Organic matter addition experiments have shown oxygen consumption by the microbial community to be greater in carbonate compared to silicate sands (Wild et al. 2005a). Additionally, nitrogen fixation rates are higher in the carbonate compared to silicate sands, which may be attributed to different diazotrophic communities residing in these environments (Bednarz et al. 2015). Overall, carbon and nutrient cycling in these reefs appear to be impacted by sand type.

Sand-associated microbes were also found to vary in the Gulf of Aqaba with sediment depth. While surface (0–2 cm), middle (2–6 cm), and deeper (6–12 cm) sands show similar bacterial community diversity, there are taxonomic shifts related to sediment depth (Schöttner et al. 2011). Sequences affiliated with the *Rhodobacteraceae* family were recovered from more aerobic surface layers, and generally more *Acidobacteriales* were identified in the middle and deeper sediments, likely due to the minimal oxygen conditions (Schöttner et al. 2011). Bacteria belonging to these same families were recovered from studies of other reef sand habitats (Gaidos et al. 2011). However, sequences affiliated with the archaea and especially the ammonia oxidizer *Nitrosopumilus maritimus* appear particularly abundant in reef sands (Gaidos et al. 2011), but methodologies employed thus far in the Red Sea sediment studies have not included the archaea.

Particulate organic matter is produced by many reef biota including fish through excretion as well as the release of mucus from corals. Studies have demonstrated that organism-derived organic matter is an important driver of reef sediment microbial dynamics. One study in the Red Sea linked picoplankton to sediment microbial dynamics and demonstrated that *Synechococcus* accumulate on coral mucus strings, thereby acting as a carrier for picoplankton biomass to the sediments (Naumann et al. 2009). In other worldwide reefs, this mucus has been shown to provide significant carbon, nitrogen, and phosphorous to the sediments (Wild et al. 2004, 2005b). In fact, a comparison of sedimentary oxygen uptake in lagoon sands at different depths revealed an increase with depth that was attributed to the delivery of organic particles from the adjacent reef (Wild et al. 2009).

Lastly, studies in the Gulf of Aqaba have also shown seasonal related impacts on the microbial community structure in reef sediments. Seasonal impacts are most notable in carbonate sediments, which are likely related to the winter shallow mixing and upwelling of higher nutrient waters (Schöttner et al. 2011). The seasonal variability also extends to rates of nitrogen fixation, which is elevated in carbonate sands during the spring and summer months and possibly attributed to phototrophic diazotrophs (Bednarz et al. 2015). Overall, the sediment microbial community appears to provide an important and sensitive record of the overlying water column and reef organismal processes (Roik et al. 2016).

Enhanced knowledge of these reef sediment microbial community members and dynamics within the greater Red Sea basin, with particular attention to the archaea, could improve understanding of the role of these organisms within the broader reef ecosystem.

4.3 Microbial Associations with Reef Animals

4.3.1 Coral Associated Microbes

4.3.1.1 Diversity of Coral Associated Microbes

Corals associate with many eukaryotic and prokaryotic microorganisms. These metaorganisms, comprised of the animal host, symbiotic dinoflagellates of the family Symbiodiniaceae (LaJeunesse et al. 2018), bacteria, viruses, archaea, and fungi, have collectively been termed the coral holobiont (Knowlton and Rohwer 2003). Of all the organisms in the coral holobiont, bacteria are one of the most diverse and abundant components. Initial molecular studies by Rohwer et al. (2002) revealed a much higher bacterial diversity associated with corals than previously thought. They estimated 6000 bacterial ribotypes in three Caribbean corals; far greater than earlier estimates using bacterial culturing techniques. The most common bacterial class were the Gammaproteobacteria, although there were substantial differences across coral species. More recent studies also tend to show a predominance of Gammaproteobacteria in the coral microbiome and estimate the total number of associated bacterial species to be several hundred to several thousand (Ceh et al. 2012; Roder et al. 2014a, b; Sunagawa et al. 2010). In the Red Sea, Lee et al. (2012) used pyrosequencing to study the bacterial community associated with three hard corals and two soft corals. They estimated up to a 1000 bacterial taxa from a single coral, the majority of which belonged to the Proteobacteria. Other studies of Red Sea *Ctenactis crassa* and *Herpolithia limax* (Apprill et al. 2013), *Fungia granulosa* (Röthig et al. 2016), *Acropora hemprichii* (Jessen et al. 2013; Ziegler et al. 2016), *Pocillopora verrucosa* (Pogoreutz et al. 2017, 2018; Ziegler et al. 2016), *Stylophora pistillata* (Bayer et al. 2013; Neave et al. 2017b), and *Ctenactis echinata* (Roder et al. 2015) also estimated a few hundred to approximately a thousand coral-associated bacterial species, similar to reefs elsewhere.

Although coral microbiomes are typically diverse, the abundance of microbial species can be highly skewed toward few very abundant species and many rare species. In fact, several studies from the Red Sea have found a coral microbiome dominated by only a handful of bacterial groups. In the central Red Sea, two operational taxonomic units (OTUs), classified to the genera *Endozoicomonas* (Gammaproteobacteria) and *Burkholderia* (Betaproteobacteria), accounted for up to 90% of the bacterial abundance in *Stylophora pistillata*, and *Endozoicomonas* alone comprised more than 60% of the bacteria associated with *Acropora humilis* (Bayer et al. 2013), *Pocillopora damicornis* (Bayer et al. 2013; Pogoreutz et al. 2017, 2018), *Acropora hemprichii* (Jessen et al. 2013), and *Ctenactis echinata* (Roder et al. 2015). Lee et al. (2012) also found that Red Sea corals were dominated by a few abundant bacteria, although the most abundant bacteria varied and were different across coral species. These studies raise the possibility of a 'core' microbiome, i.e. bacterial taxa that consistently associate with a particular coral species. Current Red Sea coral studies implicate *Endozoicomonas* as a potential core microbiome member for several corals, which has support from other reef locations (see Neave et al. 2016 and references therein). However, integrated and global comparative studies are required to fully explore the concept of a core coral microbiome and to identify putative bacterial core microbiome members, as recently conducted by Neave et al. (2017b). In this global study across 7 major geographical regions and 28 reefs, the authors assayed microbial communities associated with *Stylophora pistillata* and *Pocillopora verrucosa* and found that, indeed, *Endozoicomonas* were present in corals globally. Interestingly, the *Endozoicomonas* genotypes associated with *S. pistillata* changed in different regions, while *Endozoicomonas* genotypes in *P. verrucosa* were similar across global scales, suggesting that symbiont selection may be linked to host reproductive strategy. Still, despite the wide distribution of *Endozoicomonas*, a definite functional role remains elusive. Comparative analyses based on available genomes of *Endozoicomonas* symbionts suggest that *Endozoicomonas* participate in host-associated protein and carbohydrate transport and cycling, and show that *Endozoicomonas* harbor a high degree of genomic plasticity due to the large proportion of transposable elements residing in their genomes (Neave et al. 2016, 2017a).

Coral-associated bacteria are generally dissimilar from bacteria in the surrounding seawater (Roder et al. 2014a, 2014b; Rohwer et al. 2002; Sunagawa et al. 2010). This has been shown for central and northern Red Sea reefs (Apprill et al. 2013; Jessen et al. 2013; Pogoreutz et al. 2017, 2018; Roder et al. 2015; Röthig et al. 2016, 2017a, b; Ziegler et al. 2016) and reefs in the Red Sea Gulf of Aqaba (Meron et al. 2011). Less well understood is how the bacterial community is structured in different compartments of the coral animal, i.e. the mucus layer, tissue cavities, and skeleton. One of the few studies designed to address this question examined *Acropora eurystoma* from the Red Sea under different pH conditions (Meron et al. 2011). Though they found subtle differences between mucus, tissue, and skeletal compartments, bacterial diversity was relatively similar and the effects were overshadowed by the pH treatments. Other studies have compared the bacteria associated with coral mucus across sites (Hadaidi et al. 2017). Kooperman et al. (2007)

found that microbes in the mucus of the Red Sea coral *Fungia granulosa* were significantly more diverse in the environment than when the coral was kept in aquaria, implying a role for seawater bacteria in the colonization of coral mucus (Sunagawa et al. 2010). On the other hand, mucus of the Red Sea corals *Fungia scutaria* and *Platygyra lamellina* from the same sites contained species-specific bacterial communities (Lampert et al. 2008). These studies suggest that bacteria colonize coral from the seawater, but cannot infect coral species indiscriminately.

4.3.1.2 Factors Structuring Coral-Associated Bacterial Communities

The factors structuring coral-associated microbial communities have been difficult to ascertain. Early reports by Rohwer et al. (2002) found that the same coral species tended to have similar bacterial communities even in distant locations, and different species had different bacterial communities. Others have found a dynamic microbial community that fluctuates depending on environmental factors, such as rainfall (Chen et al. 2011). The stability of coral-bacteria associations is likely dependent on several internal and external factors. Internally, the coral may have mechanisms for the acquisition or exclusion of certain bacterial members, and bacteria may be transmitted vertically from parent to offspring (Sharp et al. 2007). Beneficial bacteria that the coral preferentially acquires may be part of a core microbiome that is less likely to vary by location and more likely to vary by coral species. External factors affecting coral-microbe communities include salinity, pH, temperature, oxygen levels, light intensity, nutrients, and chemicals.

In the Red Sea, strong north-south environmental gradients may have more influence on bacterial communities compared to other reefs. Indeed, Apprill et al. (2013) found that reef location had a major influence on the coral bacterial community for several northern and central Red Sea reefs. Across smaller spatial scales (~50 km in the central Red Sea), Lee et al. (2012) also found that the abundance of coral-associated microbes varied, particularly for members of *Vibrio*, *Pseudoalteromonas*, *Serratia*, *Stenotrophomonas*, *Pseudomonas*, and *Achromobacter*. This variability was best correlated with salinity, depth, and temperature. Interestingly however, the authors highlighted two genera, *Chloracidobacterium* and *Endozoicomonas*, that varied with coral host rather than by location, suggesting that these particular bacteria have a more intimate relationship with the coral. Roder et al. (2015) found that when the coral *Ctenactis echinata* was abundant and grew in its preferred Red Sea habitat, its microbiome was dominated by *Endozoicomonas*. In less preferred habitats, the *C. echinata* microbiome had less structure and became more diverse, suggesting that microbiome composition is also influenced by the suitability of the host habitat.

Anthropogenic impacts may also affect the structure of coral microbiomes. Meron et al. (2011) found increases in the disease-associated bacteria *Vibrionaceae* and *Alteromonadaceae* in the Red Sea coral *Acropora eurystoma* after a pH reduction from 8.2 to 7.3, although they found that Gammaproteobacteria and Cyanobacteria maintained their abundance. In response to increasing nutrient concentrations, Jessen et al. (2013) identified several bacteria associated with *Acropora hemprichii* that increased in abundance, but also noted that *Endozoicomonas* bacteria were consistently abundant regardless of treatment. The results of these manipulation studies have also been supported by recent field observations. For example, near the large Red Sea city of Jeddah, the corals *Pocillopora verrucosa* and *Acropora hemprichii* had an altered microbiome, with increases in the opportunistic bacteria *Vibrionaceae* and *Rhodobacteraceae* (Ziegler et al. 2016). Another study by Röthig et al. (2016) in the central Red Sea showed that the microbiome of *Fungia granulosa* changed in response to long-term salinity exposure. In this case, the bacterial shift changed the functional repertoire of the microbiome, possibly helping the coral host adapt to the saline environment. Overall, environmental gradients in the Red Sea appear to influence many coral-microbe relationships, although certain associations, that may be important for the coral, seem to be consistently maintained.

4.3.1.3 Microbes and Coral Disease

Coral health is intimately linked to the multiple microbial partners within the holobiont and disruption of any of the partners can compromise the health of the coral animal causing disease (Bourne et al. 2009). Coral disease has severely altered coral reefs in the Caribbean (Gladfelter 1982; Porter et al. 2001; Sutherland et al. 2004) and is affecting reefs across the Indo-Pacific (Aeby 2005; Raymundo et al. 2003; Vargas-Ángel 2009; Willis et al. 2004). Comparatively few studies have been conducted on coral disease on reefs in the Red Sea, but of the studies that have been published, three types of tissue loss diseases (black band disease, skeletal eroding band, white syndrome) and one report of growth anomalies have been documented and these findings are summarized below. In addition, crustose coralline algae diseases were reported recently (Aeby et al. 2017). Crustose coralline algae (CCA) are an important component of coral reef ecosystems as major contributors to reef productivity, cementing reefs together (Roik et al. 2016).

Black band disease (BBD) is caused by a microbial consortium, visually dominated by filamentous cyanobacteria, that creates the characteristic black band (Richardson 2004). Other constituents of the BBD lesion include sulfide-oxidizing bacteria (*Beggiatoa* sp.), sulfate-reducing bacteria that include members of the *Desulfovibrio* genus, and numerous heterotrophic bacteria (Barneah et al. 2007; Cooney et al. 2002; Frias-Lopez et al. 2002). The sulfate-reducing bacteria are responsible for the highly concentrated sulfide and anoxic conditions underneath the BBD mat that is lethal to coral tissue (Richardson 1996; Richardson et al. 2009). In the Red Sea, BBD was first reported from reefs off the coast of Saudi Arabia in the 1980s where 7 of 33 survey sites were found with BBD infections (Antonius 1985, 1988). An unusual outbreak of BBD was reported from the northern tip of the Gulf of Aqaba in 1996 (Al-Moghrabi 2001) and BBD was reported from the reefs of Eilat in 2000 (Loya 2004). BBD can infect numerous coral genera (Sutherland et al. 2004), but in the Red Sea is found most commonly on faviids (Antonius 1985; Zvuloni et al. 2009). As in other regions of the world, BBD infections on Red Sea reefs are seasonal with increased prevalence in warmer months (Antonius 1985; Zvuloni et al. 2009) and a higher occurrence associated with pollution (Antonius 1985). Consistent with other studies, BBD on the reefs of Eilat were found to be a polymicrobial infection visually dominated by cyanobacteria with the specific bacterial species differing from other regions (Barneah et al. 2007). BBD remains present on reefs in the Red Sea with a recent outbreak documented in 2015. Hadaidi et al. (2018) reported a high prevalence of BBD on multiple coral genera on a reef on the southern central Red Sea and found the main indicator bacteria, although distinct, to have a high similarity to BBD-associated microbes found worldwide. It has been suggested that microbial communities found in BBD may be primarily derived from the local reef environment, which could explain the regional differences (Miller & Richardson 2011).

Skeleton eroding band (SEB) is caused by a colonial heterotrich, folliculinid ciliate, which advances across colonies destroying polyps, coenosarcs, and the surface of the coenosteum (Winkler et al. 2004). SEB was first reported on reefs in the Red Sea in 1994 (Antonius 1995), and in 2001 SEB was found affecting a wide variety of branching and massive reef corals off the coast of Sinai with no evidence of seasonality in disease occurrence (Antonius & Lipscomb 2000). SEB also infects numerous coral genera on reefs in Jordan with *Acropora* and *Stylophora* species most commonly infected (Winkler et al. 2004). Winkler et al. (2004) found SEB at all four sites surveyed and found prevalence to be highest on the reef flat where 77% of the surveyed colonies had signs of the disease.

White syndrome (WS) refers to a tissue loss disease on corals with no pathogens grossly evident on lesions in the field (Bourne et al. 2015). The etiologies of white syndromes differ and have been associated with pathogenic bacteria (Ben-Haim 2003; Roder et al. 2014a; Sussman et al. 2008; Ushijima et al. 2012, 2014), chimeric parasites (Work et al. 2011), programmed cell death (Ainsworth et al. 2007), as well as a variety of other organisms (Work et al. 2012). White syndromes are some of the most virulent of the coral diseases resulting in substantial colony mortality (Bourne et al. 2015). In the Red Sea, coral diseases that fit the description of WS (white band disease & shut down reaction) were first reported in the 1980s off the coast of Saudi Arabia and Hurghada, Egypt (Antonius 1985, 1988). Antonius (1985) reported 13 genera affected by WS and experiments to transmit the disease to healthy colonies were unsuccessful. He also found that exposing WS-infected colonies to antibiotics *in situ* did not slow down or stop the disease, suggesting it may not have been due to bacterial infections. This is in contrast to black band disease, which he found was easily transferred to healthy colonies in direct contact with disease lesions and disease progression was stopped by exposure to antibiotics. Antonius (1988) found WS widespread along the coast of Saudi Arabia occurring in 31 of 33 survey sites. White syndrome (termed white plague) was first documented on corals from Eilat in 2000 on *Goniastrea* spp. and *Favia* spp. (Loya 2004). Further studies carried out on the pathogenesis of the disease by Barash et al. (2005) showed that in aquaria studies, transmission occurred from diseased fragments to non-touching healthy fragments, implicating disease transmission through the water column. Bacterial strain BA-3 combined with a filterable factor from disease lesions was sufficient to recreate the disease in the lab. Bacterial strain BA-3 was subsequently identified as *Thalassomonas loyana* (Thompson et al. 2006).

Growth anomalies (GA) appear as distinctive, protuberant masses on corals and have been reported to affect a variety of coral genera from both the Caribbean and the Indo-Pacific (Peters et al. 1986; Sutherland et al. 2004). Although the causes of GAs in corals are unknown, they are associated with reduced colony growth (Bak 1983), partial colony mortality (Irikawa et al. 2011; Peters et al. 1986; Work et al. 2007), and decreased reproduction (Irikawa et al. 2011; Work et al. 2007; Yamashiro et al. 2000). There has only been one report of growth anomalies from the Red Sea and that was on *Platygyra* sp. in Eilat (Loya 2004).

Coral reefs are declining precipitously (Bellwood et al. 2004; Hughes et al. 2003) and, in response to continued problems associated with anthropogenic overuse and global climate change, coral disease outbreaks are predicted to increase over time (Maynard et al. 2015; Ziegler et al. 2016). Recent surveys along the Red Sea coast of Saudi Arabia documented the continued presence of white syndromes, skeletal eroding band, black band disease, and growth anomalies (Aeby, Voolstra, and colleagues, pers. comm.). Baseline disease surveys, which are lacking in many areas in the Red Sea, are a critical first step in understanding coral disease, giving scientists the capacity to identify and respond to changes in disease levels through time. Research is critically needed to fill in the knowledge gaps on disease distribution, prevalence, and pathogenesis for reefs of the Red Sea (Fig. 4.2).

Fig. 4.2 Coral diseases observed in the Red Sea during a survey in 2015 by Aeby, Voolstra, and colleagues. (**a**) BBD on a *Platygyra* coral at Al Lith, central Red Sea; (**b**) SEB of *Pocillopora* at Yanbu, northern Red Sea; (**c**) WS of *Acropora* in Al Fahal reef, central Red Sea; (**d**) GA of *Acropora* in the offshore Al Mashpah reef, central Red Sea. Arrows indicate disease lesions. (Photo credits: Greta Aeby)

4.3.2 Sponge Associated Microbes

Similar to corals, marine sponges are associated with a remarkable diversity of microorganisms, including archaea (Preston et al. 1996), bacteria (Hentschel et al. 2002), cyanobacteria (Thacker & Starnes 2003), algae (Vacelet 1982), dinoflagellates (Garson et al. 1998), and fungi (Maldonado et al. 2005). The interactions between sponges and their associated microbes are diverse, ranging from mutualistic to parasitic, but in most cases the specific functional relationships are unknown. However, it has been shown that distantly related sponges from around the world appear to share a substantial proportion of their microbiota (Hentschel et al. 2002), some of which constitute sponge-specific divergent bacterial lineages, such as the candidate phylum Poribacteria, which contains members that are nearly exclusively found in marine sponges (Fieseler et al. 2004).

An estimated 240 sponge species were recorded in the Red Sea (Radwan et al. 2010) and have mainly been analyzed for natural products and bioactive compounds (O'Rourke et al. 2016, 2018). To further shed light on diversity and conservation of microbial communities from sponges of the Red Sea, a recent study by Lee et al. (2011) used 16S pyrosequencing to explore the diversity of bacteria and archaea associated with the sponges *Hyrtios erectus*, *Stylissa carteri*, and *Xestospongia testudinaria*. Similar to the bacterial diversity found in corals, the study revealed sponge-associated bacterial communities on the order of hundreds to thousands of bacterial taxa covering 26 bacterial phyla, further extending the presence of the 17 formally described phyla and candidate divisions (Simister et al. 2012). This study also confirmed that sponge-associated microbial communities were highly consistent within sponge species, but variable between sponge species. Additionally, the study by Lee et al. (2011) revealed the presence of potentially hundreds of archaeal species, which exceeds previously reported diversity estimates for members of this kingdom associated with sponges. A study by Moitinho-Silva et al. (2014) analyzed two of the same sponges, i.e. *Xestospongia testudinaria* and *Stylissa carteri*, from the Red Sea via pyrosequencing and confirmed the predicted diversity, but could not find the abundance of archaea, attributed to the use of different primers and bioinformatic pipelines. Further, a study by Giles et al. (2013) studied *Stylissa carteri* and *Crella cyathophora* from the Red

Sea and focused on their characteristic as being putative 'low microbial abundance' (LMA) sponges that, in contrast to 'high microbial abundance' (HMA) sponges, host significantly fewer microorganisms (Moitinho-Silva et al. 2017). Although the LMA/HMA concept might differ regionally and is not a consistent trait of particular sponge species, comparative analysis of the two sponge species from the Red Sea to *C. vaginalis* and *N. digitalis* from the Caribbean and *Raspailia topsenti* from South Pacific showed consistent differences to HMA sponges: all LMA sponges displayed lower phylum-level diversity (up to only five bacterial phyla per sponge) as well as lower sponge-specific clusters than HMA sponges, with an overall low similarity of bacterial communities. To further shed light on the functions provided by these confined assemblage of bacteria, a metatranscriptomics study of *Stylissa carteri* from the Red Sea detected high expression of archaeal ammonia oxidation and photosynthetic carbon fixation by members of the genus *Synechococcus* (Moitinho-Silva et al. 2014). Additionally, functions related to stress response and membrane transporters as well as functions related to methylotrophy were among the most highly expressed by *S. carteri* microbial symbionts. The applicability of metatranscriptomics to explore otherwise inaccessible bacterial symbionts of corals and sponges promises a better understanding of the ecologically relevant functions carried out by the diverse set of microbial partners, which will allow a comparative analyses beyond taxonomic similarities. The most recent efforts by Ryu et al. (2016) focused on sequencing and analyzing the hologenomes of *Stylissa carteri* (a putative LMA sponge) and *Xestospongia testudinaria* (a putative HMA sponge). In their analyses, the authors found that *S. carteri* in comparison to *X. testudinaria* harbors an expanded repertoire of immunological domains in line with a more diverse microbiome, providing insight into the genomic underpinnings underlying host-symbiont coevolution.

4.3.3 Reef Fish Associated Microbes

4.3.3.1 Overview

One area often neglected from microbial studies on coral reefs relates to coral reef fishes. Although the host-associated microbiota are almost synonymously associated with intestinal tract microbiota (i.e., the microbial community of the digestive tracts), a number of studies have also investigated other niches, such as skin or gill microbiota of fishes (Bowman & Nowak 2004; Larsen et al. 2013; Mitchell & Rodger 2011; Wang et al. 2010). Nevertheless, the gut microbiota is by far the best studied, and arguably the most important because of its unparalleled diversity and complexity with 10^{11-12} cells mL^{-1} totalling over 100 trillion microbes (Ley et al. 2006). Gut microbiota studies are pervasive in humans and closely related mammals, from which important discoveries have been made (Backhed et al. 2005; Ley et al. 2008; Mazmanian et al. 2008; Qin et al. 2010). The need for a better understanding of the fish gut microbiota is apparent, especially given that fish represent the most ancient (approximately 600 million years old) as well as the most diverse (over 28,000 described species) lineage of vertebrates (Nelson 2006). The limited literature available on fish primarily focuses on deciphering their nutritional and behavioural ecology, as well as their presumed impact on the environment (Clements et al. 2014), with comparatively few studies on the enteric microbes. Those that do focus on the latter primarily employed conventional techniques (cultivation or clone libraries) at suboptimal sequencing depths (Huber et al. 2004; Kim et al. 2007; Nayak 2010; Sullam et al. 2012; Ward et al. 2009). Until recently, the few next-generation sequencing (NGS) studies available focused on fish model organisms (e.g. zebrafish (Roeselers et al. 2011)) or commercially viable fishes (Gajardo et al. 2017; Lyons et al. 2017; Wilkins et al. 2016; Wong et al. 2013; Zarkasi et al. 2014), but not on coral reef fishes (but see below).

4.3.3.2 Studies on Coral Reef Fish Microbiota

The apparent lack of research on fish gut microbiota is especially true for coral reef fishes, where little is known despite their unparalleled species diversity and population density (Roberts et al. 2002). We know much more about the jaw mechanism of reef fishes (Konow et al. 2008), grazing rates (Carpenter 1986) and their influence on coral cover (Bell & Galzin 1984), than their gut microbiota, which explains the lack of a general consensus on the predominant microbes. Nevertheless, Clements et al. (2007) reported on the high abundance of *Firmicutes* in silver drummer (*Kyphosu sydneyanusi*), butterfish (*Odax pullus*), and marblefish (*Aplodactylus arctidens*). Further, Fidopiastis et al. (2006) studied the abundance of Proteobacteria in zebraperch (*Hermosilla ozurea*) and Smriga et al. (2010) described the dominance of Proteobacteria (Vibrionaceae) in parrotfish (*Chlorurus sordidus*) and snapper (*Lutjanus bohar*), while Firmicutes were the most prominent for surgeonfish (*Acanthurus nigricans*). More recently, an increasing number of studies are beginning to employ NGS technologies to investigate reef fish gut microbiota. For instance, an elasmobranch gut microbiota study by Givens et al. (2015) also included several species of reef-associated fish, notably the spinner shark (*Carcharhinus brevipinna*). Elasmobranch gut microbiota may bear special importance to reveal the evolutionary origin of fish-gut symbiont associations, given their status as basal Chondrichthyes (bony fish). However, these fishes are not commonly considered resident coral reef fish *per se*, as they are only found occasionally on or near coral reefs, in comparison to other studies (Nielsen et al. 2017; Parris et al. 2016; Tarnecki et al. 2017). Tarnecki et al. (2017) studied the faeces of Red Snapper (*Lutjanus campechanus*) as a proxy for its gut microbiota, which showed dominance of Proteobacteria (*Pseudoalteromonas* and *Photobacterium*). Parris et al. (2016) reported on the gut microbiota of pre- and post-reef settlement damselfish (Pomacentridae) and cardinalfish (Apogonidae) in Australia. Although both pre- and post-settlement fishes were abundantly associated with Proteobacteria, the composition at a finer taxonomic resolution differed, namely high abundance of Endozoicomonaceae and Shewanellaceae in pre- and Vibrionaceae and Pasteurellaceae in post-settlement fishes, respectively. Most recently, Nielsen et al. (2017) characterized the gut microbiota of rabbitfish (*Siganus fuscescens*) from the Great Barrier Reef. The study focused on a single ecologically important species, differentiating the gut contents and the walls as well as midgut and hindgut. Overall, taxa belonging to Firmicutes, Bacteoidetes, and Deltaproteobacteria were abundant, yet there was a clear distinction in the microbial community between different gut components, highlighting the selection of specific groups of bacteria at different gut locations.

4.3.3.3 Studies in the Red Sea

The paucity of research regarding reef fish microbial communities is also reflected in the Red Sea, with only 6 of 41 studies on reef-associated bacteria from the region related to reef fish (Berumen et al. 2013). All of these studies were conducted in the Gulf of Aqaba, a small stretch of sea in the northern most Red Sea. Two of these were on infectious microbes on aquaculture fishes (Diamant 2001; Diamant et al. 2000), while four dealt with *Epulopiscium* spp., an enigmatic enteric symbiont of surgeonfishes with unusual size, polyploidy, and mode of reproduction (Angert et al. 1993; Bresler et al. 1998; Clements and Bullivant 1991; Fishelson 1999). With the exception of Miyake et al. (2015) (discussed in detail below), no NGS studies on the gut microbiota of fishes from the central Red Sea are available at the time of writing.

4.3.3.4 Gut Microbiota of Red Sea Reef Fishes

In the study by Miyake et al. (2015), the authors looked at the gut microbiota of 12 species of coral reef fishes from the central Red Sea reef using 16S rRNA gene amplicon sequencing and found that, despite the presence of inter- and intra-species variations, bacteria in the phylum Firmicutes were dominant in most of these herbivorous fishes, while Proteobacteria increased in abundance for carnivorous/detritivorous fishes. In particular, herbivorous surgeonfishes were dominated by *Epulopiscium* spp., as previously reported (Clements et al. 1989). In contrast to laboratory- or aquaculture-raised fishes (Roeselers et al. 2011; Wong et al. 2013), reef fishes only shared modest numbers of bacterial taxa, even between individuals from a given species (1–9% except *Siganus stellatus* who shared 19%). However, the small number of these 'core' taxa often accounted for the majority of sequences (up to 85%), indicating that the dominant taxa are likely to be shared amongst individuals from a given species from the same location. Furthermore, microbiota compositions clustered by host diet, much like what is reported for terrestrial vertebrates (Ley et al. 2008). Because of the complexity of fish diets, a simple classification (e.g., into herbivores, carnivores, and omnivores) was insufficient to explain the clustering. Instead, three clear clusters consisted of (1) macroalgivores (fish that feed primarily on macroscopic brown algae), (2) microalgivores (fish that feed primarily on turfing and filamentous red and green algae), and (3) non-herbivores (i.e., detritivores, omnivores, and carnivores). The non-herbivore cluster is likely to split further according to different diets, but the study sampled insufficient members from the group to statistically resolve this putative relationship. Interestingly, the characteristics of the abundant taxa from two herbivorous clusters were clearly different from the non-herbivorous cluster. BLAST analysis of the abundant taxa in both microalgivore and macroalgivore clusters indicated that the majority of bacterial taxa are closely related to bacteria previously reported from the gut environment (and often from fish), while more non-gut bacteria were prominent for the non-herbivorous cluster. This raises the hypothesis that more autochthonous (indigenous) microbes persist in the former. This is in line with the notion that herbivorous gut microbiota are unique because they play an important role in the digestion and assimilation of structural and storage plant components that are resistant to host-produced digestive cocktails (Cantarel et al. 2012; Clements et al. 2014; Hehemann et al. 2012; Mackie 1997; Van Soest 1982). Of potential importance to coral reef ecology is the persistence of bacteria related to coral diseases in the gut of non-herbivorous fishes. Although this is likely to be a relic of coral ingestion by some fish species (as the study included multiple corallivores), it raises the possibility that some coral reef fishes may effectively function as carriers of (harmful or beneficial) coral microbes. Of note, it has been shown that coral disease prevalence was significantly negatively correlated with surrounding fish taxonomic diversity (Raymundo et al. 2010). In the future, the consideration of reef fish as coral disease-related microbe reservoirs and vectors may become an important component when assessing coral disease, in addition to moving focus to all compartments of the coral holobiont to understand disease processes (Daniels et al. 2015).

4.3.3.5 Regional Specificity in the Red Sea Gut Microbiota

In the absence of comparable studies from reefs around the world, one question that remains unanswered is the uniqueness of the fish gut microbes from the Red Sea. Available data on *Epulopiscium* spp. in surgeonfishes show that they are ubiquitous in herbivorous surgeonfishes around the world (Clements et al. 1989), which was also confirmed for some

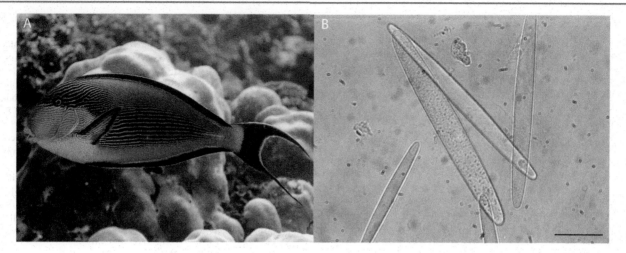

Fig. 4.3 (**a**) *Acanthurus sohal* (Sohal surgeonfish) endemic to the Red Sea and its vicinity (Sea of Oman and Persian/Arabian Gulf) and (**b**) its associated gut bacteria *Epulopiscium fishelsoni* morphotype A. The scale bar denotes 50 μm. (Image credits: Till Röthig/Anna Roik/Sou Miyake)

bacterial morphoypes at the 16S rRNA gene sequence level (Angert et al. 1993; Flint et al. 2005; Miyake et al. 2016). There may be some geographic differentiation at a sub-clade level, as observed between clades A1 and A2 (Miyake et al. 2016); similar to 'ecotypes' described for some oligotrophic ocean bacteria (Thompson et al. 2013). However, any such assertion remains inconclusive given the limited number of sequences (of different clades) available from other reefs.

Because *Epulopiscium* spp. were originally discovered in the gut of a brown surgeonfish (*Acanthurus nigrofuscus*) from the Red Sea (Fishelson et al. 1985), the remaining section describes some of the very unique features of this enigmatic bacterium (Fig. 4.3). *Epulopiscium* spp. is a general term given to *E. fishelsoni*, a giant cigar-shaped bacterium (now re-named as clades A1 and A2 due to phylogenetic differentiation), and a group of related giant bacteria found in the gut of surgeonfishes. These bacteria seem to bypass the physical limit of diffusion for unicellular organisms by growing up to ~700 μm (i.e., 0.7 mm), large enough to be observed with the naked eye. The biovolume of these bacteria can reach in excess of 350,000 μm^3 (Bresler et al. 1998), over 10^9 times larger than the most abundant free-living marine bacterium from the SAR 11 clade (Rappe et al. 2002; Schulz and Jorgensen 2001). Because of the enormous size, they were initially identified as protists until ultrastructural characterization rectified them to be bacteria (Clements and Bullivant 1991), confirmed by 16S rRNA gene analysis (Angert et al. 1993). Remarkably, some clades reproduce viviparously, where fully-functional daughter cells emerge from a mother cell (Montgomery and Pollak 1988), and perhaps related to this mode of reproduction and cell size is the fact that mature *Epulopiscium* cells can be extremely polyploid (up to 50,000–120,000 copies of single-copy marker genes) (Mendell et al. 2008). Different *Epulopiscium*-like giant bacteria of various morphology (morphotypes A-J) have been discovered and described (Clements et al. 1989). Recently, Miyake et al. (2016) studied the phylogenetic diversity and cophylogeny of *Epulopiscium* spp. in surgeonfishes from the central Red Sea, highlighting their large phylogenetic diversity that may have resulted from host-symbiont coevolution by specialisation according to the host gut condition.

Taken together, similar to other reef microbe studies in the Red Sea, further in-depth community characterizations from a wider range of fishes are warranted, in partiuclar employing NGS. Such studies should be carefully designed to consider both biotic and abiotic factors that may influence the gut microbiota (as discussed by Clements et al. (2014)). In parallel to the assessment of the microbes present, a focus on hypothesis-driven work should investigate the causative nature of the association, including functional studies linking hosts and their microbiota with the environment and the ecosystem.

4.4 Conclusions

The Red Sea presents a unique environment with a naturally very warm and saline water body in an opposing North-South gradient (north: colder, higher salinity; south: warmer, lower salinity) where many concepts of our current understanding of structure and function of microbial communities can be tested and validated. Despite the differences in prevailing environmental conditions, the microbial patterns and communities of corals, sponges, and fishes seem conserved, although fine-scale differences exist and warrant further investigation. In particular, the unique environment of the Red Sea suggests spatial adaptation that putatively gave rise to endemic species or 'ecotypes', which await further and exciting discoveries.

References

Aeby G (2005) Outbreak of coral disease in the Northwestern Hawaiian Islands. Coral Reefs 24:481–481

Aeby GS, Shore-Maggio A, Jensen T, Voolstra CR (2017) First record of crustose coralline algae diseases in the Red Sea. Bull Mar Sci 93:985–986

Ainsworth T, Kramasky-Winter E, Loya Y, Hoegh-Guldberg O, Fine M (2007) Coral disease diagnostics: what's between a plague and a band? Appl Environ Microbiol 73:981–992

Al-Moghrabi S (2001) Unusual black band disease (BBD) outbreak in the northern tip of the Gulf of Aqaba (Jordan). Coral Reefs 19:330–331

Alongi DM, Trott LA, Pfitzner J (2007) Deposition, mineralization, and storage of carbon and nitrogen in sediments of the far northern and northern Great Barrier Reef shelf. Cont Shelf Res 27:2595–2622

Angert ER, Clements KD, Pace NR (1993) The largest bacterium. Nature 362:239–241

Antonius A (1985) Coral diseases in the Indo-Pacific: a first record. Mar Ecol 6:197–218

Antonius A (1988) Distribution and dynamics of coral diseases in the Eastern Red Sea. Proc 6th Int Coral Reef Symp:293–298

Antonius A (1995) Sinai coral reef health survey I: first spot checks. Rep. Ras Mohamed Nat. Park Serv, Sinai, Egypt

Antonius AA, Lipscomb D (2000) First protozoan coral-killer identified in the Indo-Pacific. Atoll Res Bull 481:1–21

Apprill A, Hughen K, Mincer T (2013) Major similarities in the bacterial communities associated with lesioned and healthy Fungiidae corals. Environ Microbiol 15:2063–2072

Arotsker L, Siboni N, Ben-Dov E, Kramarsky-Winter E, Loya Y, Kushmaro A (2009) Vibrio sp. as a potentially important member of the Black Band Disease (BBD) consortium in Favia sp. corals. FEMS Microbiol Ecol 70:515–524

Azam F, Fenchel T, Field JG, Gray JS, Meyer-Reil LA, Thingstad F (1983) The ecological role of water-column microbes in the sea. Mar Ecol 10:257–263

Backhed F, Ley RE, Sonnenburg JL, Peterson DA, Gordon JI (2005) Host-bacterial mutualism in the human intestine. Science 307:1915–1920

Bak R (1983) Neoplasia, regeneration and growth in the reef-building coral *Acropora palmata*. Mar Biol 77:221–227

Bang C, Dagan T, Deines P, Dubilier N, Duschl WJ, Fraune S, Hentschel U, Hirt H, Hülter N, Lachnit T, Picazo D, Pita L, Pogoreutz C, Rädecker N, Saad MM, Schmitz RA, Schulenburg H, Voolstra CR, Weiland-Bräuer N, Ziegler M, Bosch TCG (2018) Metaorganisms in extreme environments: do microbes play a role in organismal adaptation? Zoology 127:1–19

Baranova O (2015) World ocean atlas 2005

Barash Y, Sulam R, Loya Y, Rosenberg E (2005) Bacterial Strain BA-3 and a filterable factor cause a white plague-like disease in corals from the Eilat coral reef. Aquat Microb Ecol 40:183–189

Barneah O, Ben-Dov E, Kramarsky-Winter E, Kushmaro A (2007) Characterization of black band disease in Red Sea stony corals. Environ Microbiol 9:1995–2006

Bayer T, Neave MJ, Alsheikh-Hussain A, Aranda M, Yum LK, Mincer T, Hughen K, Apprill A, Voolstra CR (2013) The microbiome of the Red Sea coral *Stylophora pistillata* is dominated by tissue-associated *Endozoicomonas* bacteria. Appl Environ Microbiol 79:4759–4762

Bednarz VN, van Hoytema N, Cardini U, Naumann MS, Al-Rshaidat MM, Wild C (2015) Dinitrogen fixation and primary productivity by carbonate and silicate reef sand communities of the Northern Red Sea. Mar Ecol Prog Ser 527:47–57

Bell J, Galzin R (1984) Influence of live coral cover on coral-reef fish communities. Mar Ecol Prog Ser 15:265–274

Bellwood DR, Hughes TP, Folke C, Nyström M (2004) Confronting the coral reef crisis. Nature 429:827–833

Ben-Haim Y (2003) Vibrio coralliilyticus sp. nov., a temperature-dependent pathogen of the coral *Pocillopora damicornis*. Int J Syst Evol Microbiol 53:309–315

Berumen ML, Hoey AS, Bass WH, Bouwmeester J, Catania D, Cochran JEM, Khalil MT, Miyake S, Mughal MR, Spaet JLY, Saenz-Agudelo P (2013) The status of coral reef ecology research in the Red Sea. Coral Reefs 32:737–748

Biller SJ, Berube PM, Lindell D, Chisholm SW (2015) Prochlorococcus: the structure and function of collective diversity. Nat Rev Microbiol 13:13–27

Boelen P, Post AF, Veldhuis MJW, Buma AGJ (2002) Diel patterns of UVBR-induced DNA damage in picoplankton size fractions from the Gulf of Aqaba, Red Sea. Microb Ecol 44:164–174

Bourne DG, Garren M, Work TM, Rosenberg E, Smith GW, Harvell CD (2009) Microbial disease and the coral holobiont. Trends Microbiol 17:554–562

Bourne DG, Ainsworth TD, Willis BL (2015) White syndromes of indo-Pacific corals. In: Diseases of coral. Wiley, Hoboken, pp 300–315

Bowman JP, Nowak B (2004) Salmonid gill bacteria and their relationship to amoebic gill disease. J Fish Dis 27:483–492

Bresler V, Montgomery WL, Fishelson L, Pollak PE (1998) Gigantism in a bacterium, Epulopiscium fishelsoni, correlates with complex patterns in arrangement, quantity, and segregation of DNA. J Bacteriol 180:5601–5611

Cantarel BL, Lombard V, Henrissat B (2012) Complex carbohydrate utilization by the healthy human microbiome. PLoS One 7:e28742

Cardenas A, Neave MJ, Haroon MF, Pogoreutz C, Radecker N, Wild C, Gardes A, Voolstra CR (2017) Excess labile carbon promotes the expression of virulence factors in coral reef bacterioplankton. ISME J 12(1):59–76

Carpenter RC (1986) Partitioning herbivory and its effects on coral reef algal communities. Ecol Monogr 56:345–364

Ceh J, Raina JB, Soo RM, van Keulen M, Bourne DG (2012) Coral-bacterial communities before and after a coral mass spawning event on Ningaloo Reef. PLoS One 7:e36920

Chen C-P, Tseng C-H, Chen CA, Tang S-L (2011) The dynamics of microbial partnerships in the coral *Isopora palifera*. ISME J 5:728–740

Clements KD, Bullivant S (1991) An unusual symbiont from the gut of surgeonfishes may be the largest known prokaryote. J Bacteriol 173:5359–5362

Clements KD, Sutton DC, Choat JH (1989) Occurrence and characteristics of unusual protistan symbionts from surgeonfishes (Acanthuridae) of the great barrier-reef, Australia. Mar Biol 102:403–412

Clements KD, Pasch IBY, Moran D, Turner SJ (2007) Clostridia dominate 16S rRNA gene libraries prepared from the hindgut of temperate marine herbivorous fishes. Mar Biol 150:1431–1440

Clements KD, Angert ER, Montgomery WL, Choat JH (2014) Intestinal microbiota in fishes: what's known and what's not. Mol Ecol 95:1891–1898

Colorn A, Diamant A, Eldar A, Kvitt H, Zlotkin A (2002) Streptococcus iniae infections in Red Sea cage-cultured and wild fishes. Dis Aquat Org 49:165–170

Cooney RP, Pantos O, Le Tissier MD, Barer MR, Bythell JC (2002) Characterization of the bacterial consortium associated with black band disease in coral using molecular microbiological techniques. Environ Microbiol 4:401–413

Daniels C, Baumgarten S, Yum LK, Michell CT, Bayer T, Arif C, Roder C, Weil E, Voolstra CR (2015) Metatranscriptome analysis of the reef-buidling coral *Orbicella faveolata* indicates holobiont response to coral disease. Front Mar Sci 2

Diamant A (2001) Cross-infections between marine cage-cultured stocks and wild fish in the northern Red Sea: is the environment at risk? Risk Analysis in Aquatic Animal Health, Proceedings, 202–208

Diamant A, Banet A, Ucko M, Colorni A, Knibb W, Kvitt H (2000) Mycobacteriosis in wild rabbitfish *Siganus rivulatus* associated with cage farming in the Gulf of Eilat, Red Sea. Dis Aquat Org 39:211–219

Ducklow HW (1990) The biomass, production and fate of bacteria in coral reefs. In: Dubinsky Z (ed) Ecosystems of the world 25: coral reefs. Elsevier Science Publishing, Amsterdam, pp 265–289

Fidopiastis PM, Bezdek DJ, Horn MH, Kandel JS (2006) Characterizing the resident, fermentative microbial consortium in the hindgut of the temperate-zone herbivorous fish, *Hermosilla azurea* (Teleostei: Kyphosidae). Mar Biol 148:631–642

Fieseler L, Horn M, Wagner M, Hentschel U (2004) Discovery of the novel candidate phylum "Poribacteria" in marine sponges. Appl Environ Microbiol 70:3724–3732

Fishelson L (1999) Polymorphism in gigantobacterial symbionts in the guts of surgeonfish (Acanthuridae: Teleostei). Mar Biol 133:345–351

Fishelson L, Montgomery WL, Myrberg AA (1985) A unique symbiosis in the gut of tropical herbivorous surgeonfish (Acanthuridae, Teleostei) from the Red-Sea. Science 229:49–51

Flint JF, Drzymalski D, Montgomery WL, Southam G, Angert ER (2005) Nocturnal production of endospores in natural populations of epulopiscium-like surgeonfish symbionts. J Bacteriol 187:7460–7470

Frias-Lopez J, Zerkle AL, Bonheyo GT, Fouke BW (2002) Partitioning of bacterial communities between seawater and healthy, black band diseased, and dead coral surfaces. Appl Environ Microbiol 68:2214–2228

Fuller NJ, West NJ, Marie D, Yallop M, Rivlin T, Post AF, Scanlan DJ (2005) Dynamics of community structure and phosphate status of picocyanobacterial populations in the Gulf of Aqaba, Red Sea. Limnol Oceanogr 50:363–375

Furby KA, Apprill A, Cervino JM, Ossolinski JE, Hughen KA (2014) Incidence of lesions on Fungiidae corals in the eastern Red Sea is related to water temperature and coastal pollution. Mar Environ Res 98:29–38

Gaidos E, Rusch A, Ilardo M (2011) Ribosomal tag pyrosequencing of DNA and RNA from benthic coral reef microbiota: community spatial structure, rare members and nitrogen-cycling guilds. Environ Microbiol 13:1138–1152

Gajardo K, Jaramillo-Torres A, Kortner TM, Merrifield DL, Tinsley J, Bakke AM, Krogdahl A (2017) Alternative protein sources in the diet modulate microbiota and functionality in the distal intestine of Atlantic Salmon (*Salmo salar*). Appl Environ Microbiol 83

Garson MJ, Flowers AE, Webb RI, Charan RD, McCaffrey EJ (1998) A sponge/dinoflagellate association in the haplosclerid sponge *Haliclona* sp.: cellular origin of cytotoxic alkaloids by percoll density gradient fractionation. Cell Tissue Res 293:365–373

Giles EC, Kamke J, Moitinho-Silva L et al (2013) Bacterial community profiles in low microbial abundance sponges. FEMS Microbiol Ecol 83:232–241

Givens CE, Ransom B, Bano N, Hollibaugh JT (2015) Comparison of the gut microbiomes of 12 bony fish and 3 shark species. Mar Ecol Prog Ser 518:209–223

Gladfelter WB (1982) White-band disease in *Acropora palmata*: implications for the structure and growth of shallow reefs. Bull Mar Sci 32:639–643

Hadaidi G, Röthig T, Yum LK, Ziegler M, Arif C, Roder C, Burt J, Voolstra CR (2017) Stable mucus-associated bacterial communities in bleached and healthy corals of *Porites lobata* from the Arabian Seas. Sci Rep 7:45362

Hadaidi G, Ziegler M, Shore-Maggio A, Jensen T, Aeby G, Voolstra CR (2018) Ecological and molecular characterization of a coral black band disease outbreak in the Red Sea during a bleaching event. PeerJ 6:e5169

Hehemann JH, Kelly AG, Pudlo NA, Martens EC, Boraston AB (2012) Bacteria of the human gut microbiome catabolize red seaweed glycans with carbohydrate-active enzyme updates from extrinsic microbes. Proc Natl Acad Sci U S A 109:19786–19791

Hentschel U, Hopke J, Horn M, Friedrich AB, Wagner M, Hacker J, Moore BS (2002) Molecular evidence for a uniform microbial community in sponges from different oceans. Appl Environ Microbiol 68:4431–4440

Huber I, Spanggaard B, Appel KF, Rossen L, Nielsen T, Gram L (2004) Phylogenetic analysis and in situ identification of the intestinal microbial community of rainbow trout (*Oncorhynchus mykiss*, Walbaum). J Appl Microbiol 96:117–132

Hughes TP, Baird AH, Bellwood DR, Card M, Connolly SR, Folke C, Grosberg R, Hoegh-Guldberg O, Jackson J, Kleypas J (2003) Climate change, human impacts, and the resilience of coral reefs. Science 301:929–933

Irikawa A, Casareto BE, Suzuki Y, Agostini S, Hidaka M, van Woesik R (2011) Growth anomalies on *Acropora cytherea* corals. Mar Pollut Bull 62:1702–1707

Jessen C, Villa Lizcano JF, Bayer T, Roder C, Aranda M, Wild C, Voolstra CR (2013) In-situ effects of eutrophication and overfishing on physiology and bacterial diversity of the Red Sea coral *Acropora hemprichii*. PLoS One 8:e62091

Kelly LW, Williams GJ, Barott KL, Carlson CA, Dinsdale EA, Edwards RA, Haas AF, Haynes M, Lim YW, McDole T, Nelson CE, Sala E, Sandin SA, Smith JE, Vermeij MJA, Youle M, Rohwer F (2014) Local genomic adaptation of coral reef-associated microbiomes to gradients of natural variability and anthropogenic stressors. Proc Natl Acad Sci 111:10227–10232

Kim DH, Brunt J, Austin B (2007) Microbial diversity of intestinal contents and mucus in rainbow trout (*Oncorhynchus mykiss*). J Appl Microbiol 102:1654–1664

Klinker J, Reiss Z, Kropach C, Levanon I, Harpaz H, Halicz E, Assaf G (1976) Observations on the circulation pattern in the Gulf of Elat (Aqaba), Red Sea. Isr J Earth Sci 25:85–103

Knowlton N, Rohwer F (2003) Multispecies microbial mutualisms on coral reefs: the host as a habitat. Am Nat 162:S51–S62

Konow N, Bellwood DR, Wainwright PC, Kerr AM (2008) Evolution of novel jaw joints promote trophic diversity in coral reef fishes. Biol J Linn Soc 93:545–555

Kooperman N, Ben-Dov E, Kramarsky-Winter E, Barak Z, Kushmaro A (2007) Coral mucus-associated bacterial communities from natural and aquarium environments. FEMS Microbiol Lett 276:106–113

LaJeunesse TC, Parkinson JE, Gabrielson PW, Jeong HJ, Reimer JD, Voolstra CR, Santos SR (2018) Systematic revision of Symbiodiniaceae highlights the antiquity and diversity of coral Endosymbionts. Curr Biol 28:2570–2580. e2576

Lampert Y, Kelman D, Nitzan Y, Dubinsky Z, Behar A, Hill RT (2008) Phylogenetic diversity of bacteria associated with the mucus of Red Sea corals. FEMS Microbiol Ecol 64:187–198

Larsen A, Tao Z, Bullard SA, Arias CR (2013) Diversity of the skin microbiota of fishes: evidence for host species specificity. FEMS Microbiol Ecol 85:483–494

Lee OO, Wang Y, Yang J, Lafi FF, Al-Suwailem A, Qian PY (2011) Pyrosequencing reveals highly diverse and species-specific microbial communities in sponges from the Red Sea. ISME J 5:650–664

Lee OO, Yang J, Bougouffa S, Wang Y, Batang Z, Tian R, Al-Suwailem A, Qian PY (2012) Spatial and species variations in bacterial communities associated with corals from the Red Sea as revealed by pyrosequencing. Appl Environ Microbiol 78:7173–7184

Ley RE, Peterson DA, Gordon JI (2006) Ecological and evolutionary forces shaping microbial diversity in the human intestine. Cell 124:837–848

Ley RE, Hamady M, Lozupone C, Turnbaugh PJ, Ramey RR, Bircher JS, Schlegel ML, Tucker TA, Schrenzel MD, Knight R, Gordon JI (2008) Evolution of mammals and their gut microbes. Science 320:1647–1651

Lindell D (2005) Expression of the nitrogen stress response gene ntcA reveals nitrogen sufficient Synechococcus populations in the oligotrophic northern Red Sea. Limnol Oceanogr 50:1932

Loya Y (2004) The coral reefs of Eilat—past, present and future: three decades of coral community structure studies. In: Coral health and disease. Springer, Berlin, pp 1–34

Lyons PP, Turnbull JF, Dawson KA, Crumlish M (2017) Phylogenetic and functional characterization of the distal intestinal microbiome of rainbow trout *Oncorhynchus mykiss* from both farm and aquarium settings. J Appl Microbiol 122:347–363

Mackie RI (1997) Gut environment and evolution of mutualistic fermentative digestion. In: Gastrointestinal microbiology, pp 13–35

Maldonado M, Cortadellas N, Trillas MI, Rützler K (2005) Endosymbiotic yeast maternally transmitted in a marine sponge. Biol Bull 209:94–106

Maynard J, Van Hooidonk R, Eakin CM, Puotinen M, Garren M, Williams G, Heron SF, Lamb J, Weil E, Willis B (2015) Projections of climate conditions that increase coral disease susceptibility and pathogen abundance and virulence. Nat Clim Chang 5:688–694

Mazmanian SK, Round JL, Kasper DL (2008) A microbial symbiosis factor prevents intestinal inflammatory disease. Nature 453:620–625

McFall-Ngai M, Hadfield MG, Bosch TCG, Carey HV, Domazet-Lošo T, Douglas AE, Dubilier N, Eberl G, Fukami T, Gilbert SF, Hentschel U, King N, Kjelleberg S, Knoll AH, Kremer N, Mazmanian SK, Metcalf JL, Nealson K, Pierce NE, Rawls JF, Reid A, Ruby EG, Rumpho M, Sanders JG, Tautz D, Wernegreen JJ (2013) Animals in a bacterial world, a new imperative for the life sciences. Proc Natl Acad Sci 110:3229–3236

McMahon KW, Thorrold SR, Houghton LA, Berumen ML (2015) Tracing carbon flow through coral reef food webs using a compound-specific stable isotope approach. Oecologia 180:809–821

McNally SP, Parsons RJ, Santoro AE, Apprill A (2016) Multifaceted impacts of the stony coral *Porites astreoides* on picoplankton abundance and community composition. Limnol Oceanogr 62:217–234

McNally SP, Parsons RJ, Santoro AE, Apprill A (2017) Multifaceted impacts of the stony coral *Porites astreoides* on picoplankton abundance and community composition. Limnol Oceanogr 62:217–234

Meron D, Atias E, Iasur Kruh L, Elifantz H, Minz D, Fine M, Banin E (2011) The impact of reduced pH on the microbial community of the coral *Acropora eurystoma*. ISME J 5:51–60

Miller AW, Richardson LL (2011) A meta-analysis of 16S rRNA gene clone libraries from the polymicrobial black band disease of corals. FEMS Microbiol Ecol 75:231–241

Mitchell SO, Rodger HD (2011) A review of infectious gill disease in marine salmonid fish. J Fish Dis 34:411–432

Miyake S, Ngugi DK, Stingl U (2015) Diet strongly influences the gut microbiota of surgeonfishes. Mol Ecol 24:656–672

Miyake S, Ngugi DK, Stingl U (2016) Phylogenetic diversity, distribution, and cophylogeny of giant bacteria (Epulopiscium) with their surgeonfish hosts in the Red Sea. Front Microbiol 7

Moitinho-Silva L, Bayer K, Cannistraci CV, Giles EC, Ryu T, Seridi L, Ravasi T, Hentschel U (2014) Specificity and transcriptional activity of microbiota associated with low and high microbial abundance sponges from the Red Sea. Mol Ecol 23:1348–1363

Moitinho-Silva L, Steinert G, Nielsen S, Hardoim CCP, Wu YC, McCormack GP, Lopez-Legentil S, Marchant R, Webster N, Thomas T, Hentschel U (2017) Predicting the HMA-LMA status in marine sponges by machine learning. Front Microbiol 8:752

Montgomery WL, Pollak PE (1988) Epulopiscium-Fishelsoni Ng, N-Sp, a protist of uncertain taxonomic affinities from the gut of an herbivorous reef fish. J Protozool 35:565–569

Naumann MS, Richter C, El-Zibdah M, Wild C (2009) Coral mucus as an efficient trap for picoplanktonic cyanobacteria: implications for pelagic-benthic coupling in the reef ecosystem. Mar Ecol Prog Ser 385:65–76

Nayak SK (2010) Role of gastrointestinal microbiota in fish. Aquac Res 41:1553–1573

Nealson KH (1997) Sediment bacteria: who's there, what are they doing, and what's new? Annu Rev Earth Planet Sci 25:403–434

Neave MJ, Apprill A, Ferrier-Pagès C, Voolstra CR (2016) Diversity and function of prevalent symbiotic marine bacteria in the genus *Endozoicomonas*. Appl Microbiol Biotechnol 100:8315–8324

Neave M, Michell C, Apprill A, Voolstra CR (2017a) *Endozoicomonas* genomes reveal functional adaptation and plasticity in bacterial strains symbiotically associated with diverse marine hosts. Sci Rep 7:40579

Neave MJ, Rachmawati R, Xun L, Michell CT, Bourne DG, Apprill A, Voolstra CR (2017b) Differential specificity between closely related corals and abundant *Endozoicomonas* endosymbionts across global scales. ISME J 11:186–200

Nelson JS (2006) Fishes of the world. Wiley, Hoboken

Nelson CE, Alldredge AL, McCliment EA, Amaral-Zettler LA, Carlson CA (2011) Depleted dissolved organic carbon and distinct bacterial communities in the water column of a rapid-flushing coral reef ecosystem. ISME J 5:1374–1387

Nielsen S, Wilkes Walburn J, Verges A, Thomas T, Egan S (2017) Microbiome patterns across the gastrointestinal tract of the rabbitfish *Siganus fuscescens*. PeerJ 5:e3317

O'Rourke A, Kremb S, Bader T, Helfer M, Schmitt-Kopplin P, Gerwick W, Brack-Werner R, Voolstra C (2016) Alkaloids from the sponge *Stylissa carteri* present prospective scaffolds for the inhibition of Human Immunodeficiency virus 1 (HIV-1). Mar Drugs 14:28

O'Rourke A, Kremb S, Duggan B, Sioud S, Kharbatia N, Raji M, Emwas A-H, Gerwick W, Voolstra C (2018) Identification of a 3-Alkylpyridinium compound from the Red Sea sponge Amphimedon chloros with in vitro inhibitory activity against the West Nile virus NS3 protease. Molecules 23:1472

Parris DJ, Brooker RM, Morgan MA, Dixson DL, Stewart FJ (2016) Whole gut microbiome composition of damselfish and cardinalfish before and after reef settlement. PeerJ 4:e2412

Partensky F, Blanchot J, and Vaulot D (1999) Differential distribution and ecology of *Prochlorococcus* and *Synechococcus* in oceanic waters: a review. Bull Inst océanogr NS19:457–475

Peters EC, Halas JC, McCarty HB (1986) Calicoblastic neoplasms in *Acropora palmata*, with a review of reports on anomalies of growth and form in corals. J Natl Cancer Inst 76:895–912

Pogoreutz C, Rädecker N, Cárdenas A, Gärdes A, Voolstra CR, Wild C (2017) Sugar enrichment provides evidence for a role of nitrogen fixation in coral bleaching. Glob Chang Biol 23:3838–3848

Pogoreutz C, Rädecker N, Cárdenas A, Gärdes A, Wild C, Voolstra CR (2018) Dominance of *Endozoicomonas* bacteria throughout coral bleaching and mortality suggests structural inflexibility of the *Pocillopora verrucosa* microbiome. Ecol Evol 8:2240–2252

Porter JW, Dustan P, Jaap WC, Patterson KL, Kosmynin V, Meier OW, Patterson ME, Parsons M (2001) Patterns of spread of coral disease in the Florida keys. In: The ecology and etiology of newly emerging marine diseases. Springer, pp 1–24

Preston CM, Wu KY, Molinski TF, DeLong EF (1996) A psychrophilic crenarchaeon inhabits a marine sponge: Cenarchaeum symbiosum gen. nov., sp. nov. Proc Natl Acad Sci U S A 93:6241–6246

Qian P-Y, Wang Y, Lee OO, Lau SCK, Yang J, Lafi FF, Al-Suwailem A, Wong TYH (2011) Vertical stratification of microbial communities in the Red Sea revealed by 16S rDNA pyrosequencing. ISME J 5:507–518

Qin JJ, Li RQ, Raes J, Arumugam M, Burgdorf KS, Manichanh C, Nielsen T, Pons N, Levenez F, Yamada T, Mende DR, Li JH, Xu JM, Li SC, Li DF, Cao JJ, Wang B, Liang HQ, Zheng HS, Xie YL, Tap J,

Lepage P, Bertalan M, Batto JM, Hansen T, Le Paslier D, Linneberg A, Nielsen HB, Pelletier E, Renault P, Sicheritz-Ponten T, Turner K, Zhu HM, Yu C, Li ST, Jian M, Zhou Y, Li YR, Zhang XQ, Li SG, Qin N, Yang HM, Wang J, Brunak S, Dore J, Guarner F, Kristiansen K, Pedersen O, Parkhill J, Weissenbach J, Bork P, Ehrlich SD, Consortium M (2010) A human gut microbial gene catalogue established by metagenomic sequencing. Nature 464:59–U70

Rädecker N, Pogoreutz C, Voolstra CR, Wiedenmann J, Wild C (2015) Nitrogen cycling in corals: the key to understanding holobiont functioning? Trends Microbiol 23:490–497

Radwan M, Hanora A, Zan J, Mohamed NM, Abo-Elmatty DM, Abou-El-Ela SH, Hill RT (2010) Bacterial community analyses of two Red Sea sponges. Mar Biotechnol (NY) 12:350–360

Rappe MS, Connon SA, Vergin KL, Giovannoni SJ (2002) Cultivation of the ubiquitous SAR11 marine bacterioplankton clade. Nature 418:630–633

Rasheed M, Badran MI, Huettel M (2003) Influence of sediment permeability and mineral composition on organic matter degradation in three sediments from the Gulf of Aqaba, Red Sea. Estuar Coast Shelf Sci 57:369–384

Raymundo LJ, Harvell CD, Reynolds TL (2003) Porites ulcerative white spot disease: description, prevalence, and host range of a new coral disease affecting Indo-Pacific reefs. Dis Aquat Org 56:95–104

Raymundo LJ, Halford AR, Maypa AP, Kerr AM (2010) Functionally diverse reef-fish communities ameliorate coral disease. Proc Natl Acad Sci 107:514–514

Ribes M, Coma R, Atkinson MJ, Kinzie RAI (2003) Particle removal by coral reef communities: picoplankton is a major source of nitrogen. Mar Ecol Prog Ser 257:13–23

Richardson L (1996) Horizontal and vertical migration patterns of *Phorrnidium corallyticum* and *Beggiatoa* spp. associated with black-band disease of corals. Microb Ecol 32:323–335

Richardson LL (2004) Black band disease. In: Coral health and disease. Springer, Berlin, pp 325–336

Richardson LL, Miller AW, Broderick E, Kaczmarsky L, Gantar M, Sekar R (2009) Sulfide, microcystin, and the etiology of black band disease. Dis Aquat Org 87:79

Roberts CM, McClean CJ, Veron JEN, Hawkins JP, Allen GR, McAllister DE, Mittermeier CG, Schueler FW, Spalding M, Wells F, Vynne C, Werner TB (2002) Marine biodiversity hotspots and conservation priorities for tropical reefs. Science 295:1280–1284

Roder C, Arif C, Bayer T, Aranda M, Daniels C, Shibl A, Chavanich S, Voolstra CR (2014a) Bacterial profiling of white plague disease in a comparative coral species framework. ISME J 8:31–39

Roder C, Arif C, Daniels C, Weil E, Voolstra CR (2014b) Bacterial profiling of white plague disease across corals and oceans indicates a conserved and distinct disease microbiome. Mol Ecol 23:965–974

Roder C, Bayer T, Aranda M, Kruse M, Voolstra CR (2015) Microbiome structure of the fungid coral *Ctenactis echinata* aligns with environmental differences. Mol Ecol 24:3501–3511

Roeselers G, Mittge EK, Stephens WZ, Parichy DM, Cavanaugh CM, Guillemin K, Rawls JF (2011) Evidence for a core gut microbiota in the zebrafish. ISME J 5:1595–1608

Rohwer F, Seguritan V, Azam F, Knowlton N (2002) Diversity and distribution of coral-associated bacteria. Mar Ecol Prog Ser 243:1–10

Roik A, Röthig T, Roder C, Ziegler M, Kremb SG, Voolstra CR (2016) Year-long monitoring of Physico-chemical and biological variables provide a comparative baseline of coral reef functioning in the Central Red Sea. PLoS One 11:e0163939

Röthig T, Ochsenkühn MA, Roik A, van der Merwe R, Voolstra CR (2016) Long-term salinity tolerance is accompanied by major restructuring of the coral bacterial microbiome. Mol Ecol 25:1308–1323

Röthig T, Roik A, Yum LK, Voolstra CR (2017a) Distinct bacterial microbiomes associate with the Deep-Sea coral *Eguchipsammia fistula* from the Red Sea and from aquaria settings. Front Mar Sci 4:259

Röthig T, Yum LK, Kremb SG, Roik A, Voolstra CR (2017b) Microbial community composition of deep-sea corals from the Red Sea provides insight into functional adaption to a unique environment. Sci Rep 7:44714

Ryu T, Seridi L, Moitinho-Silva L, Oates M, Liew YJ, Mavromatis C, Wang X, Haywood A, Lafi FF, Kupresanin M, Sougrat R, Alzahrani MA, Giles E, Ghosheh Y, Schunter C, Baumgarten S, Berumen ML, Gao X, Aranda M, Foret S, Gough J, Voolstra CR, Hentschel U, Ravasi T (2016) Hologenome analysis of two marine sponges with different microbiomes. BMC Genomics 17:1–11

Schöttner S, Pfitzner B, Grünke S, Rasheed M, Wild C, Ramette A (2011) Drivers of bacterial diversity dynamics in permeable carbonate and silicate coral reef sands from the Red Sea. Environ Microbiol 13:1815–1826

Schulz HN, Jorgensen BB (2001) Big bacteria. Annu Rev Microbiol 55:105–137

Sharp KH, Eam B, Faulkner DJ, Haygood MG (2007) Vertical transmission of diverse microbes in the tropical sponge Corticium sp. Appl Environ Microbiol 73:622–629

Sieburth JM, Smetacek V, Lenz J (1978) Pelagic ecosystem structure: heterotrophic compartments of the plankton and their relationship to plankton size fractions. Limnol Oceanogr 23:1256–1263

Simister R, Taylor MW, Tsai P, Webster N (2012) Sponge-microbe associations survive high nutrients and temperatures. PLoS One 7:e52220

Smriga S, Sandin SA, Azam F (2010) Abundance, diversity, and activity of microbial assemblages associated with coral reef fish guts and feces. FEMS Microbiol Ecol 73:31–42

Sullam KE, Essinger SD, Lozupone CA, O'Connor MP, Rosen GL, Knight R, Kilham SS, Russell JA (2012) Environmental and ecological factors that shape the gut bacterial communities of fish: a meta-analysis. Mol Ecol 21:3363–3378

Sunagawa S, Woodley CM, Medina M (2010) Threatened corals provide underexplored microbial habitats. PLoS One 5:e9554

Sussman M, Willis BL, Victor S, Bourne DG (2008) Coral pathogens identified for white syndrome (WS) epizootics in the indo-Pacific. PLoS One 3:e2393

Sutherland KP, Porter JW, Torres C (2004) Disease and immunity in Caribbean and indo-Pacific zooxanthellate corals. Mar Ecol Prog Ser 266:265–272

Tarnecki AM, Burgos FA, Ray CL, Arias CR (2017) Fish intestinal microbiome: diversity and symbiosis unravelled by metagenomics. J Appl Microbiol

Thacker RW, Starnes S (2003) Host specificity of the symbiotic cyanobacterium Oscillatoria spongeliae in marine sponges, Dysidea spp. Mar Biol 142:643–648

Theis KR, Dheilly NM, Klassen JL, Brucker RM, Baines JF, Bosch TCG, Cryan JF, Gilbert SF, Goodnight CJ, Lloyd EA, Sapp J, Vandenkoornhuyse P, Zilber-Rosenberg I, Rosenberg E, Bordenstein SR (2016) Getting the Hologenome concept right: an eco-evolutionary framework for hosts and their microbiomes. mSystems 1(2): e00028-16

Thompson F, Barash Y, Sawabe T, Sharon G, Swings J, Rosenberg E (2006) *Thalassomonas loyana* sp. nov., a causative agent of the white plague-like disease of corals on the Eilat coral reef. Int J Syst Evol Microbiol 56:365–368

Thompson LR, Field C, Romanuk T, Kamanda Ngugi D, Siam R, El Dorry H, Stingl U (2013) Patterns of ecological specialization among microbial populations in the Red Sea and diverse oligotrophic marine environments. Ecol Evol 3:1780–1797

Ushijima B, Smith A, Aeby GS, Callahan SM (2012) *Vibrio owensii* induces the tissue loss disease Montipora white syndrome in the Hawaiian reef coral *Montipora capitata*. PLoS One 7:e46717

Ushijima B, Videau P, Burger AH, Shore-Maggio A, Runyon CM, Sudek M, Aeby GS, Callahan SM (2014) *Vibrio coralliilyticus* strain OCN008 is an etiological agent of acute Montipora white syndrome. Appl Environ Microbiol 80:2102–2109

Vacelet J (1982) Algal-sponge symbioses in the coral reefs of New Caledonia: a morphological study. In: 4th international coral reef. University of the Philippines, S. Manila, Philippines, pp 713–719

Van Soest PJ (1982) Nutritional ecology of the ruminants. Cornell University Press 2:11–45

Vargas-Ángel B (2009) Coral health and disease assessment in the US Pacific remote island areas. Bull Mar Sci 84:211–227

Wang W, Zhou Z, He S, Liu Y, Cao Y, Shi P, Yao B, Ringø E (2010) Identification of the adherent microbiota on the gills and skin of poly-cultured gibel carp (*Carassius auratus* gibelio) and bluntnose black bream (*Megalobrama amblycephala* Yih). Aquac Res 41

Ward NL, Steven B, Penn K, Methe BA, Detrich WH (2009) Characterization of the intestinal microbiota of two Antarctic notothenioid fish species. Extremophiles 13:679–685

Wild C, Huettel M, Klueter A, Kremb SG, Rasheed MYM, Jorgensen BB (2004) Coral mucus functions as an energy carrier and particle trap in the reef ecosystem. Nature 428:66–70

Wild C, Rasheed M, Jantzen C, Cook P, Struck U, Huettel M, Boetius A (2005a) Benthic metabolism and degradation of natural particulate organic matter in carbonate and silicate reef sands of the northern Red Sea. Mar Ecol Prog Ser 298:69–78

Wild C, Woyt H, Huettel M (2005b) Influence of coral mucus on nutrient fluxes in carbonate sands. Mar Ecol Prog Ser 287:87–98

Wild C, Naumann MS, Haas A, Struck U, Mayer FW, Rasheed MY, Huettel M (2009) Coral sand O2 uptake and pelagic–benthic coupling in a subtropical fringing reef, Aqaba, Red Sea. Aquat Biol 6:133–142

Wilkins LG, Fumagalli L, Wedekind C (2016) Effects of host genetics and environment on egg-associated microbiotas in brown trout (*Salmo trutta*). Mol Ecol 25:4930–4945

Willis BL, Page CA, Dinsdale EA (2004) Coral disease on the great barrier reef. In: Coral health and disease. Springer, Berlin, pp 69–104

Winkler R, Antonius A, Abigail Renegar D (2004) The skeleton eroding band disease on coral reefs of Aqaba, Red Sea. Mar Ecol 25:129–144

Wong S, Waldrop T, Summerfelt S, Davidson J, Barrows F, Kenney PB, Welch T, Wiens GD, Snekvik K, Rawls JF, Good C (2013) Aquacultured rainbow trout (*Oncorhynchus mykiss*) possess a large Core intestinal microbiota that is resistant to variation in diet and rearing density. Appl Environ Microbiol 79:4974–4984

Work TM, Aeby GS, Coles SL (2007) Distribution and morphology of growth anomalies in Acropora from the Indo-Pacific. Dis Aquat Organ 78:255

Work TM, Forsman ZH, Szabó Z, Lewis TD, Aeby GS, Toonen RJ (2011) Inter-specific coral chimerism: genetically distinct multicellular structures associated with tissue loss in *Montipora capitata*. PLoS One 6:e22869

Work TM, Russell R, Aeby GS (2012) Tissue loss (white syndrome) in the coral *Montipora capitata* is a dynamic disease with multiple host responses and potential causes. Proc R Soc Lond B Biol Sci. https://doi.org/10.1098/rspb.2012.1827

Yahel G, Post AF, Fabricius K, Marie D, Vaulot D, Genin A (1998) Phytoplankton distribution and grazing near coral reefs. Limnol Oceanogr 43:551–563

Yamashiro H, Yamamoto M, van Woesik R (2000) Tumor formation on the coral *Montipora informis*. Dis Aquat Org 41:211–217

Yeo SK, Huggett MJ, Eiler A, Rappé MS (2013) Coastal bacterioplankton community dynamics in response to a natural disturbance. PLoS One 8:e56207

Zarkasi KZ, Abell GC, Taylor RS, Neuman C, Hatje E, Tamplin ML, Katouli M, Bowman JP (2014) Pyrosequencing-based characterization of gastrointestinal bacteria of Atlantic salmon (*Salmo salar* L.) within a commercial mariculture system. J Appl Microbiol 117:18–27

Zhang F, Blasiak LC, Karolin JO, Powell RJ, Geddes CD, Hill RT (2015) Phosphorus sequestration in the form of polyphosphate by microbial symbionts in marine sponges. Proc Natl Acad Sci 112:4381–4386

Ziegler M, Roik A, Porter A, Zubier K, Mudarris MS, Ormond R, Voolstra CR (2016) Coral microbial community dynamics in response to anthropogenic impacts near a major city in the Central Red Sea. Mar Pollut Bull 105:629–640

Ziegler M, Seneca FO, Yum LK, Palumbi SR, Voolstra CR (2017) Bacterial community dynamics are linked to patterns of coral heat tolerance. Nat Commun 8:14213

Zvuloni A, Artzy-Randrup Y, Stone L, Kramarsky-Winter E, Barkan R, Loya Y (2009) Spatio-temporal transmission patterns of black-band disease in a coral community. PLoS One 4:e4993

Symbiodiniaceae Diversity in Red Sea Coral Reefs & Coral Bleaching

Maren Ziegler, Chatchanit Arif, and Christian R. Voolstra

Abstract

This chapter introduces Symbiodiniaceae, the diverse group of dinoflagellate microalgae, that form an obligate symbiosis with corals and other coral reef organisms. The Symbiodiniaceae cells reside within the coral tissue, their photosynthesis fuels the productivity and diversity of coral reef ecosystem, and the breakdown of this symbiosis leads to coral bleaching and may entail the death of the host. Here, we summarize Symbiodiniaceae taxonomy and phylogeny and the molecular tools that are used to study Symbiodiniaceae diversity in the Red Sea. We provide an overview over all described Symbiodiniaceae species and discuss the functional diversity within this phylogenetically diverse group as well as the implications of this diversity for coral-Symbiodiniaceae pairings and ecological niche partitioning in coral reef ecosystems. We review host-Symbiodiniaceae associations of 57 host genera in the Red Sea and discuss the emerging patterns in light of their wider biogeographic distribution. Last, we summarize how climate change-induced thermal anomalies have repeatedly led to coral bleaching and mortality in the Red Sea and how they threaten these reef ecosystems, otherwise thought to be comparatively resilient. We conclude with a perspective of important topics for Symbiodiniaceae research in the Red Sea that have the potential to contribute to a broader understanding of the basis of thermotolerance in this fragile symbiosis.

Keywords

Symbiodiniaceae diversity · Biogeography · Host-symbiont association · Symbiosis · Molecular tools · ITS2 · Coral bleaching

5.1 Introduction

Coral reefs harbor the largest biodiversity of all marine ecosystems (Connell 1978; Roberts et al. 2002). Scleractinian or hermatypic, reef-forming, corals are primarily adapted to live in the light-flooded zone of warm tropical and subtropical oceans (Kleypas et al. 1999). Despite the oligotrophic conditions prevalent in these waters, coral reefs belong to the most productive ecosystems (Connell 1978; Patton et al. 1977; Roberts et al. 2002). The key to the success of hermatypic corals is the association with autotrophic dinoflagellates of the family Symbiodiniaceae in an obligate symbiosis (Muscatine and Porter 1977). More generally, a great variety of coral reef invertebrate taxa has been found to host Symbiodiniaceae symbionts, such as soft corals (Octocorallia) (Barneah et al. 2004; Benayahu et al. 1989; Goulet and Coffroth 2003), sponges (Porifera) (Carlos et al. 1999; Vicente 1990), flat worms (Plathyelminthes) (Barneah et al. 2007), soritid Foraminifera (Leutenegger 1984; Müller-Merz and Lee 1976; Pochon et al. 2010), and molluscs (Mollusca) such as nudibranchs and tridacnid giant clams (Belda-Baillie et al. 1999; Burghardt et al. 2005; Jeffrey and Haxo 1968; Taylor 1968; Ziegler et al. 2014a).

The Symbiodiniaceae cells are located in the endodermal tissue of their coral hosts where they are found in membrane-bound modified lysosomes, the symbiosomes (Fig. 5.1; Trench 1979; Wakefield and Kempf 2001). The spatial proximity of this endosymbiotic association facilitates a system

M. Ziegler (✉)
Red Sea Research Center, Division of Biological and Environmental Science and Engineering, King Abdullah University of Science and Technology, Thuwal, Saudi Arabia

Department of Animal Ecology and Systematics, Justus Liebig University Giessen, Giessen, Germany
e-mail: maren.ziegler@bio.uni-giessen.de

C. Arif · C. R. Voolstra
Red Sea Research Center, Division of Biological and Environmental Science and Engineering,
King Abdullah University of Science and Technology,
Thuwal, Saudi Arabia

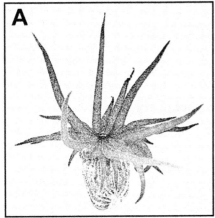 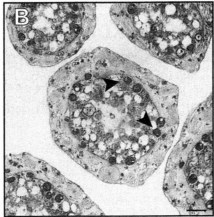 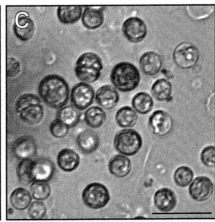

Fig. 5.1 Symbiodiniaceae cells are abundant in endodermal tissues of Cnidarians. (**A**) Fluorescence image of the Cnidarian model organism Aiptasia (strain CC7); Symbiodiniaceae cells are visualized through chlorophyll fluorescence in dark grey/black, whereas the anemone's body is translucent; (**B**) stained tissue cross-section of polyps of the soft coral *Bayerxenia* with Symbiodiniaceae cells (arrowheads) located in the endodermal tissue, which borders the cell-free mesogloea and is surrounded by ectodermal tissue; (**C**) Symbiodiniaceae cells of strain SSB01 (genus *Breviolum*, ITS2 type **B1**) in culture. Scale bars = 20 μm. Photocredit: (**A** & **C**) Fabia Simona, KAUST, (**B**) Maren Ziegler, KAUST

of tight recycling of nutrients and metabolic products (Muscatine and Porter 1977; Rädecker et al. 2015; Tanaka et al. 2006; Trench 1979). The Symbiodiniaceae cells receive protection from damaging ultraviolet radiation (UVR) (Banaszak and Trench 1995a, 1995b) and are provided with carbon dioxide (CO_2) from the coral host, which they utilize for their highly efficient photosynthesis (Falkowski et al. 1984; Muscatine and Porter 1977; Muscatine et al. 1989). The photosynthates in turn, are passed on to the coral host, typically as compounds of low-molecular weight such as glucose, glycerol, and amino acids (Burriesci et al. 2012; Markell and Trench 1993; Trench 1993). In a healthy coral, photosynthesis can cover almost the entire energy demand of the coral host (Muscatine 1990; Muscatine et al. 1984; Muscatine and Porter 1977). Supported by high photosynthetic production rates of their endosymbionts, corals secrete calcium carbonate skeletons that give rise to the large three-dimensional coral reef structures that in turn provide the habitat complexity to support a large diversity of species (Kawaguti and Sakumoto 1948; Pearse and Muscatine 1971).

The Red Sea represents a unique and rather extreme environment with thriving coral reef landscapes. Because of its long-term thermal regime at the upper limits of coral reef occurrence, it represents a suitable location to explore the perspectives of host-Symbiodiniaceae associations under climate change and to study their adaptation and acclimatization mechanisms (Hume et al. 2016). Caused by long geographic isolation and repeated extinction events, Red Sea coral reef communities are characterized by a larger proportion of endemic species than previously assumed (DiBattista et al. 2013); amongst them many Symbiodiniaceae-bearing host taxa, including e.g., octocorals (Fabricius and Alderslade 2001), scleractinian corals (Terraneo et al. 2014; Veron et al. 2015), and tridacnid clams (Richter et al. 2008). The evolutionary trajectories of these host species affect the rates and patterns of diversification of the associated symbionts (Thornhill et al. 2014), but comprehensive data on the evolutionary history of the host-Symbiodiniaceae system in the Red Sea is wanting.

5.2 Symbiodiniaceae Taxonomy and Phylogeny– Challenges in Diversity Analyses

The term 'zooxanthellae' (Brandt 1881) is commonly used to refer to dinoflagellate endosymbionts of the family Symbiodiniaceae in the order Suessiales (LaJeunesse et al. 2018). However, this term originally includes any golden-brown ('xanthos') algae of diatom and dinoflagellate origin living in symbioses with animals ('zoo') (Blank and Trench 1985, 1986; Trench 1979). The family Symbiodiniaceae (Fensome et al. 1993) was recently revised including a formal description of 7 genera (i.e. *Symbiodinium*, formerly clade A; *Breviolum*, formerly clade B; *Cladocopium*, formerly clade C; *Durusdinium*, formerly clade D; *Effrenium*, formerly clade E; *Fugacium*, formerly clade F; *Gerakladium*, formerly clade G) and the identification of further 8 lineages that require taxonomic classification (LaJeunesse et al. 2018). *Symbiodinium microadriaticum* LaJeunesse 2017 was the first Symbiodiniaceae species to be formally described by Freudenthal (1962). This original description was later found to be invalid because of the lack of a holotype that was only designated in 2017 (LaJeunesse 2017). *S. microadriaticum* was isolated from the scyphozoid upside-down jellyfish *Cassiopea xamachana* Bigelow, 1982 in the Bahamas by

McLaughlin and Zahl (1957, 1959) and by David A. Schoenberg from the same host in Florida in 1977 (LaJeunesse 2017). Because the first isolate of the species was lost, the second isolate, also known under culture strain number CCMP 2462/rt-061, was used to designate a species holotype (LaJeunesse 2017); the genome sequence of this species (strain CCMP2467) became recently available (Aranda et al. 2016). Since the original description, *S. microadriacticum* has been encountered in a range of hosts throughout the Red Sea (LaJeunesse 2001; Sawall et al. 2014; Ziegler et al. 2014b), as well as in other oceans (Correa and Baker 2009; LaJeunesse 2002; Reimer et al. 2007; Stat and Gates 2008; Stat et al. 2009).

Initially, taxonomic studies on these microalgae were hampered by the lack of distinguishing morphological attributes in symbiosis and further aggravated by the difficulty to maintain them in culture (Freudenthal 1962; Taylor 1969). Until today, only members of *Cladocopium* can be morphologically diagnosed and distinguished from other genera in the family Symbiodiniaceae (LaJeunesse et al. 2018). Consequently, *S. microadriaticum* was at first perceived as the exclusive panmictic symbiotic dinoflagellate species in cnidarians (Kevin et al. 1969; Taylor 1968, 1969), until studies on morphology, physiology, and biochemistry of cultured isolates revealed distinct ecological features and led to the description of several novel species in the family Symbiodiniaceae (Table 5.1; Banaszak et al. 1993; Blank and Trench 1985; Chang et al. 1983; Schoenberg and Trench 1980a, 1980b, 1980c; Trench and Blank 1987).

Until today, the establishment of cultures of different Symbiodiniaceae species remains a challenge (Krueger and Gates 2012; Santos et al. 2001). Hence, the advancement of Symbiodiniaceae taxonomy and phylogeny was driven by molecular techniques. Sequence analyses of the ribosomal small subunit (SSU) 18S rDNA revealed high phylogenetic divergence between Symbiodiniaceae lineages up to that between taxonomic orders of non-symbiotic dinoflagellates (Rowan and Powers 1992). The investigation of additional DNA marker regions, such as the ribosomal large subunit (LSU) 28S rDNA together with 18S rDNA from a wide array of invertebrate hosts corroborated these findings and prompted the division into 9 phylogenetic clades, designated A to I (Coffroth and Santos 2005; Loh et al. 2001; Pochon and Gates 2010; Pochon et al. 2004; Rodriguez-Lanetty et al. 2001; Rowan 1998; Stat et al. 2006), and later, a further subdivision into 15 genus-level lineages (LaJeunesse et al. 2018). But not all Symbiodiniaceae species are specific in their association with scleractinian corals. For example, to date members belonging to clades H and I have only been encountered in association with benthic Foraminifera (Pochon and Gates 2010; Pochon and Pawlowski 2006; Pochon et al. 2001), while scleractinian corals most commonly associate with Symbiodiniaceae of the genera *Symbiodinium*, *Breviolum*, *Cladocopium*, *Durusdinium* (formerly clade A to D), and occasionally with the genera *Fugacium* and *Gerakladium* (formerly clade F lineage Fr5 and clade G) as well as the undescribed genus represented by clade F lineage Fr2 (Baker 2003; Coffroth and Santos 2005; Rowan 1998).

The distinction into evolutionary subgeneric lineages has received further support from sequence analyses of the chloroplast LSU 23S (Santos et al. 2002) and the mitochondrial cytochrome c oxidase subunit 1 (COI) markers (Stern et al. 2010; Takabayashi et al. 2004). However, these investigations have also underlined limitations of such coarse taxonomic approaches, highlighting the importance of addressing discrete evolutionary units (i.e., species) at higher resolution. Analyses using the fast-evolving, non-coding internal transcribed spacer (ITS) regions of rDNA promised to fill this gap and drove the sub-division into so called phylotypes (hereafter referred to as 'types'), designated by the clade followed by an alphanumeric identifier (e.g., A1, C1, C2, etc.; Baillie et al. 2000; LaJeunesse 2001; van Oppen et al. 2001). Until today, hundreds of different (ITS2) Symbiodiniaceae types have been discovered, but to date only 25 of them have formally been described as biological species (Table 5.1). While evidence attests reasonable validity to the ITS2 marker for estimating Symbiodiniaceae species diversity in the majority of cases (Sampayo et al. 2009; Thornhill et al. 2007), recent research efforts have been aimed at developing more specific ITS2 primer pairs (Hume et al. 2018b) and a novel analytical framework (http://symportal.org; Hume et al. 2019) to delineate Symbiodiniaceae species diversity. In addition, combinations of alternative molecular markers including genes from all 3 compartments, i.e., chloroplast (cp23S, psbA), mitochondrion (COI, cob), and nucleus (nr28S, elf2), as well as microsatellites, are being analyzed to resolve species relationships (Lajeunesse et al. 2012; LaJeunesse and Thornhill 2011; LaJeunesse et al. 2014; Pochon et al. 2012, 2014).

Overall, the most commonly used method for determination of Symbiodiniaceae types has been denaturing gradient gel electrophoresis (DGGE) of polymerase chain reaction (PCR) amplified ITS2 sequences, and approximately half (46%) of all studies conducted in the Red Sea used this technique. A side effect of DGGE analyses on multicopy loci harboring intragenomic variation, such as ITS2, is the occurrence of heteroduplexes, which are mismatched DNA strands from different ITS2 copies within a sample. Although this is sometimes seen as a disadvantage of DGGE ITS2 analyses, heteroduplexes have successfully been used as a diagnostic feature that can increase DGGE resolution (Myers et al. 1985). One further constraint of the DGGE technique is that the detection limit of different Symbiodiniaceae types within mixed samples varies between clades and commonly ranges between 1 to 10%, which results in an underestimation of the total diver-

Table 5.1 List of formally described genera and species belonging to the family Symbiodiniaceae and "*nomina nuda*" (published specific epithets without formal diagnosis)

Symbiodiniaceae genera and species	Authors[a]	References	Isolated from	clade	ITS2 type[b]	Region	Country / type locality
Symbiodinium	Gert Hansen & Daugbjerg	Hansen and Daugbjerg (2009)					
S. microadriaticum	LaJeunesse	Freudenthal (1962), Trench and Blank (1987), Blank and Huss (1989), Kevin et al. (1969), LaJeunesse (2017), Lee et al. (2015), and Loeblich and Sherley (1979)	Cassiopea xamachana	A	A1	Caribbean	Florida Keys
"S. cariborum"			Condylactis gigantea	A	A.1.1	Caribbean	Jamaica
"S. microadriaticum var. condylactis"			C. xamachana /Cassiopea frondosa	A	A.1.1	Caribbean	Florida Keys/Jamaica
S. pilosum	Trench & Blank	Trench (2000) and Trench and Blank (1987)	Zoanthus sociatus	A	A2	Caribbean	Jamaica
"S. corculorum"			Corculum cardissa	A	A2	West Pacific	Palau
"S. meandrinae"			Meandrina meandrites	A	A2	Caribbean	Jamaica
"S. corculorum"			Corculum cardissa	A			
S. natans[d]	Hansen & Daugbjerg	Hansen and Daugbjerg (2009)	free-living, planktonic	A	A3	Northeast Atlantic	Canary Islands
"S. fitti"			Acropora palmata, Tridacna maxima	A	A3	Caribbean/Indopacific	
S. tridacnidorum	Lee, Jeong, Kang & LaJeunesse	Lee et al. (2015)	Hippopus hippopus, Tridacna gigas	A	A3	Indopacific	Great Barrier Reef, Palau
S. linucheae	(Trench and Thinh) LaJeunesse	LaJeunesse (2017) and Trench and Thinh (1995)	Linuche unguiculata	A	A4	Western Atlantic	Bermuda
S. necroappetens	LaJeunesse, Lee, Knowlton & Jeong	LaJeunesse et al. (2015)	Orbicella annularis	A	A13 (A1.1)	Caribbean	Jamaica
Breviolum	J.E.Parkinson & LaJeunesse	LaJeunesse et al. (2018)					
B. minutum[d]	(LaJeunesse, Parkinson & Reimer) J.E.Parkinson & LaJeunesse	LaJeunesse et al. (2012)	Aiptasia sp.	B	B1	Caribbean	Florida Keys
B. antillogorgium	(Parkinson, Coffroth & LaJeunesse) J.E.Parkinson & LaJeunesse	Parkinson et al. (2015)	Antillogorgia bipinnata	B	B1	Caribbean	Florida Keys
B. pseudominutum	(Parkinson, Coffroth & LaJeunesse) J.E.Parkinson & LaJeunesse	Parkinson et al. (2015)	Oculina diffusa	B	B1	Western Atlantic	Bermuda
"B. pulchrorum"			Aiptasia pulchella	B	B1	Central Pacific	Hawaii
"B. bermudense"			Aiptasia tagetes	B	B1	Western Atlantic	Bermuda
B. dendrogyrum	A.M. Lewis, A.N. Chan & LaJeunesse	Lewis et al. (2018)	Dendrogyra cylindrus	B	B1-1k	Caribbean	Curacao
B. faviinorum	A.M. Lewis & LaJeunesse	Lewis et al. (2018)	Diploria labyrinthiformis and other Faviidae, and the genera Isophyllia and Orbicella	B	B1-14-14a-24	Caribbean	Curacao

Species	Authority	Reference	Host	Clade	ITS2 type	Ocean	Region	Location
B. meandrinium	A.M. Lewis & LaJeunesse	Lewis et al. (2018)	Meandrina meandrites and other Meansdrinidae and the genus Orbicella	B	B1-20	Caribbean		Curacao
B. psygmophilum	(LaJeunesse, Parkinson & Reimer) J.E.Parkinson & LaJeunesse	Lajeunesse et al. (2012)	Oculina diffusa	B	B2	Western Atlantic		Bermuda
"B. muscatinei"				B	B4	East Pacific		USA
B. endomadracis	(Parkinson, Coffroth & LaJeunesse) J.E.Parkinson & LaJeunesse	Parkinson et al. (2015)	Anthopleura elegantissima Madracis spp.	B	B7	Caribbean		Curacao
B. aenigmaticum	(Parkinson, Coffroth & LaJeunesse) J.E.Parkinson & LaJeunesse	Parkinson et al. (2015)	Porites astreoides	B	close to B23	Caribbean		Florida Keys
Cladocopium	LaJeunesse & H.J. Jeong	LaJeunesse et al. (2018)						
C. goreaui[d]	LaJeunesse & H.J. Jeong	LaJeunesse et al. (2018), Trench (2000), and Trench and Blank (1987)	Heteractis/Rhodactis lucida	C	C1	Caribbean		Jamaica
C. thermophilum	(Hume, D'Angelo, Smith, Stevens, Burt & Wiedenmann) LaJeunesse & H.J. Jeong	Hume et al. (2015), and Hume et al. (2018a)	Porites lobata	C	C3, C3gulf[c]	Persian Gulf		United Arab Emirates, Abu Dhabi
Durusdinium	LaJeunesse	LaJeunesse et al. (2018)						
D. glynnii	(Wham & LaJeunesse) LaJeunesse	Wham et al. (2017)	Pocillopora type I, Seriatopora, Montipora	D	D1, D1-4-6	Entire Pacific		Palau
D. trenchii[d]	(LaJeunesse) LaJeunesse	LaJeunesse et al. (2014)	various Scleractinia	D	D1-4 (D1a)	West Pacific		Japan
D. eurythalpos	(LaJeunesse & Chen) LaJeunesse	LaJeunesse et al. (2014)	Oulastrea crispata	D	D8, D8-12, D12-13, D13	West Pacific		Taiwan
D. boreum	(LaJeunesse & Chen) LaJeunesse	LaJeunesse et al. (2014)	Oulastrea crispata	D	D15	West Pacific		Taiwan
Effrenium	LaJeunesse & H.J. Jeong	LaJeunesse et al. (2018)						
S. voratum[d]	(Jeong, Lee, Kang & LaJeunesse) LaJeunesse & H.J. Jeong	Jeong et al. (2014)	free-living to symbiotic	E		Pacific Ocean / Mediterranean		Korea
"S. californium"			Anthopleura elegantissima	E	E1	Central Pacific		Hawaii
Fugacium	LaJeunesse	LaJeunesse et al. (2018)						
F. kawagutii[d]	LaJeunesse	LaJeunesse et al. (2018), Trench (2000), and Trench and Blank (1987)	Montipora verrucosa	F	F1	Central Pacific		Hawaii
Gerakladium	LaJeunesse	LaJeunesse et al. (2018)						
G. endoclionum[d]	(Ramsby & LaJeunesse) LaJeunesse	Ramsby et al. (2017)	Cliona orientalis	G		Indopacific		Great Barrier Reef
G. spongiolum	(M.S. Hill & LaJeunesse) LaJeunesse	Ramsby et al. (2017)	Cliona varians	G		Caribbean		Florida Keys

[a] For species of newly erected genera, the original species authors are given in parentheses and the author of the new genus is given at the end in accordance with taxonomic nomenclature
[b] Please note that different species of Symbiodiniaceae may have an identical main ITS2 sequence. Former designated ITS2 types are listed in parentheses.
[c] Minor sequence variant not resolved by DGGE.
[d] Type species of the genus

sity (LaJeunesse et al. 2008; Thornhill et al. 2006b). Bacterial cloning, on the other hand, which was used in about one fifth (21%) of Red Sea studies, overestimates the diversity, because it retrieves a high number of intragenomic ITS2 variants, alongside the intergenomic variability within a sample (Arif et al. 2014; Thornhill et al. 2007). The remaining third (29%) of studies from the Red Sea used restriction fragment length polymorphisms (RFLPs) of 18S rDNA, which was used in Symbiodiniaceae molecular research early on (Rowan and Powers 1991a, 1991b). So far, only one study (corresponding to 4%) used high-resolution, high-throughput next-generation sequencing (NGS), yielding a high number of ITS2 sequence reads, thus capturing a high proportion of the diversity in mixed Symbiodiniaceae assemblages and providing information on the relative abundance of distinct sequence variants within a sample (Ziegler et al. 2017).

5.3 Functional Diversity of Different Host-Symbiodiniaceae Pairings

Symbiodiniaceae species can be attributed specific physiological and biochemical properties, which reflect their adaptation to distinct environments. These adaptations translate into different properties for the associated coral host, for example by increasing growth rates in coral recruits depending on the Symbiodiniaceae type (Little et al. 2004). Consequently, the ability to associate with different Symbiodiniaceae types is an important factor influencing a coral species' distribution range (Rodriguez-Lanetty et al. 2001), metabolic performance (Cooper et al. 2011b), and stress tolerance (Abrego et al. 2008; Berkelmans and van Oppen 2006; Howells et al. 2012).

Between coral species, the niche partitioning in host-Symbiodiniaceae associations is most commonly observed along depth-mediated gradients of light and temperature, where it is an important variable explaining depth zonation. For example, photosynthetic properties of *Durusdinium* type D1 symbionts in *Pocillopora verrucosa* Ellis & Solander, 1786 dominating shallow habitats between 0 – 6 m were distinct from those of *Cladocopium* type C1c in *Pavona gigantea* Verrill, 1869, occurring in deeper water from 6 – 14 m (Iglesias-Prieto et al. 2004). These host-specific symbionts were adapted to different light regimes, and host-symbiont fidelity contributed to vertical niche partitioning between the 2 coral species (Iglesias-Prieto et al. 2004). Observations of four scleractinian genera over a large depth gradient in the central Red Sea (Ziegler et al. 2015a) and within the genus *Agaricia* in the Caribbean (Bongaerts et al. 2013) support the concept of host-specific Symbiodiniaceae association as one of the drivers of depth-niche partitioning between taxa.

A possible determinant of host-symbiont specificity is the mode of symbiont acquisition. In brooding and some broadcast spawning corals, Symbiodiniaceae cells are directly passed on to the offspring vertically (Trench 1987). In contrast, corals with horizontal symbiont transmission have symbiont-free gametes and each generation has to acquire symbionts from the environment *de novo* (Trench 1987). While vertical symbiont transmission avoids the risk associated with having to find new symbiont partners, as is the case with horizontal transmission, the resulting tight co-evolution may also limit the flexibility of the host to associate with a wide (phylogenetic) range of symbionts. In fact, vertical symbiont transmission promotes the evolution of specialist symbionts (LaJeunesse et al. 2004a). In contrast, each generation in horizontally transmitting coral species can potentially yield new host-symbiont combinations and the initial uptake of Symbiodiniaceae is relatively flexible (Abrego et al. 2009; Coffroth et al. 2001; Gómez-Cabrera et al. 2008; Little et al. 2004; Voolstra et al. 2009), although it may be limited by the symbionts' cell size (Biquand et al. 2017). Such flexibility may be particularly important with regard to range expansions (Grupstra et al. 2017) and global climate change (Decelle et al. 2018). However, studies addressing the connection between different reproductive strategies and host-symbiont specificity remain inconclusive, and hence, the issue remains a matter of debate (Barneah et al. 2004; LaJeunesse et al. 2004a, 2004b; Rodriguez-Lanetty et al. 2004; Stat et al. 2008; Thornhill et al. 2006a; van Oppen 2004).

Symbiont generalist coral species are characterized by more flexible Symbiodiniaceae associations (Baker 2003). In these generalist corals, the distribution of Symbiodiniaceae can vary with irradiance levels within a single colony, and in fact most of these coral colonies harbor more than one Symbiodiniaceae genus and/or type at the same time, often in uneven proportions (Mieog et al. 2007; Silverstein et al. 2012). This was first observed in *Orbicella annularis* Ellis & Solander, 1786 and *Orbicella faveolata* Ellis & Solander, 1786, that harbored members of the genera *Symbiodinium* and *Breviolum* in sun-exposed and *Cladocopium* in shaded parts of the colonies (Rowan et al. 1997). Similar patterns were later found in other coral species (Ulstrup and Van Oppen 2003). However, spatial differences in association within a single coral colony do not seem to be a universal phenomenon, as e.g. within colonies of *Pocillopora* symbiont types are distributed uniformly (LaJeunesse et al. 2008; Pettay et al. 2011). More generally, it is assumed that in the majority of cases only a single Symbiodiniaceae taxon is predominant in an individual coral (Goulet and Coffroth 2003; Thornhill et al. 2009; Pettay et al. 2011; Baums et al. 2014).

Stratification of symbionts within generalist species also exists between colonies along environmental gradients. For example, some corals from the genera *Madracis* (Frade et al. 2008) and *Orbicella* (Rowan and Knowlton 1995) associate with different Symbiodiniaceae in shallow and deep water. In the Red Sea, it was recently demonstrated for *Porites lutea* that the symbiont community of a single coral host species is variable across depth, cross-shelf location, and sampling times (Ziegler et al. 2015b). These finding contradict the concept of high symbiont specificity in *Porites* (Ziegler et al. 2015b) and highlight the need for more comprehensive sampling efforts to study the diversity of host-Symbiodiniaceae associations, in particular because this relationship is directly compromised by the consequences of global climate change.

5.4 Symbiodiniaceae Diversity in the Red Sea

Overall, 24 studies reported host-Symbiodiniaceae associations in the Red Sea, spanning 57 host genera belonging to 23 families and 8 orders that were associated with a total of 65 Symbiodiniaceae types from 5 genera (*Symbiodinium*, formerly clade A; *Breviolum*, formerly clade B; *Cladocopium*, formerly clade C; *Durusdinium*, formerly clade D; *Fugacium*, formerly clade F / lineage Fr5; and representatives of clade F lineages Fr2 and Fr4 with yet undescribed genera). Members of the genus *Cladocopium* dominated the endosymbiont assemblages throughout the Red Sea (Fig. 5.2). The majority of host genera (49/57, 86%) were associated with members of *Cladocopium* at least once and a total of 45 *Cladocopium* ITS2 types were recorded (Table 5.2).

The most common ITS2 types were C1 and C41, present in 23 and 21 genera across all Red Sea regions, respectively. In contrast, other *Cladocopium* types displayed more specific associations with their host organisms. For example, although considered a generalist type, *Cladocopium* C3 was found in only 4 genera (*Montipora, Pachyseris, Pocillopora,* and *Xenia*), and C38 was limited to *Montipora*, C161 and C162 to *Stylophora*, C163 to *Seriatopora*, C39 to Agaricidae (*Gardineroseris, Leptoseris, Pachyseris, Pavona*), and C65 to Alcyoniidae (*Lobophytum, Sarcophyton, Sinularia*) (Table 5.2).

The genus *Symbiodinium* (formerly clade A) was found in 14 host genera, and it occurred in almost even proportions along the Red Sea coast (Fig. 5.2). The overall third most abundant type after *Cladocopium* C1 and C41 was *Symbiodinium* A1, but its occurrence was limited to the genus *Montipora* and the family Pocilloporidae (*Pocillopora, Seriatopora,* and *Stylophora*), whose members belonged to the most frequently sampled taxa.

The proportion of host genera found to harbor the genus *Durusdinium* (formerly clade D) increased from 2 (6% of sampled genera) in the north, 9 (30%) in the central north, 16 (57%) in the central Red Sea to 18 and 3 (each representing 75% of sampled genera) in the central south and the southern Red Sea, respectively. More specifically, the genera *Acropora, Astreopora, Diploastrea, Gardineroseris, Pavona, Pocillopora,* and *Porites* changed from Symbiodiniaceae assemblages consisting of the genera *Symbiodinium* and/or *Cladocopium* to (additionally) containing *Durusdinium* towards the southern localities of their respective distributions (Table 5.2). Other genera, such as *Echinopora, Montipora,* and *Stylophora* were found to associate with *Durusdinium* at some localities throughout their range.

The genus *Fugacium* (formerly clade F / lineage Fr5) was found in association with the foraminiferan genus *Amphisorus* in the northern Red Sea. The clade F lineage Fr4 was found in association with the foraminiferan genus *Sorites* in the northern Red Sea, and the clade F lineage Fr2 was found in association with both Foraminifera in the northern and with the coral *Stylophora* in the central Red Sea. The genus *Breviolum* (formerly clade B), uncommon to the Indopacific region, was recorded once in association with *P. verrucosa* (Ziegler et al. 2014b).

5.5 Biogeographic Patterns in Symbiodiniaceae Diversity and Host-Symbiont Associations

The presence of 65 Symbiodiniaceae types encountered in 57 host genera compares favorably with diversity estimates from surveys in other locations. For instance, LaJeunesse et al. (2004b, 2010) sampled a comparable, mixed host assemblage consisting of 58 genera in the Andaman Sea (Thailand) and observed only 37 Symbiodiniaceae types. Similarly, 50 host genera in the Caribbean contained 35 Symbiodiniaceae types (LaJeunesse et al. 2003) and higher numbers of host genera sampled in the Western Indian Ocean (70) and the central Great Barrier Reef (GBR) (72) yielded 47 and 33 Symbiodiniaceae types, respectively (LaJeunesse et al. 2004b, 2010), highlighting the high relative diversity of Symbiodiniaceae in the Red Sea.

The distribution and occurrence of Symbiodiniaceae from the different genera and lineages varies across biogeographic regions. In the IndoPacific, the two main Symbiodiniaceae genera associated with hard corals are *Cladocopium* and *Durusdinium*. Hard coral-symbiont assemblages in the Red Sea share the dominance of *Cladocopium* and the occurrence of *Durusdinium* with those in the IndoPacific, however, they are distinct with regard to the presence of *Symbiodinium* symbionts in Pocilloporidae and few other species. The genus *Symbiodinium* is rarely found in hard corals of the IndoPacific region, while it is common in the Atlantic Ocean. The presence and large diversity of the genus *Breviolum* in the Caribbean and North Atlantic in turn separates these Symbiodiniaceae assemblages from those in the Red Sea. In the Caribbean, *Breviolum, Cladocopium, Symbiodinium,* and *Durusdinium*, in descending order of prevalence, dominate Symbiodiniaceae assemblages in hard corals, which are considered to be more diverse in relation to the number of

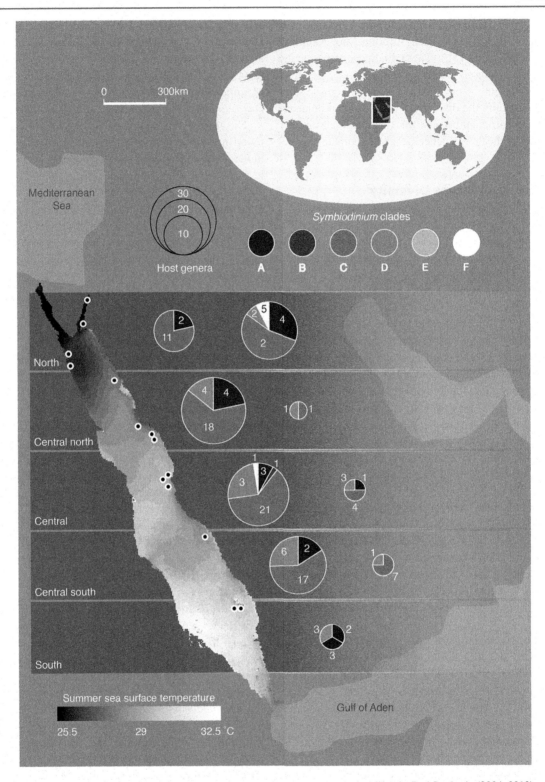

Fig. 5.2 Distribution of Symbiodiniaceae lineages (formerly clades) in scleractinian corals (left pie charts) and other host organisms (right pie charts) sampled along the coast of the Red Sea. Symbiodiniacean clades correspond to the recently described genera: former clade A, *Symbiodinium*; former clade B, *Breviolum*; former clade C, *Cladocopium*; former clade D, *Durusdinium*; former clade F lineage Fr5, *Fugacium*; and representatives of clade F lineages Fr2 and Fr4 with yet undescribed genera. Average summer sea surface temperatures are depicted for the Red Sea basin (2004–2013) and sampling sites are demarcated by black circles. Scale bar denotes distances across the Red Sea. Sizes of pie charts denote the number of host genera sampled in each region of the Red Sea (north, central north, central, central south, and south, respectively) and the numbers in the pie charts represent the number of ITS2 phylotypes encountered per clade at a location. (Data based on references listed in Table 5.2)

Table 5.2 Invertebrate host genera found in association with Symbiodiniaceae ITS2 types along the coast of five Red Sea regions (N = north, CN = central north, C = central, CS = central south, S = south)

	Genus	Red Sea regions (underlined) and Symbiodiniaceae ITS2 types [a]	References
Scleractinia	Acanthastrea	CN: C1, C41	Ziegler et al. (2017)
	Acropora	N: A, C41, C74; CN: A, C, C1, C41; C: C1, C41, D1, D17, D1-4; CS: C1, C41, C41a; S: C41a, D1-4	Baker et al. (2005), Baker et al. (2004), Barneah et al. (2004), Jessen et al. (2013), Pochon et al. (2006), Santos and LaJeunesse (2006), and Ziegler et al. (2017)
	Astreopora	CN: A, C, C41; C: C1, D1; CS: C41, D1, D1-4	Baker et al. (2005) and Ziegler et al. (2017)
	Cladocora	CS: D1-4, D5	Ziegler et al. (2017)
	Coscinarea	C: C1	Ziegler et al. (2017)
	Cyphastrea	CN: C, C1, C1b, C41; CS: C1, C41	Ziegler et al. (2017)
	Diploastrea	CN: C; C: C41, D1, D1-4; CS: D1, D6, D10	Baker et al. (2005), Baker et al. (2004), and Ziegler et al. (2017)
	Echinophyllia	CS: A1, C22	Ziegler et al. (2017)
	Echinopora	CN: C, C1, C41, D; C: D1, D1-4; CS: C1, C41	Baker et al. (2005) and Ziegler et al. (2017)
	Favia	N: C; CN: C, C1, C41; C: C1, C41; CS: C1, C1b, C41	Baker et al. (2005), Baker et al. (2004), Karako-Lampert et al. (2004), and Ziegler et al. (2017)
	Favites	CN: C; C: C41	Baker et al. (2005) and Ziegler et al. (2017)
	Fungia	N: C; CN: C, C1; C: C1, C1b; CS: C1, C1b, C41	Baker et al. (2005), Karako-Lampert et al. (2004), and Ziegler et al. (2017)
	Galaxea	CN: C, D; C: C1, C1b; CS: C1, C41	Baker et al. (2005) and Ziegler et al. (2017)
	Gardineroseris	CN: C, C39; C: C1, C39, D1, D1-4	Baker et al. (2005) and Ziegler et al. (2017)
	Goniastrea	CN: C1, C41, D1-4; C: C15, C41; CS: C15, D1, D1-4	Ziegler et al. (2017)
	Goniopora	N: C	Karako-Lampert et al. (2004)
	Hydnophora	CN: C; C: D1	Baker et al. (2005)
	Leptoria	C: C1, C1mm, C15, C39, C63, D1-4	Baker et al. (2005), Baker et al. (2004), and Ziegler et al. (2017)
	Leptoseris	CS: C1, C41	Ziegler et al. (2015a)
	Lobophyllia	CN: C41	Ziegler et al. (2017)
	Montastrea	N: C; CN: A, A1, C, C1b, C3, C3w, C38, D1-4; C: C3, C38; CS: A1, C3, C38, C41, D1, D1-4, D6	Baker et al. (2005), Baker et al. (2004), Karako-Lampert et al. (2004), and Ziegler et al. (2017)
	Mussa	CN: C	Baker et al. (2005)
	Mycedium	C: C1, C1b	Ziegler et al. (2017)
	Pachyseris	CN: A; C: C1, C1mm, C3, C39	Baker et al. (2005) and Ziegler et al. (2015a)
	Pavona	CN: C, C1, C15, C39, C41; C: C1, C39; CS: C1, C1b, C39, C83, C116, D1	Baker et al. (2005), Baker et al. (2004), and Ziegler et al. (2017)
	Plerogyra	CN: D	Baker et al. (2005)
	Plesiastrea	N: C41	Barneah et al. (2007)
	Pocillopora	N: A1, C, C1, C1ª, C1b, C1nn, C1oo; CN: A, A1, A1b, A1c, C, C3, C3w, C15, C19; C: A1, A1c, A21, B1, C#, C1#; C15, C98, C116, D1-4; CS: A1, A1c, C1oo, C21, C41; S: A1, A21, C1ª, C19, D6, D17	Baker et al. (2005), Baker et al. (2004), Karako-Lampert et al. (2004), LaJeunesse et al. (2009), Santos and LaJeunesse (2006), Sawall et al. (2014), Ziegler et al. (2017), Ziegler et al. (2014b), and Ziegler et al. (2015b)
	Podabacia	CN: C; C: C1, C39	Baker et al. (2005) and Ziegler et al. (2015a)
	Porites	CN: A, C, C15; C: C15, C15n, C15o, C15p, C97, C99, C116, D1-4; CS: C22, D1, D1-4, D6	Baker et al. (2005), Baker et al. (2004), Ziegler et al. (2017), Ziegler et al. (2015a), and Ziegler et al. (2015b)
	Seriatopora	N: C, C3nt; CN: A, C, C163, C163a, C163b; C: A1, C163a, C163b; CS: A1; S: A1	Baker et al. (2005), Karako-Lampert et al. (2004), Nir et al. (2011), Santos and LaJeunesse (2006), and Ziegler et al. (2017)

(continued)

Table 5.2 (continued)

	Genus	Red Sea regions (underlined) and Symbiodiniaceae ITS2 types [a]	References
	Stephanocoenia	CN: C1	Ziegler et al. (2017)
	Stylophora	N: A, A1, C, C72; CN: A, A1, C, C21, C160, C161, C161a, C162a; C: A1, C1#, C116, Fr2-2; CS: A1, C19, C162, D1-4; S: C19	Baker et al. (2005), Baker et al. (2004), Barneah et al. (2007), Karako-Lampert et al. (2004), LaJeunesse et al. (2009), LaJeunesse (2001), Lampert-Karako et al. (2008), Santos and LaJeunesse (2006), and Ziegler et al. (2017)
	Symphyllia	CS: C41	Ziegler et al. (2017)
	Turbinaria	N: C, C1, C1n, C41; CN: N: C; C: C1; CS: C1, C1b, C41	Baker et al. (2005), Barneah et al. (2007), Karako-Lampert et al. (2004), and Ziegler et al. (2017)
Alcyonacea	*Anthelia*	N: C	Barneah et al. (2004)
	Cladiella	N: C	Barneah et al. (2004)
	Lithophyton	N: A, A10	Barneah et al. (2004) and LaJeunesse et al. (2009)
	Lobophytum	C: C65; CS: C65	Ziegler et al. (2017)
	Nephthea	N: A; C: A10	Barneah et al. (2004) and Ziegler et al. (2017)
	Paralemnalia	N: C	Barneah et al. (2004)
	Rhytisma	N: C	Barneah et al. (2004)
	Sarcophyton	N: C; CS: C65	Barneah et al. (2004) and Ziegler et al. (2017)
	Sinularia	N: C; C: C65; CS: C1, C1b, C65	Barneah et al. (2004) and Ziegler et al. (2017)
	Stereonephthya	N: A, A9	Barneah et al. (2007) and Barneah et al. (2004)
	Xenia	N: C; CN: C3; C: C3, C41, C115a, D1-4, D3, D5; CS: C3, C3.7, C3n, D3	Barneah et al. (2004) and Ziegler et al. (2017)
	Heteroxenia	N: C	Barneah et al. (2004) and Goulet et al. (2008b)
others	*Discosoma*	N: C1, D1-4	Kuguru et al. (2008)
	Rhodactis	N: C, C1, D, D1-4	Kuguru et al. (2008) and Kuguru et al. (2007)
	Tridacna	N: A, C	Richter et al. (2008)
	Millepora	N: A	Karako-Lampert et al. (2004) and Pochon et al. (2001)
	Amphisorus	N: C, Fr2, Fr5	Pawlowski et al. (2001) and Pochon et al. (2001)
	Marginopora	N: C	Pochon et al. (2001)
	Sorites	N: C, F2, Fr2, Fr4	Pawlowski et al. (2001), Pochon et al. (2006), and Pochon et al. (2001)
	Cassiopea	N: A1	LaJeunesse (2001)
	Waminoa	N: A11	Barneah et al. (2007)

Symbiodiniaceae clades correspond to the recently described genera: former clade A, *Symbiodinium*; former clade B, *Breviolum*; former clade C, *Cladocopium*; former clade D, *Durusdinium*; former clade F lineage Fr5, *Fugacium*; and representatives of clade F lineages Fr2 and Fr4 with yet undescribed genera

[a]Symbiodiniaceae detection methods used: DGGE (Barneah et al. 2007; Jessen et al. 2013; LaJeunesse 2001, 2005; Santos and LaJeunesse 2006; Sawall et al. 2014; Ziegler et al. 2014b; Ziegler et al. 2015a; Ziegler et al. 2015b), cloning & sequencing (Kuguru et al. 2008; Nir et al. 2011; Pawlowski et al. 2001; Pochon et al. 2006; Richter et al. 2008), RFLP (Baker et al. 2004; Barneah et al. 2004; Goulet et al. 2008b; Karako-Lampert et al. 2004; Kuguru et al. 2007; Lampert-Karako et al. 2008; Pochon et al. 2001), NGS (Ziegler et al. 2017)

host species than their IndoPacific counterparts (LaJeunesse et al. 2003). Data presented herein further points towards the Red Sea as a hot spot of Symbiodiniaceae diversity.

Cladocopium types C1 and C3 are believed to be at the basis of a series of regional adaptive radiation events in this genus in the miocene-pleistocene transition (LaJeunesse 2005; Thornhill et al. 2014). Repeated radiation led to high diversity within the genus *Cladocopium* (also reflected in its name) compared to the other Symbiodiniaceae genera and lineages (LaJeunesse et al. 2004b), and this pattern was also apparent in the Red Sea, with 69% of all types belonging to *Cladocopium*. The ancestral and globally occurring *Cladocopium* types C1 and C3 were also found in the Red Sea. While C1 represented the most widespread *Cladocopium* type in the Red Sea, *Cladocopium* C3 was rather uncommon, as opposed to other regions, where both types mostly occur in co-dominance (LaJeunesse et al. 2003). A different type, *Cladocopium* C41 so far only reported from Red Sea waters (LaJeunesse 2005; Pochon et al. 2001; Ziegler et al. 2017) was almost as ubiquitous as C1. Its sequence similarity with C1, from which it is separated by a single base pair difference in the ITS2 region, suggests a diversification event, specific to the Red Sea.

Some Symbiodiniaceae types are strongly associated with certain host taxa over large geographic ranges. For instance, the association of *Cladocopium* type C65 with Alcyoniidae extends beyond the Red Sea to other locations in the Western Indian Ocean (LaJeunesse et al. 2010) and along the GBR (Goulet et al. 2008a; LaJeunesse 2005; LaJeunesse et al. 2004b). Furthermore, *Seriatopora hystrix* associates with *Cladocopium* C3nt in both the Red Sea and the GBR. But while its association in the GBR is limited to symbionts from the genus *Cladocopium* (Bongaerts et al. 2010; LaJeunesse et al. 2003; Sampayo et al. 2007; Stat et al. 2008), it is additionally associated with symbionts from the genus *Symbiodinium* in the Red Sea. *Porites* in turn, though widely regarded a symbiont specialist coral for *Cladocopium* C3 in the Persian Gulf (Hume et al. 2013) and C15 in the IndoPacific (see Franklin et al. 2012), was found to be associated with a wide range of Symbiodiniaceae from the genera *Symbiodinium*, *Cladodopium*, and *Durusdinium* along the Red Sea coast. The diversity encountered in this genus in the Red Sea equals that found in Caribbean *Porites* (Finney et al. 2010; Green et al. 2010; LaJeunesse 2002, 2005), suggesting local, species-specific adaptive events driving host-symbiont specificity.

Patterns of variable host-symbiont association have previously been related to latitudinal gradients of temperature and inorganic nutrients (LaJeunesse 2002, 2005; LaJeunesse et al. 2004b; Loh et al. 2001; Macdonald et al. 2008). For instance, comparable to the reports of shifting Symbiodiniaceae communities to *Durusdinium* dominance for several host genera in warmer regions of the Red Sea (*Acropora*, *Astreopora*, *Diploastrea*, *Gardineroseris*, *Pavona*, *Pocillopora*, and *Porites*), *Acropora tenuis* in Australia associated with *Cladocopium* in the south (C3) and central region (C1, C3), i.e., the more temperate environments of the GBR, and with *Cladocopium* (C1) and *Durusdinium* (D1) in the northern warmer parts (LaJeunesse et al. 2004b; LaJeunesse et al. 2003). Latitudinal shifts in the association between types within a genus were observed in *P. verrucosa*, which associated with *Symbiodinium* A1 throughout its distribution range and with *Symbiodinium* A21 at the most southern location of the Red Sea (Sawall et al. 2014). These latitudinal shifts of Symbiodiniaceae assemblages in the Red Sea towards higher proportions of *Durusdinium* and *Symbiodinium* type A21 symbionts is also apparent along cross-shelf gradients, as evidenced by their presence in warmer, nutrient enriched and more turbid nearshore reefs, while being absent from colder offshore reefs in the central Red Sea (Ziegler et al. 2015b). Cross-shelf and habitat specific shifts between the genera *Cladocopium* and *Durusdinium* also occurred in coral communities in the GBR (Cooper et al. 2011a; Ulstrup and Van Oppen 2003), Palau (Fabricius et al. 2004), and Indonesia (Hennige et al. 2010). Moreover, corals harboring *Durusdinium* bleached less compared to those harboring other Symbiodiniaceaen lineages (Baker et al. 2004; Berkelmans and van Oppen 2006). Taken together, these observations suggest a competitive advantage for the respective coral host when associated with symbionts from the genus *Durusdinium* under warmer and nutrient enriched environmental settings (Berkelmans and van Oppen 2006; Stat and Gates 2011). However, not all host-symbiont combinations show this effect (Abrego et al. 2008), and the recently described 'heat-loving' *Cladocopium thermophilum* Hume, D'Angelo, Smith, Stevens, Burt & Wiedenmann, 2018 does not belong to the assumed heat-tolerant genus *Durusdinium* (Hume et al. 2015). This indicates that thermotolerance is not associated with a specific Symbiodiniaceaen lineage, but rather a type or species-specific trade that can be found in some members of all Symbiodiniaceaen lineages (Swain et al. 2017).

5.6 Coral Bleaching and Symbiosis Breakdown

Coral bleaching is the dissociation of the coral-Symbiodiniaceae symbiosis, during which the coral host loses large proportions of Symbiodiniaceae cells leading to the white, i.e., bleached appearance (Hoegh-Guldberg 1999). Global climate change induced sea surface warming and increased frequency and severity of temperature anomalies are the main causes for mass bleaching and global coral die-off events (Hoegh-Guldberg et al. 2007) that are becoming more frequent and devastating as evidenced by the third global coral bleaching event that affected coral reefs across the globe and devastated roughly one third (29%) of the coral reefs in the northern GBR (Hughes et al. 2018a; Hughes et al. 2018b). As most corals live very close to their upper thermal limit (Jokiel and Coles 1990), they are already susceptible to only small deviations from the long-term minima and maxima of temperature and other environmental factors (Kleypas et al. 1999). The effects of thermal stress can be aggravated when combined with eutrophication or imbalance of inorganic nutrients (Rädecker et al. 2015; Wiedenmann et al. 2012) and/or high solar irradiance (Fitt et al. 2001; Lesser 1996), which can also trigger bleaching on its own (Glynn 1993, 1996; Hoegh-Guldberg and Smith 1989; Krämer et al. 2013). Depending on the severity and duration of the stressor, corals can recover from bleaching events (e.g. Connell 1997). However, often coral bleaching leads to the death of the host, entailing mass mortalities and degradation of whole reefscapes (Hughes et al. 2018b; Sheppard 2003).

Several processes in the photosynthetic apparatus of Symbiodiniaceae are affected during coral bleaching. The breakdown of photosynthetic pathways and the continued absorption of light energy in photosystem II eventually exceed the capacity for non-photochemical quenching (Gorbunov et al. 2001; Wooldridge 2009) and ultimately lead to the production of reactive oxygen species (ROS) (Lesser 1996; Suggett et al. 2008; Tchernov et al. 2004; Warner et al. 1999). The different ROS impair and damage the structure and function of proteins, lipids, and DNA (Fey et al. 2005; Hideg et al. 1995; Martindale and Holbrook 2002; Smith et al. 2005). Consequently, ROS transgress to the coral host, causing further oxidative stress that is linked to the onset of coral bleaching (Lesser 1996; Smith et al. 2005) and initiation of apoptotic pathways in the host (Tchernov et al. 2011). Notably, recent studies suggest that other factors besides photodamage-induced ROS must be considered to explain observed bleaching phenomena (Tolleter et al. 2013; Diaz et al. 2016; Pogoreutz et al. 2017). As such, the elucidation of the cellular mechanisms underlying bleaching is an area of active investigation and critical to the design of meaningful interventions and mitigation strategies. Generally, coral species vary in their bleaching tolerance, with branching growth forms, as prevalent in e.g. Acroporids and Pocilloporids, displaying higher susceptibility than massive or encrusting species (Baird et al. 2009; Fitt et al. 2001; Loya et al. 2001; Stimson et al. 2002) and these trends are also apparent in bleaching events reported from the Red Sea (Table 5.3).

Similar to the flexible formation of host-symbiont relationships as a means to broaden the ecological niche, a resembling process has been formulated as a response to stress within individual colonies. The so-called 'adaptive bleaching hypothesis' was proposed as a mechanism through which the coral host can shift Symbiodiniaceae types to enhance its resilience to environmental changes (Buddemeier and Fautin 1993). This process is based on replacement of one Symbiodiniaceae type by another less abundant type ('shuffling') or by intake of exogenous Symbiodiniaceae from the environment ('switching') (Baker 2003). Background shuffling, i.e., changing proportions of Symbiodiniaceae types present in low abundances, may be a common phenomenon (McGinley et al. 2012) and the emergence of rare, less-abundant background Symbiodiniaceae can play a role during acute stress events (Boulotte et al. 2016; Lee et al. 2016), or as a source of adaptive potential over evolutionary time scales (Hume et al. 2016). But generally, many Symbiodiniaceae communities are stable over time (Thornhill et al. 2006a; Thornhill et al. 2006b), even during exposure to changing environmental conditions (Bongaerts et al. 2011), or when recovering from stress events (Goulet and Coffroth 2003; McGinley et al. 2012). Unfortunately, to date, there is a lack of information on the dynamics of the host-Symbiodiniaceae association during bleaching events in the Red Sea.

5.7 Coral Bleaching in the Red Sea

Coral reef ecosystems in the Red Sea thrive in warm seawater temperatures that exceed the tolerable limits of reef corals elsewhere (Kleypas et al. 1999). Caused by a selective bottleneck in the southern Red Sea possibly selecting for heat-resistant populations, the cooler northern part of the Red Sea is proposed to be a refuge for corals under global warming (Fine et al. 2013; Osman et al. 2018). In line with this, bleaching thresholds for corals in the Red Sea are higher than for most locations in the GBR, Indo-Pacific, and the Caribbean (Berkelmans 2002; Osman et al. 2018). In the last global report on the status of coral reefs in 2008, an estimated 82% of reefs in the Red Sea were classified at low risk (Wilkinson 2008). However, growth rates of *Diploastrea heliopora* have been declining since 1998 as a response to increased water temperatures, indicating that global warming also poses a major threat for Red Sea coral reefs (Cantin et al. 2010).

During the 1990s the Red Sea started to experience frequent SST anomalies and abrupt warming (Raitsos et al. 2011), and this period coincides with the earliest reports of coral bleaching in the Red Sea during the global coral bleaching event in 1998 (Table 5.3). Earlier *in situ* bleaching data from the region is wanting and large-scale surveys only started to take place under the umbrella of the Regional Organization for the Conservation of the Environment of the Red Sea and Gulf of Aden (PERSGA) after its foundation in 1995. However, using coral cores to estimate growth rates, a study by Cantin et al. (2010) suggests a possible thermal

5 Symbiodiniaceae Diversity in Red Sea Coral Reefs & Coral Bleaching

Table 5.3 Record of bleaching events in the Red Sea with details on severity and affected coral taxa

Year	Red Sea region	Country	Location	Bleaching severity	Coral taxa affected and comments	References
1996/1997	south	Yemen	Al Khawkhah	High	Indirect observation of deceased colonies of *Goniastrea*, *Montipora*, *Porites* between survey years	Turak et al. (2007)
1998	central north, north	Saudi Arabia	Yanbu, Al-Wajh to Gulf of Aqaba	Low–medium	Low incidence, patchy bleaching, ~10%	DeVantier et al. (2005)
1998	central	Saudi Arabia	Rabigh to Yanbu	Medium–high	Severe bleaching on shallow reefs <6 m, up to 90% affected in Rabigh area, most affected, common taxa: *Acanthastrea*, *Acropora*, *Dipsastraea*, *Galaxea*, *Gardineroseris*, *Goniastrea*, *Leptastrea*, *Merulina*, *Millepora*, *Pocillopora*, *Porites*, *Stylophora*, *Tubastraea*	DeVantier et al. (2005)
1998	south	Eritrea	Massawa, Green Island	Medium–high	Water temperatures up to 40 °C, shallow and deep bleaching with subsequent mortality in shallow and recovery in deeper locations	ReefBase, NOAA, C. Wilkinson
1998?	south	Sudan	Dungonab Bay	–	Surveys in 2002 outside the bay found many dead colonies (*Galaxea* amongst others) and reefs in poor health state, possibly linked to previous bleaching	PERSGA (2003)
2002/2003	north	Israel	Gulf of Aqaba	–	sporadic bleaching of *Montipora*	Loya (2004)
2007	north / central north	Egypt, Jordan, Sudan	Fringing reefs	–	Low tides exposed reef flats with subsequent bleaching and mortality	Kotb et al. (2008)
2007	central north	Egypt	Rocky Island	–	Localized SST anomaly led to bleaching down to 20 m depth	Kotb et al. (2008)
2010	central	Saudi Arabia	Thuwal	Low–high	Cross-shelf gradient of bleaching with highest incidence nearshore of: Acroporidae, Agariciidae, Faviidae, Fungiidae, Merulinidae, Pectiniidae, Pocilloporidae, Poritidae, Siderastreidae, Oculinidae, Mussidae, Dendrophylliidae	Furby et al. (2013)
2010	central	Saudi Arabia	Thuwal	Variable	Spatial bleaching pattern of *Stylophora*; offshore reefs and nearshore protected areas unbleached	Piñeda et al. (2013)
2010	central	Saudi Arabia	Thuwal	High	Bleaching of all anemones in the area: *Heteractis*, *Entacmaea*, *Stichodactyla*	Hobbs et al. (2013)
2010/2011	north	Israel	Gulf of Aqaba	Medium	Mesophotic bleaching of *Stylophora* between 40 – 63 m, regular seasonal phenomenon?	Nir et al. (2014)
2010?	north	Egypt	Hurghada, Safaga, El Quesier	Low–medium	Based on the report, the year of bleaching is unclear. Medium bleaching: *Acropora*, *Montipora*, *Stylophora* Low bleaching: *Echinopora*, *Favia*, *Fungia*, *Galaxea*, *Pavona*, *Platygra*, *Pocillopora*	Ammar et al. (2012)
2015	central	Saudi Arabia	Thuwal	Low–high	Low bleaching at offshore and midshore locations, high bleaching nearshore	Monroe et al. (2018); Roik et al. (2015)
2015	central south	Saudi Arabia	Al Lith	High	All shallow corals affected nearshore to offshore, partial bleaching down to 20 – 30 m	Osman et al. (2018)

anomaly in the central Red Sea as early as 1941/1942, when growth rates intermittently decreased by 44%.

The first recorded coral mortality in the Red Sea, which was later attributed to a bleaching event, dates back to Khawkhah (Yemen) where areas of large *Goniastrea retiformis*, *Montipora* spp., and *Porites* spp. succumbed to bleaching between 1996 and 1997 (Turak et al. 2007). While this seemed to be a local event, the first large-scale coral bleaching in the Red Sea was reported 1 year later during the 1998 global coral bleaching. A census of the central to northern Saudi Arabian Red Sea coast during summer and fall of 1998 found 10% of surveyed reefs to be affected by bleaching (DeVantier et al. 2005). The highest incidence occurred on shallow reefs (<6 m) between Rabigh and Yanbu, at water temperatures above 31 °C, which is 2 °C above the mean monthly average. In the area around Rabigh bleaching affected hard, soft, and fire corals with recently dead and bleached colonies accounting for up to 90% of the total coral cover (DeVantier et al. 2005). The most affected genera included *Acropora*, *Dipsastraea*, *Galaxea*, *Goniastrea*, *Millepora*, *Pocillopora*, and *Porites* (DeVantier et al. 2005). The reefs outside Dungonab Bay on the western shore of the Red Sea (Sudan) at roughly the same latitude were possibly affected by this event too, as was suggested from their poor health state during surveys in 2002 (PERSGA 2003). At the same time, the northern Red Sea (i.e., coral reefs in Egypt) (Kotb et al. 2004), the areas to the north of Yanbu, and areas with coastal upwelling (DeVantier et al. 2005) were largely unaffected.

Although reefs in the northern Red Sea largely escaped the 1998 bleaching event, coral cover was still declining in some regions between 1998 and 2004 (Kotb et al. 2004). These changes were attributed to local anthropogenic stressors, such as coastal development, pollution, and tourism related activities (Kotb et al. 2004), as well as sporadic coral bleaching of *Montipora* spp. in the Gulf of Aqaba in the summers of 2002 and 2003 (Loya 2004). After 1998, coral recovery along the Sudanese and Yemeni coastlines varied from almost no recovery to areas with high levels of recruitment and regrowth until 2007 (Klaus et al. 2008; Kotb et al. 2004). In March 2007 extremely low tides exposed reef flats along large stretches of coast in Egypt, Jordan, and Sudan leading to extensive coral bleaching and mortality (Kotb et al. 2008). In October of the same year a localized warm water event caused further coral bleaching down to 20 m depth on the offshore archipelago of 'Rocky Island' in south Egypt (Kotb et al. 2008).

The next record of coral bleaching dates back to 2010, where coral reefs in the central Saudi Arabian Red Sea near Thuwal were exposed to up to 11 degree heating weeks (Furby et al. 2013). Bleaching increased with proximity to shore and in shallow areas, where the majority of corals was affected; Oculinidae and Agariciidae being the worst impacted with up to 80 – 100% bleaching (Furby et al. 2013). Moreover, all anemone species bleached during the bleaching event, incl. *Heteractis magnifica*, *Entacmaea quadricolor*, *Stichodactyla haddoni* (Hobbs et al. 2013). Nearshore reefs experienced subsequent mortality of many taxa, while midshore and offshore reefs recovered to levels close to before the bleaching (Furby et al. 2013). A study investigating spatial patterns of bleaching in *Stylophora pistillata* largely supported the observations by Furby et al. (2013) and highlighted fine scale bleaching patterns with nearshore protected corals being less affected than those from the exposed side of the same reef, stating that: "Corals from the mildest and the most extreme thermal environments escape mortality" (Piñeda et al. 2013). At the same time *S. pistillata* at mesophotic depth (40 - 63 m) in the Gulf of Aqaba underwent repeated seasonal coral bleaching and recovery during the summers of 2010 and 2011 (Nir et al. 2014), questioning the role of deep reefs as coral refuges at least in this location (Fine et al. 2013; Glynn 1996). One more study published in 2012 reports on coral bleaching along the central Egyptian coast from northern Hurghada to El Quseer mostly affecting corals from the genera *Stylophora*, *Galaxea*, *Acropora*, and *Montipora*, but it is unclear when this bleaching was observed (Ammar et al. 2012).

The following El Niño-Southern Oscillation (ENSO) event during 2015 again hit coral reefs in the central and southern Red Sea (Monroe et al. 2018; Osman et al. 2018; Roik et al. 2015). Preliminary surveys along the Saudi Arabian coast showed coral bleaching in the central south around 20°N, with reefs up to 60 km offshore and down to >20 m being severely affected (Osman et al. 2018). The reefs around Thuwal (22°N) followed cross-shelf bleaching patterns comparable to the reports from 2010 (Furby et al. 2013; Monroe et al. 2018), while reefs in the Yanbu area (24°N) and those north of Yanbu seemed largely unaffected (Osman et al. 2018). Overall, bleaching susceptibility of coral genera throughout the Red Sea was comparable to other geographic provinces, with fast-growing branching Acroporids and Pocilloporids being affected fastest and least likely to recover. However, at the most impacted locations bleaching was a mass phenomenon that affected all coral species.

As highlighted by the increasing numbers of bleaching reports, coral reefs in the Red Sea are impacted by global climate change. Although phase shifts from coral-dominated to algal-dominated habitats have not been reported from the Red Sea yet, comparative surveys spanning the entire Red Sea coast over 2 decades indicate increasing coral community homogenization, loss of rare coral species, and a general decline in coral colony sizes (Riegl et al. 2012). Further, the 2010 bleaching event resulted in almost complete local extinction of certain taxa in some reefs and declines in diversity and coral cover in many reefs (Furby et al. 2013). Of note, bleaching is not the only cause of coral decline in the Red Sea. Heavy construction activities around urban areas along the coast, oil spills, landfilling, pollutant discharge, and effluents from desalination centers continue to threaten coastal ecosystems in the Red Sea, but observations that assess the effect of these stressors are rare (Ziegler et al. 2016) and largely anecdotal.

5.8 Outlook: What Can We Learn from Red Sea Algal Symbionts in Regard to 'Future Oceans'?

Coral host-Symbiodiniaceae associations provide the foundation of reef ecosystems and studying their relationships is the key to understanding the implications of changing environmental conditions on coral reef functioning. One of the main challenges in Symbiodiniaceae research to date is the scarcity of properly described Symbiodiniaceae species and the difficulties in assigning evolutionarily and ecologically distinct lineages as species. Although the ITS2 marker has long been used for this purpose and reasonable validity is attested to the resolution of Symbiodiniaceae ITS2 types as species (Sampayo et al. 2009; Thornhill et al. 2007), a multi-copy genetic marker such as ITS2 poses various challenges for diversity analyses (Arif et al. 2014; LaJeunesse and Thornhill 2011). At the same time, and in combination with high throughput next-generation sequencing approaches, such intragenomic diversity may be used to resolve between symbiont taxa at a level far surpassing previous approaches (Hume et al. 2019). In addition, efforts in developing alternative molecular markers will benefit from Symbiodiniaceae genomes as an available resource (Aranda et al. 2016; Lin et al. 2015; Shoguchi et al. 2013). Another line of research to overcome these challenges is the establishment of cultured isolates of Symbiodiniaceae from the Red Sea to further address Symbiodiniaceae species' physiological and biochemical properties.

Despite the long research tradition in the northern Red Sea, specifically in the Gulf of Aqaba, large parts of the Red Sea remain difficult to study due to limited access. This is highlighted by the limited number of studies on Symbiodiniaceae diversity, their distribution and sampling periods, but also the general lack of ecological data, such as bleaching observations from the Red Sea. Thus, we advocate extended sampling efforts on both coasts along the entire Red Sea coast to enhance the understanding of Symbiodiniaceae assemblage patterns in this understudied, but globally important region. The large diversity of Symbiodiniaceae communities in the Red Sea offers a unique opportunity to study the ecological performance of distinct host-symbiont combinations and shuffling/switching events in relation to changing environmental conditions, but only few studies have begun to do so (Sawall et al. 2014; Ziegler et al. 2015b, 2018). Furthermore, the presence of an apparently endemic, but regionally common type such as *Cladocopium* C41 suggests regional adaptation and speciation processes (Ziegler et al. 2017). This offers the opportunity to investigate the origin of symbionts and adaptation to local conditions (but see Hume et al. 2016).

Distribution patterns of Symbiodiniaceae in the Red Sea support the putative role of members of the genus *Durusdinium* in thermally challenged environments that were previously observed elsewhere (Fabricius et al. 2004; Hennige et al. 2010). First, *Durusdinium* occurs in warm nearshore locations in the Red Sea, and second, it becomes more ubiquitous among host species in the warmer southern Red Sea. Beyond validating and extending the thermal tolerance of *Durusdinium* geographically, the Red Sea provides a good place to identify other heat resistant symbiont types. One of these may be found in *Symbiodinium* A21, which followed a similar pattern of occurrence to that of *Durusdinium*. These observations further warrant comparative investigations into the molecular, biochemical, and physiological basis underlying thermotolerance of Symbiodiniaceae. The application of functional genomic tools will aid in elucidating the molecular underpinnings of resilience to the extreme environmental conditions in the Red Sea and thus contribute to a broader understanding of the impacts of climate change on coral reef ecosystems on a global scale.

Author Declaration Parts of subchapters 1, 2, 3, and 6 were part of the first author's PhD thesis.

References

Abrego D, Ulstrup KE, Willis BL, van Oppen MJH (2008) Species-specific interactions between algal endosymbionts and coral hosts define their bleaching response to heat and light stress. Proc R Soc B Biol Sci 275:2273–2282

Abrego D, Van Oppen MJH, Willis BL (2009) Onset of algal endosymbiont specificity varies among closely related species of Acropora corals during early ontogeny. Mol Ecol 18:3532–3543

Ammar MSA, Obuid-Allah AH, Al-Hammady MAM (2012) Corals differential susceptibilities to bleaching along the Red Sea coast, Egypt, Proc Soc Indon Biodiv Intl Conf, London, pp 255–263

Aranda M, Li Y, Liew YJ, Baumgarten S, Simakov O, Wilson MC, Piel J, Ashoor H, Bougouffa S, Bajic VB, Ryu T, Ravasi T, Bayer T, Micklem G, Kim H, Bhak J, Lajeunesse TC, Voolstra CR (2016) Genomes of coral dinoflagellate symbionts highlight evolutionary adaptations conducive to a symbiotic lifestyle. Sci Rep 6:39734

Arif C, Daniels C, Bayer T, Banguera-Hinestroza E, Barbrook A, Howe CJ, LaJeunesse TC, Voolstra CR (2014) Assessing *Symbiodinium* diversity in scleractinian corals via next-generation sequencing-based genotyping of the ITS2 rDNA region. Mol Ecol 23:4418–4433

Baillie BK, Belda-Baillie CA, Maruyama T (2000) Conspecificity and indo-Pacific distribution of *Symbiodinium* genotypes (Dinophyceae) from giant clams. J Phycol 36:1153–1161

Baird AH, Bhagooli R, Ralph PJ, Takahashi S (2009) Coral bleaching: the role of the host. Trends Ecol Evol 24:16–20

Baker AC (2003) Flexibility and specificity in coral-algal symbiosis: diversity, ecology, and biogeography of *Symbiodinium*. Annu Rev Ecol Evol Syst 34:661–689

Baker AC, Starger CJ, McClanahan TR, Glynn PW (2004) Corals' adaptive response to climate change. Nature 430:741–741

Baker A, Jones SH IV, Lee TS (2005) Symbiont diversity in Arabian corals and its relation to patterns of contemporary and historical environmental stress. In: Abuzinada AH, Joubert E, Krupp F (eds) The extent and impact of coral bleaching in the Arabian region. National Commission for Wildlife Conservation and Development, Riyadh, pp 24–36

Banaszak AT, Trench RK (1995a) Effects of ultraviolet (UV) radiation on marine microalgal-invertebrate symbioses. I. Response of the algal symbionts in culture an in hospite. J Exp Mar Biol Ecol 194:213–232

Banaszak AT, Trench RK (1995b) Effects of ultraviolet (UV) radiation on marine microalgal-invertebrate symbioses. II. The synthesis of mycosporine-like amino acids in response to exposure to UV in *Anthopleura elegantissima* and *Cassiopeia xamachana*. J Exp Mar Biol Ecol 194:233–250

Banaszak AT, Iglesias-Prieto R, Trench RK (1993) *Scrippsiella velelae* sp. nov. (Peridiniales) and *Gloeodinium viscum* sp. nov. (Phytodiniales), dinoflagellate symbionts of 2 hydrozoans (Cnidaria). J Phycol 29:517–528

Barneah O, Weis VM, Perez S, Benayahu Y (2004) Diversity of dinoflagellate symbionts in Red Sea soft corals: mode of symbiont acquisition matters. Mar Ecol Prog Ser 275:89–95

Barneah O, Brickner I, Hooge M, Weis VM, Lajeunesse TC, Benayahu Y (2007) Three party symbiosis: acoelomorph worms, corals and unicellular algal symbionts in Eilat (Red Sea). Mar Biol 151:1215–1223

Baums IB, Devlin-Durante MK, LaJeunesse TC (2014) New insights into the dynamics between reef corals and their associated dinoflagellate endosymbionts from population genetic studies. Mol Ecol 23:4203–4215

Belda-Baillie CA, Sison M, Silvestre V, Villamor K, Monje V, Gomez ED, Baillie BK (1999) Evidence for changing symbiotic algae in juvenile tridacnids. J Exp Mar Biol Ecol 241:207–221

Benayahu Y, Achituv Y, Berner T (1989) Metamorphosis of an octocoral primary polyp and its infection by algal symbionts. Symbiosis 7:159–169

Berkelmans R (2002) Time-integrated thermal bleaching thresholds of reefs and their variation on the Great Barrier Reef. Mar Ecol Prog Ser 229:73–82

Berkelmans R, van Oppen MJH (2006) The role of zooxanthellae in the thermal tolerance of corals: a 'nugget of hope' for coral reefs in an era of climate change. Proc R Soc B Biol Sci 273:2305–2312

Biquand E, Okubo N, Aihara Y, Rolland V, Hayward DC, Hatta M, Minagawa J, Maruyama T, Takahashi S (2017) Acceptable symbiont cell size differs among cnidarian species and may limit symbiont diversity. ISME J 11:1702

Blank RJ, Huss VAR (1989) DNA divergency and speciation in *Symbiodinium* (Dinophyceae). Plant Syst Evol 163:153–163

Blank RJ, Trench RK (1985) Speciation and symbiotic dinoflagellates. Science 229:656–658

Blank RJ, Trench RK (1986) Nomenclature of endosymbiotic dinoflagellates. Taxon 35:286–294

Bongaerts P, Riginos C, Ridgway T, Sampayo EM, van Oppen MJH, Englebert N, Vermeulen F, Hoegh-Guldberg O (2010) Genetic divergence across habitats in the widespread coral *Seriatopora hystrix* and its associated *Symbiodinium*. PLoS One 5:e10871

Bongaerts P, Riginos C, Hay K, van Oppen M, Hoegh-Guldberg O, Dove S (2011) Adaptive divergence in a scleractinian coral: physiological adaptation of *Seriatopora hystrix* to shallow and deep reef habitats. BMC Evol Biol 11:303

Bongaerts P, Frade P, Ogier J, Hay K, van Bleijswijk J, Englebert N, Vermeij M, Bak R, Visser P, Hoegh-Guldberg O (2013) Sharing the slope: depth partitioning of agariciid corals and associated *Symbiodinium* across shallow and mesophotic habitats (2-60m) on a Caribbean reef. BMC Evol Biol 13:205

Boulotte NM, Dalton SJ, Carroll AG, Harrison PL, Putnam HM, Peplow LM, van Oppen MJH (2016) Exploring the *Symbiodinium* rare biosphere provides evidence for symbiont switching in reef-building corals. ISME J 10:2693–2701

Brandt K (1881) Über das Zusammenleben von Thieren und Algen. Verh Physiologischer Ges:22–26

Buddemeier RW, Fautin DG (1993) Coral bleaching as an adaptive mechanism - a testable hypothesis. Bioscience 43:320–326

Burghardt I, Evertsen J, Johnsen G, Wägele H (2005) Solar powered seaslugs - mutualistic symbiosis of aeolid nudibranchia (Mollusca, Gastropoda, Opisthobranchia) with *Symbiodinium*. Symbiosis 38:227–250

Burriesci MS, Raab TK, Pringle JR (2012) Evidence that glucose is the major transferred metabolite in dinoflagellate–cnidarian symbiosis. J Exp Biol 215:3467–3477

Cantin NE, Cohen AL, Karnauskas KB, Tarrant AM, McCorkle DC (2010) Ocean warming slows coral growth in the central Red Sea. Science 329:322–325

Carlos AA, Baillie BK, Kawachi M, Maruyama T (1999) Phylogenetic position of *Symbiodinium* (Dinophyceae) isolates from tridacnids (Bivalvia), cardiids (Bivalvia), a sponge (Porifera), a soft coral (Anthozoa), and a free-living strain. J Phycol 35:1054–1062

Chang SS, Prézelin BB, Trench RK (1983) Mechanisms of photoadaptation in three strains of the symbiotic dinoflagellate *Symbiodinium microadriaticum*. Mar Biol 76:219–229

Coffroth MA, Santos SR (2005) Genetic diversity of symbiotic dinoflagellates in the genus *Symbiodinium*. Protist 156:19–34

Coffroth MA, Santos SR, Goulet TL (2001) Early ontogenetic expression of specificity in a cnidarian-algal symbiosis. Mar Ecol Prog Ser 222:85–96

Connell JH (1978) Diversity in tropical rain forests and coral reefs. Science 199:1302–1310

Connell JH (1997) Disturbance and recovery of coral assemblages. Coral Reefs 16:S101–S113

Cooper TF, Berkelmans R, Ulstrup KE, Weeks S, Radford B, Jones AM, Doyle J, Canto M, O'Leary RA, van Oppen MJH (2011a) Environmental factors controlling the distribution of *Symbiodinium* harboured by the coral *Acropora millepora* on the Great Barrier Reef. PLOS ONE 6:e25536

Cooper TF, Ulstrup KE, Dandan SS, Heyward AJ, Kuhl M, Muirhead A, O'Leary RA, Ziersen BEF, Van Oppen MJH (2011b) Niche specialization of reef-building corals in the mesophotic zone: metabolic trade-offs between divergent *Symbiodinium* types. Proc R Soc B Biol Sci 278:1840–1850

Correa AMS, Baker AC (2009) Understanding diversity in coral-algal symbiosis: a cluster-based approach to interpreting fine-scale genetic variation in the genus *Symbiodinium*. Coral Reefs 28:81–93

Decelle J, Carradec Q, Pochon X, Henry N, Romac S, Mahé F, Dunthorn M, Kourlaiev A, Voolstra CR, Wincker P, de Vargas C (2018) Worldwide occurrence and activity of the reef-building coral symbiont *Symbiodinium* in the Open Ocean. Curr Biol 28:3625–3633, e3

Devantier L, Turak E, Al-Shaikh K (2005) Coral bleaching in the central-northern Saudi Arabian Red Sea August - September 1998. In: Abuzinada AH, Joubert E, Krupp F (eds) The extent and impact of coral bleaching in the Arabian region. National Commission for Wildlife Conservation and Development, Riyadh, pp 75–90

Diaz JM, Hansel CM, Apprill A, Brighi C, Zhang T, Weber L, McNally S, Xun L (2016) Species-specific control of external superoxide levels by the coral holobiont during a natural bleaching event. Nat Commun 7:13801

DiBattista JD, Berumen ML, Gaither MR, Rocha LA, Eble JA, Choat JH, Craig MT, Skillings DJ, Bowen BW (2013) After continents divide: comparative phylogeography of reef fishes from the Red Sea and Indian Ocean. J Biogeogr 40:1170–1181

Fabricius K, Aldersdale P (2001) Soft corals and sea fans: a comprehensive guide to the tropical shallow water genera of the central-west Pacific, the Indian Ocean and the Red Sea. Australian Institute of Marine Science, AIMS, Townsville, Qld

Fabricius KE, Mieog JC, Colin PL, Idip D, Van Oppen MJH (2004) Identity and diversity of coral endosymbionts (zooxanthellae) from three Palauan reefs with contrasting bleaching, temperature and shading histories. Mol Ecol 13:2445–2458

Falkowski PG, Dubinsky Z, Muscatine L, Porter JW (1984) Light and the bioenergetics of a symbiotic coral. Bioscience 34:705–709

Fensome RA, Taylor FJR, Norris G, Sarjeant WAS, Wharton DI, Williams GL (1993) A classification of fossil and living dinoflagellates. Micropaleontol Press Spec Paper 7

Fey V, Wagner R, Brautigam K, Pfannschmidt T (2005) Photosynthetic redox control of nuclear gene expression. J Exp Bot 56:1491–1498

Fine M, Gildor H, Genin A (2013) A coral reef refuge in the Red Sea. Glob Change Biol 19:3640–3647

Finney JC, Pettay D, Sampayo E, Warner M, Oxenford H, Lajeunesse T (2010) The relative significance of host–habitat, depth, and geography on the ecology, endemism, and speciation of coral endosymbionts in the genus *Symbiodinium*. Microb Ecol 60:250–263

Fitt WK, Brown BE, Warner ME, Dunne RP (2001) Coral bleaching: interpretation of thermal tolerance limits and thermal thresholds in tropical corals. Coral Reefs 20:51–65

Frade PR, De Jongh F, Vermeulen F, Van Bleijswijk J, Bak RPM (2008) Variation in symbiont distribution between closely related coral species over large depth ranges. Mol Ecol 17:691–703

Franklin E, Stat M, Pochon X, Putnam H, Gates R (2012) GeoSymbio: a hybrid, cloud-based web application of global geospatial bioinformatics and ecoinformatics for *Symbiodinium*-host symbioses. Mol Ecol Resour 12:369–373

Freudenthal HD (1962) *Symbiodinium* gen. nov. and *Symbiodinium microadriaticum* sp. nov., a Zooxanthella: taxonomy, life cycle, and morphology. J Protozool 9:45–52

Furby KA, Bouwmeester J, Berumen ML (2013) Susceptibility of central Red Sea corals during a major bleaching event. Coral Reefs 32:505–513

Glynn PW (1993) Coral reef bleaching: ecological perspectives. Coral Reefs 12:1–17

Glynn PW (1996) Coral reef bleaching: facts, hypotheses and implications. Glob Chang Biol 2:495–509

Gómez-Cabrera MdC, Ortiz JC, Loh WKW, Ward S, Hoegh-Guldberg O (2008) Acquisition of symbiotic dinoflagellates (*Symbiodinium*) by juveniles of the coral *Acropora longicyathus*. Coral Reefs 27:219–226

Gorbunov MY, Kolber ZS, Lesser MP, Falkowski PG (2001) Photosynthesis and photoprotection in symbiotic corals. Limnol Oceanogr 46:75–85

Goulet TL, Coffroth MA (2003) Stability of an octocoral-algal symbiosis over time and space. Mar Ecol Prog Ser 250:117–124

Goulet TL, LaJeunesse T, Fabricius K (2008a) Symbiont specificity and bleaching susceptibility among soft corals in the 1998 Great Barrier Reef mass coral bleaching event. Mar Biol 154:795–804

Goulet TL, Simmons C, Goulet D (2008b) Worldwide biogeography of *Symbiodinium* in tropical octocorals. Mar Ecol Prog Ser 355:45–58

Green DH, Edmunds PJ, Pochon X, Gates RD (2010) The effects of substratum type on the growth, mortality, and photophysiology of juvenile corals in St. John, US Virgin Islands. J Exp Mar Biol Ecol 384:18–29

Grupstra CGB, Coma R, Ribes M, Posbic Leydet K, Parkinson JE, McDonald K, Catllà M, Voolstra CR, Hellberg ME, Coffroth MA (2017) Evidence for coral range expansion accompanied by reduced diversity of *Symbiodinium* genotypes. Coral Reefs 36:981–985

Hansen G, Daugbjerg N (2009) *Symbiodinium natans* sp. nov.: a "free-living" dinoflagellate from Tenerife (northeast Atlantic Ocean). J Phycol 45:251–263

Hennige SJ, Smith DJ, Walsh SJ, McGinley MP, Warner ME, Suggett DJ (2010) Acclimation and adaptation of scleractinian coral communities along environmental gradients within an Indonesian reef system. J Exp Mar Biol Ecol 391:143–152

Hideg E, Spetea C, Vass I (1995) EPR spectroscopy detection of active oxygen and free radicals in thylakoids exposed to photoinhibition. Acta Phytopathol Entomol Hung 30:51–57

Hobbs J-PA, Frisch AJ, Ford BM, Thums M, Saenz-Agudelo P, Furby KA, Berumen ML (2013) Taxonomic, spatial and temporal patterns of bleaching in anemones inhabited by anemonefishes. PLOS ONE 8:e70966

Hoegh-Guldberg O (1999) Climate change, coral bleaching and the future of the world's coral reefs. Mar Freshw Res 50:839–866

Hoegh-Guldberg O, Smith GJ (1989) The effect of sudden changes in temperature, light and salinity on the population density and export of zooxanthellae from the reef corals *Stylophora pistillata* Esper and *Seriatopora hystrix* Dana. J Exp Mar Biol Ecol 129:279–303

Hoegh-Guldberg O, Mumby PJ, Hooten AJ, Steneck RS, Greenfield P, Gomez E, Harvell CD, Sale PF, Edwards AJ, Caldeira K, Knowlton N, Eakin CM, Iglesias-Prieto R, Muthiga N, Bradbury RH, Dubi A, Hatziolos ME (2007) Coral reefs under rapid climate change and ocean acidification. Science 318:1737–1742

Howells EJ, Beltran VH, Larsen NW, Bay LK, Willis BL, van Oppen MJH (2012) Coral thermal tolerance shaped by local adaptation of photosymbionts. Nat Clim Chang 2:116–120

Hughes TP, Anderson KD, Connolly SR, Heron SF, Kerry JT, Lough JM, Baird AH, Baum JK, Berumen ML, Bridge TC, Claar DC, Eakin CM, Gilmour JP, Graham NAJ, Harrison H, Hobbs J-PA, Hoey AS, Hoogenboom M, Lowe RJ, McCulloch MT, Pandolfi JM, Pratchett M, Schoepf V, Torda G, Wilson SK (2018a) Spatial and temporal patterns of mass bleaching of corals in the Anthropocene. Science 359:80–83

Hughes TP, Kerry JT, Baird AH, Connolly SR, Dietzel A, Eakin CM, Heron SF, Hoey AS, Hoogenboom MO, Liu G, McWilliam MJ, Pears RJ, Pratchett MS, Skirving WJ, Stella JS, Torda G (2018b) Global warming transforms coral reef assemblages. Nature 556:492–496

Hume BCC, D'Angelo C, Burt J, Baker AC, Riegl B, Wiedenmann J (2013) Corals from the Persian/Arabian Gulf as models for thermotolerant reef-builders: Prevalence of clade C3 *Symbiodinium*, host fluorescence and ex situ temperature tolerance. Mar Pollut Bull 72:313–322

Hume BCC, D'Angelo C, Smith EG, Stevens JR, Burt J, Wiedenmann J (2015) *Symbiodinium thermophilum* sp. nov., a thermotolerant symbiotic alga prevalent in corals of the world's hottest sea, the Persian/Arabian Gulf. Sci Rep 5:8562

Hume BCC, Voolstra CR, Arif C, D'Angelo C, Burt JA, Eyal G, Loya Y, Wiedenmann J (2016) Ancestral genetic diversity associated with the rapid spread of stress-tolerant coral symbionts in response to Holocene climate change. Proc Natl Acad Sci USA 113:4416–4421

Hume BCC, D'Angelo C, Smith EG, Stevens JR, Burt JA, Wiedenmann J (2018a) Validation of the binary designation *Symbiodinium thermophilum* (Dinophyceae). J Phycol 54(5):762–764

Hume BCC, Ziegler M, Poulain J, Pochon X, Romac S, Boissin E, de Vargas C, Planes S, Wincker P, Voolstra CR (2018b) An improved primer set and amplification protocol with increased specificity and sensitivity targeting the *Symbiodinium* ITS2 region. PeerJ 6:e4816

Hume BCC, Smith EG, Ziegler M, Warrington HJM, Burt JA, LaJeunesse TC, Wiedenmann J, Voolstra CR (2019) SymPortal: a novel analytical framework and platform for coral algal symbiont next-generation sequencing ITS2 profiling. Mol Ecol Resour. https://doi.org/10.1111/1755-0998.13004

Iglesias-Prieto R, Beltran VH, Lajeunesse TC, Reyes-Bonilla H, Thome PE (2004) Different algal symbionts explain the vertical distribution of dominant reef corals in the eastern Pacific. Proc R Soc Lond Ser B Biol Sci 271:1757–1763

Jeffrey SW, Haxo FT (1968) Photosynthetic pigments of symbiotic dinoflagellates (zooxanthellae) from corals and clams. Biol Bull 135:149–165

Jeong HJ, Lee SY, Kang NS, Yoo YD, Lim AS, Lee MJ, Kim HS, Yih W, Yamashita H, Lajeunesse TC (2014) Genetics and morphology characterize the Dinoflagellate *Symbiodinium voratum*, n. sp., (Dinophyceae) as the sole representative of *Symbiodinium* clade E. J Eukaryot Microbiol 61:75–94

Jessen C, Villa Lizcano JF, Bayer T, Roder C, Aranda M, Wild C, Voolstra CR (2013) *In-situ* effects of eutrophication and overfishing on physiology and bacterial diversity of the Red Sea coral *Acropora hemprichii*. PLOS ONE 8:e62091

Jokiel PL, Coles SL (1990) Response of Hawaiian and other Indo-Pacific reef corals to elevated temperature. Coral Reefs 8:155–162

Karako-Lampert S, Katcoff DJ, Achituv Y, Dubinsky Z, Stambler N (2004) Do clades of symbiotic dinoflagellates in scleractinian corals of the Gulf of Eilat (Red Sea) differ from those of other coral reefs? J Exp Mar Biol Ecol 311:301–314

Kawaguti S, Sakumoto D (1948) The effect of light on the calcium deposition of corals. Bull Oceanogr Inst Taiwan 4:65–70

Kevin MJ, Hall WT, McLaughlin JJA, Zahl PA (1969) *Symbiodinium microadriaticum* Freudenthal, a revised taxonomic description, ultrastructure. J Phycol 5:341–350

Klaus R, Kemp J, Samoilys M, Anlauf H, El Din S, Abdalla EO, Chekchak T (2008) Ecological patterns and status of the reefs of Sudan, 11th International Coral Reef Symposium, Ft. Lauderdale, pp. 716–720

Kleypas JA, McManus JW, Menez LAB (1999) Environmental limits to coral reef development: where do we draw the line? Am Zool 39:146–159

Kotb M, Abdulaziz M, Al-Agwan Z, Alshaikh K, Al-Yami H, Banajah A, Devantier L, Eisinger M, Eltayeb M, Hassan M, Heiss G, Howe S, Kemp J, Klaus R, Krupp F, Mohamed N, Rouphael T, Turner J, Zajonz U (2004) Status of Coral Reefs in the Red Sea and Gulf of Aden in 2004. In: Wilkinson CR (ed) Status of Coral Reefs of the World, 2004. Australian Institute of Marine Science, Townsville, p 301

Kotb M, Hanafy MH, Rirache H, Matsumura S, Al-Sofyani A, Ahmed AG, Bawazir G, Al-Horani FA (2008) Status of Coral Reefs in the Red Sea and Gulf of Aden Region. In: Wilkinson CR (ed) Status of coral reefs of the world: 2008. Global Coral Reef Monitoring Network and Reef and Rainforest Research Centre, Townsville, pp 67–78

Krämer WE, Schrameyer V, Hill R, Ralph PJ, Bischof K (2013) PSII activity and pigment dynamics of *Symbiodinium* in two Indo-Pacific corals exposed to short-term high-light stress. Mar Biol 160:563–577

Krueger T, Gates RD (2012) Cultivating endosymbionts — Host environmental mimics support the survival of *Symbiodinium* C15 *ex hospite*. J Exp Mar Biol Ecol 413:169–176

Kuguru B, Winters G, Beer S, Santos SR, Chadwick NE (2007) Adaptation strategies of the corallimorpharian *Rhodactis rhodostoma* to irradiance and temperature. Mar Biol 151:1287–1298

Kuguru B, Chadwick N, Achituv Y, Zandbank K, Tchernov D (2008) Mechanisms of habitat segregation between corallimorpharians: photosynthetic parameters and *Symbiodinium* types. Mar Ecol Prog Ser 369:115–129

LaJeunesse TC (2001) Invesitgating the biodiversity, ecology and phylogeny of endosymbiontic dinoflagellates in the genus *Symbiodinium* using the ITS region: in search of a "species" level marker. J Phycol 37:866–880

LaJeunesse TC (2002) Diversity and community structure of symbiotic dinoflagellates from Caribbean coral reefs. Mar Biol 141:387–400

LaJeunesse TC (2005) "Species" radiations of symbiotic dinoflagellates in the Atlantic and Indo-Pacific since the miocene-pliocene transition. Mol Biol Evol 22:570–581

LaJeunesse TC (2017) Validation and description of *Symbiodinium microadriaticum*, the type species of *Symbiodinium* (Dinophyta). J Phycol 53:1109–1114

LaJeunesse TC, Thornhill DJ (2011) Improved resolution of reef-coral endosymbiont (*Symbiodinium*) species diversity, ecology, and evolution through psbA non-coding region genotyping. PLOS ONE 6:e29013

LaJeunesse TC, Loh WKW, van Woesik R, Hoegh-Guldberg O, Schmidt GW, Fitt WK (2003) Low symbiont diversity in southern Great Barrier Reef corals, relative to those of the Caribbean. Limnol Oceanogr 48:2046–2054

LaJeunesse T, Thornhill D, Cox E, Stanton F, Fitt W, Schmidt G (2004a) High diversity and host specificity observed among symbiotic dinoflagellates in reef coral communities from Hawaii. Coral Reefs 23:596–603

LaJeunesse TC, Bhagooli R, Hidaka M, DeVantier L, Done T, Schmidt GW, Fitt WK, Hoegh-Guldberg O (2004b) Closely related *Symbiodinium* spp. differ in relative dominance in coral reef host communities across environmental, latitudinal and biogeographic gradients. Mar Ecol Prog Ser 284:147–161

LaJeunesse TC, Bonilla HR, Warner ME, Wills M, Schmidt GW, Fitt WK (2008) Specificity and stability in high latitude eastern Pacific coral-algal symbioses. Limnol Oceanogr 53:719–727

LaJeunesse T, Loh W, Trench R (2009) Do introduced endosymbiotic dinoflagellates 'take' to new hosts? Biol Invasions 11:995–1003

LaJeunesse TC, Pettay DT, Sampayo EM, Phongsuwan N, Brown B, Obura DO, Hoegh-Guldberg O, Fitt WK (2010) Long-standing environmental conditions, geographic isolation and host–symbiont specificity influence the relative ecological dominance and genetic diversification of coral endosymbionts in the genus *Symbiodinium*. J Biogeogr 37:785–800

LaJeunesse TC, Parkinson JE, Reimer JD (2012) A genetics-based description of *Symbiodinium minutum* sp. nov. and *S. psygmophilum* sp. nov. (dinophyceae), two dinoflagellates symbiotic with Cnidaria. J Phycol 48:1380–1391

LaJeunesse TC, Wham DC, Pettay DT, Parkinson JE, Keshavmurthy S, Chen CA (2014) Ecologically differentiated stress-tolerant endosymbionts in the dinoflagellate genus *Symbiodinium* (Dinophyceae) Clade D are different species. Phycologia 53:305–319

LaJeunesse TC, Lee SY, Gil-Agudelo DL, Knowlton N, Jeong HJ (2015) *Symbiodinium necroappetens* sp. nov. (Dinophyceae): an opportunist 'zooxanthella' found in bleached and diseased tissues of Caribbean reef corals. Eur J Phycol:1–16

LaJeunesse TC, Parkinson JE, Gabrielson PW, Jeong HJ, Reimer JD, Voolstra CR, Santos SR (2018) Systematic revision of Symbiodiniaceae highlights the antiquity and diversity of coral endosymbionts. Curr Biol 28: 2570–2580, e6

Lampert-Karako S, Stambler N, Katcoff DJ, Achituv Y, Dubinsky Z, Simon-Blecher N (2008) Effects of depth and eutrophication on the zooxanthella clades of *Stylophora pistillata* from the Gulf of Eilat (Red Sea). Aquat Conserv Mar Freshwat Ecosyst 18:1039–1045

Lee SY, Jeong HJ, Kang NS, Jang TY, Jang SH, Lajeunesse TC (2015) *Symbiodinium tridacnidorum* sp. nov., a dinoflagellate common to Indo-Pacific giant clams, and a revised morphological description of *Symbiodinium microadriaticum* Freudenthal, emended Trench & Blank. Eur J Phycol 50:155–172

Lee MJ, Jeong HJ, Jang SH, Lee SY, Kang NS, Lee KH, Kim HS, Wham DC, Lajeunesse TC (2016) Most low-abundance "Background" *Symbiodinium* spp. are transitory and have minimal functional significance for symbiotic corals. Microb Ecol 71:771–783

Lesser MP (1996) Elevated temperatures and ultraviolet radiation cause oxidative stress and inhibit photosynthesis in symbiotic dinoflagellates. Limnol Oceanogr 41:271–283

Leutenegger S (1984) Symbiosis in benthic foraminifera: specificity and host adaptations. J Foraminifer Res 14:16–35

Lewis AM, Chan AN, LaJeunesse TC (2018) New species of closely related endosymbiotic dinoflagellates in the greater Caribbean have niches corresponding to host coral phylogeny. J Eukaryot Microbiol. https://doi.org/10.1111/jeu.12692

Lin S, Cheng S, Song B, Zhong X, Lin X, Li W, Li L, Zhang Y, Zhang H, Ji Z, Cai M, Zhuang Y, Shi X, Lin L, Wang L, Wang Z, Liu X, Yu S, Zeng P, Hao H, Zou Q, Chen C, Li Y, Wang Y, Xu C, Meng S, Xu X, Wang J, Yang H, Campbell DA, Sturm NR, Dagenais-Bellefeuille S, Morse D (2015) The *Symbiodinium kawagutii* genome illuminates dinoflagellate gene expression and coral symbiosis. Science 350:691–694

Little AF, van Oppen MJH, Willis BL (2004) Flexibility in algal endosymbioses shapes growth in reef corals. Science 304:1492–1494

Loeblich AR, Sherley JL (1979) Observations on the theca of the motile phase of free-living and symbiotic isolates of *Zooxanthella microadriatica* (Freudenthal) comb.nov. J Mar Biol Assoc UK 59:195–205

Loh WKW, Loi T, Carter D, Hoegh-Guldberg O (2001) Genetic variability of the symbiotic dinoflagellates from the wide ranging coral species *Seriatopora hystrix* and *Acropora longicyathus* in the Indo-West Pacific. Mar Ecol Prog Ser 222:97–107

Loya Y (2004) The coral reefs of Eilat — past, present and future: three decades of coral community structure studies. In: Rosenberg E, Loya Y (eds) Coral health and disease. Springer, Berlin Heidelberg, pp 1–34

Loya Y, Sakai K, Yamazato K, Nakano Y, Sambali H, van Woesik R (2001) Coral bleaching: the winners and the losers. Ecol Lett 4:122–131

Macdonald AH, Sampayo E, Ridgway T, Schleyer M (2008) Latitudinal symbiont zonation in *Stylophora pistillata* from southeast Africa. Mar Biol 154:209–217

Markell DA, Trench RK (1993) Macromolecules exuded by symbiotic dinoflagellates in culture: amino acids and sugar composition. J Phycol 29:64–68

Martindale JL, Holbrook NJ (2002) Cellular response to oxidative stress: signaling for suicide and survival. J Cell Physiol 192:1–15

McGinley MP, Aschaffenburg MD, Pettay DT, Smith RT, LaJeunesse TC, Warner ME (2012) *Symbiodinium* spp. in colonies of eastern Pacific *Pocillopora* spp. are highly stable despite the prevalence of low-abundance background populations. Mar Ecol Prog Ser 462:1–7

McLaughlin JJA, Zahl PA (1957) Studies in marine biology. II. In vitro culture of zooxanthellae. Exp Biol Med 95:115–120

McLaughlin JJA, Zahl PA (1959) Axenic zooxanthellae from various invertebrate hosts. Ann NY Acad Sci 77:55–72

Mieog JC, van Oppen MJH, Cantin NE, Stam WT, Olsen JL (2007) Real-time PCR reveals a high incidence of *Symbiodinium* clade D at low levels in four scleractinian corals across the Great Barrier Reef: implications for symbiont shuffling. Coral Reefs 26:449–457

Monroe A, Ziegler M, Roik A, Röthig T, Hardestine R, Emms M, Jensen T, Voolstra CR, Berumen M (2018) *In-situ* observations of coral bleaching in the central Saudi Arabian Red Sea during the 2015/2016 global coral bleaching event. PLOS ONE 13(4):e0195814

Müller-Merz E, Lee JJ (1976) Symbiosis in larger foraminiferan *Sorites marginalis-* (with notes on *Archaias* spp.). J Protozool 23:390–396

Muscatine L (1990) The role of symbiotic algae in carbon and energy flux in reef corals. In: Dubinsky Z (ed) Ecosystems of the world, Coral reefs. Elsevier, Amsterdam, pp 75–87

Muscatine L, Porter JW (1977) Reef corals: mutualistic symbioses adapted to nutrient-poor environments. Bioscience 27:454–460

Muscatine L, Falkowski PG, Porter JW, Dubinsky Z (1984) Fate of photosynthetic fixed carbon in light-adapted and shade-adapted colonies of the symbiotic coral *Stylophora pistillata*. Proc R Soc B Biol Sci 222:181–202

Muscatine L, Porter JW, Kaplan IR (1989) Resource partitioning by reef corals as determined from stable isotope composition. Mar Biol 100:185–193

Myers RM, Fischer SG, Maniatis T, Lerman LS (1985) Modification of the melting properties of duplex DNA by attachment of a GC-rich DNA sequence as determined by denaturing gradient gel electrophoresis. Nucleic Acids Res 13:3111–3129

Nir O, Gruber DF, Einbinder S, Kark S, Tchernov D (2011) Changes in scleractinian coral *Seriatopora hystrix* morphology and its endocellular *Symbiodinium* characteristics along a bathymetric gradient from shallow to mesophotic reef. Coral Reefs 30:1089–1100

Nir O, Gruber DF, Shemesh E, Glasser E, Tchernov D (2014) Seasonal mesophotic coral bleaching of *Stylophora pistillata* in the northern Red Sea. PLOS ONE 9:e84968

Osman EO, Smith DJ, Ziegler M, Kürten B, Conrad C, El-Haddad KM, Voolstra CR, Suggett DJ (2018) Thermal refugia against coral bleaching throughout the Northern Red Sea. Glob Change Biol 24:1354–1013

Parkinson JE, Coffroth MA, Lajeunesse TC (2015) New species of Clade B *Symbiodinium* (Dinophyceae) from the greater Caribbean belong to different functional guilds: *S. aenigmaticum* sp. nov., *S. antillogorgium* sp. nov., *S. endomadracis* sp. nov., and *S. pseudominutum* sp. nov. J Phycol 51:850–858

Patton JS, Abraham S, Benson AA (1977) Lipogenesis in the intact coral *Pocillopora capitata* and its isolated zooxanthellae: evidence for a light-driven carbon cycle between symbiont and host. Mar Biol 44:235–247

Pawlowski J, Holzmann M, Fahrni JF, Pochon X, Lee JJ (2001) Molecular identification of algal endosymbionts in large miliolid foraminifera: 2. Dinoflagellates. J Eukaryot Microbiol 48:368–373

Pearse VB, Muscatine L (1971) Role of symbiotic algae (zooxanthallae) in coral calcification. Biol Bull 141:350–363

PERSGA (2003) Survey of the proposed marine protected area at Dungonab Bay and Mukkawar Island, Sudan, In: Kemp J (ed) Technical Series Report. PERSGA, Jeddah, p 84

Pettay DT, Wham DC, Pinzón JH, Lajeunesse TC (2011) Genotypic diversity and spatial–temporal distribution of *Symbiodinium* clones in an abundant reef coral. Mol Ecol 20:5197–5212

Piñeda J, Starczak V, Tarrant A, Blythe J, Davis K, Farrar T, Berumen M, da Silva JCB (2013) Two spatial scales in a bleaching event: corals from the mildest and the most extreme thermal environments escape mortality. Limnol Oceanogr 58:1531–1545

Pochon X, Gates RD (2010) A new *Symbiodinium* clade (Dinophyceae) from soritid foraminifera in Hawai'i. Mol Phylogenet Evol 56:492–497

Pochon X, Pawlowski J (2006) Evolution of the soritids-*Symbiodinium* symbiosis. Symbiosis 42:77–88

Pochon X, Pawlowski J, Zaninetti L, Rowan R (2001) High genetic diversity and relative specificity among *Symbiodinium*-like endosymbiotic dinoflagellates in soritid foraminiferans. Mar Biol 139:1069–1078

Pochon X, LaJeunesse TC, Pawlowski J (2004) Biogeographic partitioning and host specialization among foraminiferan dinoflagellate symbionts (*Symbiodinium*; Dinophyta). Mar Biol 146:17–27

Pochon X, Montoya-Burgos JI, Stadelmann B, Pawlowski J (2006) Molecular phylogeny, evolutionary rates, and divergence timing of the symbiotic dinoflagellate genus *Symbiodinium*. Mol Phylogenet Evol 38:20–30

Pochon X, Stat M, Takabayashi M, Chasqui L, Chauka LJ, Logan DDK, Gates RD (2010) Comparison of endosymbiotic and free-living *Symbiodinium* (Dinophyceae) diversity in a Hawaiian reef environment. J Phycol 46:53–65

Pochon X, Putnam HM, Burki F, Gates RD (2012) Identifying and characterizing alternative molecular markers for the symbiotic and free-living dinoflagellate genus *Symbiodinium*. PLOS ONE 7:e29816

Pochon X, Putnam HM, Gates RD (2014) Multi-gene analysis of *Symbiodinium* dinoflagellates: a perspective on rarity, symbiosis, and evolution. PeerJ 2:e394

Pogoreutz C, Rädecker N, Cárdenas A, Gärdes A, Voolstra CR, Wild C (2017) Sugar enrichment provides evidence for a role of nitrogen fixation in coral bleaching. Glob Chang Biol 23:3838–3848

Rädecker N, Pogoreutz C, Voolstra CR, Wiedenmann J, Wild C (2015) Nitrogen cycling in corals: the key to understanding holobiont functioning? Trends Microbiol 23:490–497

Raitsos DE, Hoteit I, Prihartato PK, Chronis T, Triantafyllou G, Abualnaja Y (2011) Abrupt warming of the Red Sea. Geophys Res Lett 38:L14601

Ramsby BD, Hill MS, Thornhill DJ, Steenhuizen SF, Achlatis M, Lewis AM, Lajeunesse TC (2017) Sibling species of mutualistic *Symbiodinium* clade G from bioeroding sponges in the western Pacific and western Atlantic oceans. J Phycol 53:951–960

Reimer JD, Ono S, Tsukahara J, Takishita K, Maruyama T (2007) Non-seasonal clade-specificity and subclade microvariation in symbiotic dinoflagellates (*Symbiodinium* spp.) in *Zoanthus sansibaricus* (Anthozoa: Hexacorallia) at Kagoshima Bay, Japan. Phycol Res 55:58–65

Richter C, Roa-Quiaoit H, Jantzen C, Al-Zibdah M, Kochzius M (2008) Collapse of a new living species of giant clam in the Red Sea. Curr Biol 18:1349–1354

Riegl BM, Bruckner AW, Rowlands GP, Purkis SJ, Renaud P (2012) Red Sea coral reef trajectories over 2 decades suggest increasing community homogenization and decline in coral size. PLOS ONE 7:e38396

Roberts CM, McClean CJ, Veron JEN, Hawkins JP, Allen GR, McAllister DE, Mittermeier CG, Schueler FW, Spalding M, Wells F, Vynne C, Werner TB (2002) Marine biodiversity hotspots and conservation priorities for tropical reefs. Science 295:1280–1284

Rodriguez-Lanetty M, Loh W, Carter D, Hoegh-Guldberg O (2001) Latitudinal variability in symbiont specificity within the widespread scleractinian coral *Plesiastrea versipora*. Mar Biol 138:1175–1181

Rodriguez-Lanetty M, Krupp DA, Weis VM (2004) Distinct ITS types of *Symbiodinium* in Clade C correlate with cnidarian/dinoflagellate specificity during onset of symbiosis. Mar Ecol Prog Ser 275:97–102

Roik A, Röthig T, Ziegler M, Voolstra CR (2015) Coral bleaching event in the central Red Sea. Mideast Coral Reef Soc Newsl 3:3

Rowan R (1998) Diversity and ecology of zooxanthellae on coral reefs. J Phycol 34:407–417

Rowan R, Knowlton N (1995) Intraspecific diversity and ecological zonation in coral algal symbiosis. Proc Natl Acad Sci USA 92:2850–2853

Rowan R, Powers DA (1991a) A molecular genetic classification of zooxanthellae and the evolution of animal-algal symbioses. Science 251:1348–1351

Rowan R, Powers DA (1991b) Molecular genetic identification of symbiotic dinoflagellates (zooxanthellae). Mar Ecol Prog Ser 71:65–73

Rowan R, Powers DA (1992) Ribosomal RNA sequences and the diversity of symbiotic dinoflagellates (zooxanthellae). Proc Natl Acad Sci USA 89:3639–3643

Rowan R, Knowlton N, Baker A, Jara J (1997) Landscape ecology of algal symbionts creates variation in episodes of coral bleaching. Nature 388:265–269

Sampayo EM, Franceschinis L, Hoegh-Guldberg O, Dove S (2007) Niche partitioning of closely related symbiotic dinoflagellates. Mol Ecol 16:3721–3733

Sampayo EM, Dove S, LaJeunesse TC (2009) Cohesive molecular genetic data delineate species diversity in the dinoflagellate genus *Symbiodinium*. Mol Ecol 18:500–519

Santos SR, LaJeunesse T (2006) Searchable database of *Symbiodinium* diversity—geographic and ecological diversity (SD2-GED). Auburn University, http://www.auburn.edu/~santosr/sd2_ged.htm

Santos SR, Taylor DJ, Coffroth MA (2001) Genetic comparisons of freshly isolated versus cultured symbiotic dinoflagellates: implications for extrapolating to the intact symbiosis. J Phycol 37:900–912

Santos SR, Taylor DJ, Kinzie RA, Hidaka M, Sakai K, Coffroth MA (2002) Molecular phylogeny of symbiotic dinoflagellates inferred from partial chloroplast large subunit (23S)-rDNA sequences. Mol Phylogen Evol 23:97–111

Sawall Y, Al-Sofyani A, Banguera-Hinestroza E, Voolstra CR (2014) Spatio-temporal analyses of *Symbiodinium* physiology of the coral *Pocillopora verrucosa* along large-scale nutrient and temperature gradients in the Red Sea. PLOS ONE 9:e103179

Schoenberg DA, Trench RK (1980a) Genetic variation in *Symbiodinium* (=*Gymnodinium*) *microadriaticum* Freudenthal, and specificity in its symbiosis with marine invertebrates. I. Isoenzyme and soluble protein patterns of axenic cultures of *Symbiodinium microadriaticum*. Proc R Soc Lond B Biol Sci 207:405–427

Schoenberg DA, Trench RK (1980b) Genetic variation in *Symbiodinium* (=*Gymnodinium*) *microadriaticum* Freudenthal, and specificity in its symbiosis with marine invertebrates. II. Morphological variation in *Symbiodinium microadriaticum*. Proc R Soc Lond B Biol Sci 207:429–444

Schoenberg DA, Trench RK (1980c) Genetic variation in *Symbiodinium* (=*Gymnodinium*) *microadriaticum* Freudenthal, and specificity in its symbiosis with marine invertebrates. III. Specificity and infectivity of *Symbiodinium microadriaticum*. Proc R Soc Lond B Biol 207:445–460

Sheppard CRC (2003) Predicted recurrences of mass coral mortality in the Indian Ocean. Nature 425:294–297

Shoguchi E, Shinzato C, Kawashima T, Gyoja F, Mungpakdee S, Koyanagi R, Takeuchi T, Hisata K, Tanaka M, Fujiwara M, Hamada M, Seidi A, Fujie M, Usami T, Goto H, Yamasaki S, Arakaki N, Suzuki Y, Sugano S, Toyoda A, Kuroki Y, Fujiyama A, Medina M, Coffroth MA, Bhattacharya D, Satoh N (2013) Draft assembly of the *Symbiodinium minutum* nuclear genome reveals dinoflagellate gene structure. Curr Biol 23:1399–1408

Silverstein RN, Correa AMS, Baker AC (2012) Specificity is rarely absolute in coral–algal symbiosis: implications for coral response to climate change. Proc R Soc B Biol Sci 279:2609–2618

Smith DJ, Suggett DJ, Baker NR (2005) Is photoinhibition of zooxanthellae photosynthesis the primary cause of thermal bleaching in corals? Glob Chang Biol 11:1–11

Stat M, Gates R (2008) Vectored introductions of marine endosymbiotic dinoflagellates into Hawaii. Biol Invasions 10:579–583

Stat M, Gates RD (2011) Clade D *Symbiodinium* in scleractinian corals: a nugget of hope, a selfish opportunist, an ominous sign, or all of the above? J Mar Biol 2011:730715

Stat M, Carter D, Hoegh-Guldberg O (2006) The evolutionary history of *Symbiodinium* and scleractinian hosts - Symbiosis, diversity, and the effect of climate change. Perspect Plant Ecol Evol Syst 8:23–43

Stat M, Loh WKW, Hoegh-Guldberg O, Carter DA (2008) Symbiont acquisition strategy drives host-symbiont associations in the southern Great Barrier Reef. Coral Reefs 27:763–772

Stat M, Pochon X, Cowie ROM, Gates RD (2009) Specificity in communities of *Symbiodinium* in corals from Johnston Atoll. Mar Ecol Prog Ser 386:83–96

Stern RF, Horak A, Andrew RL, Coffroth M-A, Andersen RA, Küpper FC, Jameson I, Hoppenrath M, Véron B, Kasai F, Brand J, James ER, Keeling PJ (2010) Environmental barcoding reveals massive dinoflagellate diversity in marine environments. PLOS ONE 5:e13991

Stimson J, Sakai K, Sembali H (2002) Interspecific comparison of the symbiotic relationship in corals with high and low rates of bleaching-induced mortality. Coral Reefs 21:409–421

Suggett DJ, Warner ME, Smith DJ, Davey P, Hennige S, Baker NR (2008) Photosynthesis and production of hydrogen peroxide by *Symbiodinium* (Pyrrhophyta) phylotypes with different thermal tolerances. J Phycol 44:948–956

Swain TD, Chandler J, Backman V, Marcelino L (2017) Consensus thermotolerance ranking for 110 *Symbiodinium* phylotypes: an exemplar utilization of a novel iterative partial-rank aggregation tool with broad application potential. Funct Ecol 31:172–183

Takabayashi M, Santos SR, Cook CB (2004) Mitochondrial DNA phylogeny of the symbiotic dinoflagellates (*Symbiodinium*, Dinophyta). J Phycol 40:160–164

Tanaka Y, Miyajima T, Koike I, Hayashibara T, Ogawa H (2006) Translocation and conservation of organic nitrogen within the coral-zooxanthella symbiotic system of *Acropora pulchra*, as demonstrated by dual isotope-labeling techniques. J Exp Mar Biol Ecol 336:110–119

Taylor DL (1968) In situ studies on the cytochemistry and ultrastructure of a symbiotic marine dinoflagellate. J Mar Biol Assoc UK 48:349–366

Taylor DL (1969) Identity of zooxanthellae isolated from some Pacific Tridacnidae. J Phycol 5:336–340

Tchernov D, Gorbunov MY, de Vargas C, Yadav SN, Milligan AJ, Haggblom M, Falkowski PG (2004) Membrane lipids of symbiotic algae are diagnostic of sensitivity to thermal bleaching in corals. Proc Natl Acad Sci USA 101:13531–13535

Tchernov D, Kvitt H, Haramaty L, Bibby TS, Gorbunov MY, Rosenfeld H, Falkowski PG (2011) Apoptosis and the selective survival of host animals following thermal bleaching in zooxanthellate corals. Proc Natl Acad Sci USA 108:9905–9909

Terraneo TI, Berumen ML, Arrigoni R, Waheed Z, Bouwmeester J, Caragnano A, Stefani F, Benzoni F (2014) *Pachyseris inattesa* sp. n. (Cnidaria, Anthozoa, Scleractinia): a new reef coral species from the Red Sea and its phylogenetic relationships. ZooKeys 433:1–30

Thornhill D, Fitt W, Schmidt G (2006a) Highly stable symbioses among western Atlantic brooding corals. Coral Reefs 25:515–519

Thornhill DJ, Lajeunesse TC, Kemp DW, Fitt WK, Schmidt GW (2006b) Multi-year, seasonal genotypic surveys of coral-algal symbioses reveal prevalent stability or post-bleaching reversion. Mar Biol 148:711–722

Thornhill DJ, Lajeunesse TC, Santos SR (2007) Measuring rDNA diversity in eukaryotic microbial systems: how intragenomic variation, pseudogenes, and PCR artifacts confound biodiversity estimates. Mol Ecol 16:5326–5340

Thornhill DJ, Xiang Y, Fitt WK, Santos SR, Vollmer S (2009) Reef endemism, host specificity and temporal stability in populations of symbiotic dinoflagellates from two ecologically dominant Caribbean corals. PLoS One 4:e6262

Thornhill D, Lewis A, Wham D, Lajeunesse T (2014) Host specialist lineages dominate the adaptive radiation of reef coral endosymbionts. Evolution 68:352–367

Tolleter D, Seneca FO, DeNofrio JC, Krediet CJ, Palumbi SR, Pringle JR, Grossman AR (2013) Coral bleaching independent of photosynthetic activity. Curr Biol 23:1782–1786

Trench RK (1979) The cell biology of plant-animal symbiosis. Annu Rev Plant Physiol 30:485–531

Trench RK (1987) The biology of Dinoflagellates in non-parasitic symbioses. Blackwell Scientific Publications, Oxford

Trench RK (1993) Microalgal-invertebrate symbioses - a review. Endocytobiosis Cell Res 9:135–175

Trench R (2000) Validation of some currently used invalid names of dinoflagellates. J Phycol 36:972–972

Trench RK, Blank RJ (1987) *Symbiodinium microadriaticum* Freudenthal, *S. goreauii* sp. nov., *S. kawagutii* sp. nov. and *S. pilosum* sp. nov.: Gymnodinioid dinoflagellate symbionts of marine invertebrates. J Phycol 23:469–481

Trench RK, Thinh L-v (1995) *Gymnodinium linucheae* sp. nov.: the dinoflagellate symbiont of the jellyfish *Linuche unguiculata*. Eur J Phycol 30:149–154

Turak E, Brodie J, Devantier L (2007) Reef-building corals and coral communities of the Yemen Red Sea. Fauna Arab 23:1–40

Ulstrup KE, Van Oppen MJH (2003) Geographic and habitat partitioning of genetically distinct zooxanthellae (*Symbiodinium*) in *Acropora* corals on the Great Barrier Reef. Mol Ecol 12:3477–3484

van Oppen M (2004) Mode of zooxanthella transmission does not affect zooxanthella diversity in acroporid corals. Mar Biol 144:1–7

van Oppen MJH, Palstra FP, Piquet AM-T, Miller DJ (2001) Patterns of coral–dinoflagellate associations in *Acropora*: significance of local availability and physiology of *Symbiodinium* strains and host–symbiont selectivity. Proc R Soc London Ser B 268:1759–1767

Veron J, Stafford-Smith M, Devantier L, Turak E (2015) Overview of distribution patterns of zooxanthellate Scleractinia. Front Mar Sci 1

Vicente VP (1990) Response of sponges with autotrophic endosymbionts during the coral-bleaching episode in Puerto Rico. Coral Reefs 8:199–202

Voolstra CR, Schwarz JA, Schnetzer J, Sunagawa S, Desalvo MK, Szmant AM, Coffroth MA, Medina M (2009) The host transcriptome remains unaltered during the establishment of coral-algal symbioses. Mol Ecol 18:1823–1833

Wakefield TS, Kempf SC (2001) Development of host- and symbiont-specific monoclonal antibodies and confirmation of the origin of the symbiosome membrane in a cnidarian–dinoflagellate symbiosis. Biol Bull 200:127–143

Warner ME, Fitt WK, Schmidt GW (1999) Damage to photosystem II in symbiotic dinoflagellates: a determinant of coral bleaching. Proc Natl Acad Sci U S A 96:8007–8012

Wham DC, Ning G, Lajeunesse TC (2017) *Symbiodinium glynnii* sp. nov., a species of stress-tolerant symbiotic dinoflagellates from pocilloporid and montiporid corals in the Pacific Ocean. Phycologia 56:396–409

Wiedenmann J, D'Angelo C, Smith EG, Hunt AN, Legiret F-E, Postle AD, Achterberg EP (2012) Nutrient enrichment can increase the susceptibility of reef corals to bleaching. Nat Clim Chang 3:160–164

Wilkinson CR (2008) Status of Coral Reefs of the World: 2008. Global Coral Reef Monitoring Network and Reef and Rainforest Research Centre, Townsville

Wooldridge SA (2009) A new conceptual model for the warm-water breakdown of the coral–algae endosymbiosis. Mar Freshw Res 60:483–496

Ziegler M, Fitzpatrick SK, Burghardt I, Liberatore KL, Joshua Leffler A, Takacs-Vesbach C, Shepherd U (2014a) Thermal stress response in a dinoflagellate-bearing nudibranch and the octocoral on which it feeds. Coral Reefs 33:1085–1099

Ziegler M, Roder C, Büchel C, Voolstra C (2014b) Limits to physiological plasticity of the coral *Pocillopora verrucosa* from the central Red Sea. Coral Reefs 33:1115–1129

Ziegler M, Roder C, Büchel C, Voolstra CR (2015a) Mesophotic coral depth acclimatization is a function of host-specific symbiont physiology. Front Mar Sci 2

Ziegler M, Roder C, Büchel C, Voolstra CR (2015b) Niche acclimatization in Red Sea corals is dependent on flexibility of host-symbiont association. Mar Ecol Prog Ser 533:149–161

Ziegler M, Roik A, Porter A, Zubier K, Mudarris MS, Ormond R, Voolstra CR (2016) Coral microbial community dynamics in response to anthropogenic impacts near a major city in the Central Red Sea. Mar Pollut Bull 105:629–640

Ziegler M, Arif C, Burt J, Dobretsov SV, Roder C, Lajeunesse TC, Voolstra CR (2017) Biogeography and molecular diversity of coral symbionts in the genus *Symbiodinium* around the Arabian Peninsula. J Biogeogr 44:674–686

Ziegler M, Eguiluz VM, Duarte CM, Voolstra CR (2018) Rare symbionts may contribute to the resilience of coral–algal assemblages. ISME J 12:161

Sponges of the Red Sea

Michael K. Wooster, Oliver Voigt, Dirk Erpenbeck, Gert Wörheide, and Michael L. Berumen

Abstract

Sponges are found in virtually all marine habitats. The Red Sea is no exception, harboring a diverse community of sponge species. However, the state of knowledge of the Red Sea sponge fauna remains in early stages. Various taxonomic efforts have been initiated, starting with early explorers at the beginning of the nineteenth century. Subsequently, published work has focused on modern taxonomic approaches, potential bioactive molecules, microbiological associations of host sponges, and a variety of ecological topics. The majority of studies are restricted to few locations and/or small numbers of species. Overall, this collective knowledge represents a sound foundation but there remains great potential for Red Sea sponges to inform the broader context of sponge work throughout the tropics. This chapter aims to provide an overview of previous work in the region and identify fruitful areas of potential future work.

Keywords

Porifera · Biodiversity · Taxonomy · Bioactive compounds · Ecology · Microbes

6.1 Introduction

The Red Sea has long been recognized as a region of high biodiversity (Stehli and Wells 1971) and endemism (Ormond and Edwards 1987; DiBattista et al. 2016), for example, home to well over 1000 species of fishes and over 50 genera of hermatypic corals. Few comprehensive estimates of diversity are available for other taxa, but recent attempts to compile species lists estimate 635 polychaete species, 211 echinoderm species, and 79 ascidians (DiBattista et al. 2016). However, the Red Sea and Arabian region in general have been largely understudied compared to comparably biodiverse coral reef systems (Berumen et al. 2013; Vaughan and Burt 2016), and this is especially true for sponges. Further complicating an understanding of the Red Sea ecosystem, the majority of the accessible published research originates from a relatively short (~6 km) stretch of coastline in the far northern Red Sea within the Gulf of Eilat / Aqaba (hereafter Gulf of Aqaba) (e.g., Spaet et al. 2012). The Red Sea, however, is of increasing interest to scientists working on climate change due to its relatively high and variable water temperatures (from 20 °C in spring to 35 °C in summer) and high salinity (40.0 psu in the northern Red Sea; Edwards 1987), conditions that may reflect the near-future state of oceans in other parts of the world (e.g., Voolstra et al. 2015).

Sponges are integral members of benthic communities in virtually all aqueous habitats, ranging from polar seas (McClintock et al. 2005; Peters et al. 2009; Dayton et al.

M. K. Wooster (✉) · M. L. Berumen
Red Sea Research Center, Division of Biological and Environmental Science and Engineering, King Abdullah University of Science and Technology, Thuwal, Saudi Arabia
e-mail: michael.wooster@kaust.edu.sa

O. Voigt
Department of Earth and Environmental Sciences, Palaeontology and Geobiology, Ludwig-Maximilians-Universität München, Munich, Germany

D. Erpenbeck
GeoBio-Center, Ludwig-Maximilians-Universität München, Munich, Germany

Department of Earth and Environmental Sciences, Palaeontology and Geobiology, Ludwig-Maximilians-Universität München, Munich, Germany

G. Wörheide
GeoBio-Center, Ludwig-Maximilians-Universität München, Munich, Germany

Department of Earth and Environmental Sciences, Palaeontology and Geobiology, Ludwig-Maximilians-Universität München, Munich, Germany

Bayerische Staatssammlung für Paläontologie und Geologie, Munich, Germany

2013) to temperate and tropical waters (for an overview, see Bell 2008). In coral reef systems, sponges are important structural components and play important functional roles. They are efficient filter feeders, play crucial roles in carbon and nitrogen cycles in coral reef ecosystems, and exert control on plankton communities (Reiswig 1974; Pile et al. 1997; Savarese et al. 1997; Peterson et al. 2006; De Goeij et al. 2013). While many sponge species provide shelter and habitat for large numbers of invertebrates and fishes (e.g., Westinga and Hoetjes 1981; Pawlik 1983; Duffy 1992; Henkel and Pawlik 2005), they are themselves aggressive competitors for space (e.g., Targett and Schmahl 1984; Suchanek et al. 1985; Aerts 1998; Loh et al. 2015) due to the most prolific production of deterrent biochemical compounds among marine organisms (see Blunt et al. (2007) and subsequent publications in that series). They host diverse communities of symbiotic microorganisms that contribute to primary productivity and nitrification (Erwin and Thacker 2007; Southwell et al. 2008a, b; Gibson 2011) and may serve as valuable models for 'holobiont' co-evolution (Ryu et al. 2016; Thomas et al. 2016). Sponges have attracted the attention of climate change researchers as sponges may be resistant to warming seawater temperatures (Simister et al. 2012). In some locations, excavating sponges are among the primary agents of carbonate bioerosion on coral reefs (Rützler and Rieger 1973; Zundelevich et al. 2007), but other sponge species may also consolidate coral rubble and thus facilitate the settlement of coral (Wulff 1984). Most of the ecological work on sponges has been conducted in the Caribbean, where sponge communities appear to have similar community composition (i.e., relative abundances are fairly consistent) within the region and community biodiversity has been well documented (van Soest et al. 2012; Pawlik and Loh 2016).

The aim of this chapter is to provide an overview on the state of sponge research that has been conducted in the Red Sea. We also attempt to highlight areas of research that are lacking or other important knowledge gaps that may serve as a guide for future work.

6.2 Red Sea Sponge Biodiversity

Little is known about Red Sea sponge biodiversity in comparison to other regional sponge communities, for example, in Oman, Seychelles, India, and East Africa (van Soest and Beglinger 2008; Berumen et al. 2013). The earliest work on Red Sea sponges was conducted by early natural historians focused primarily on cataloging the biodiversity of the region. We identified 34 papers related to taxonomy of Red Sea sponges with 12 of them focused on larger regions with only one or two species from the Red Sea. The major source of early Red Sea sponge observations began in the late nineteenth century, with Haeckel (1870, 1872), Topsent (1892) and Keller (1889, 1891). These efforts continued well into the twentieth century with important publications from Topsent (1906), Row (1909, 1911), Burton (1926, 1952, 1959), and Lévi (1958, 1965). These studies were based on work with preserved material, which poses some challenges and creates the potential for some taxonomic confusion, as discussed later. Since 2000, 23 new species (13 from the Gulf of Aqaba and 10 from the main body of the Red Sea) have been described (Vacelet et al. 2001; Klautau and Valentine 2003; Ilan et al. 2004; Helmy et al. 2004, 2005; Gugel et al. 2011; Voigt et al. 2017; van Soest and de Voogd 2018). Only a few more recent works specifically address the biodiversity of sponges in the Red Sea (e.g. Ilan et al. 2004; Helmy and van Soest 2005; Erpenbeck et al. 2016b; Voigt et al. 2017; van Soest and de Voogd 2018), and in some cases researchers had to rely on older collection material (e.g. Klautau and Valentine 2003). One of the major impediments to continued discovery of new species was the difficulty for non-regional scientists to access the region and conduct research in Red Sea for many decades (see Berumen et al. 2013; Vaughan and Burt 2015). However, for a long period, studies focused mostly on the Gulf of Suez and Gulf of Aqaba in the north, while the central and southern Red Sea remained largely understudied (Berumen et al. 2013). Although there is some anecdotal evidence that present-day sponge communities have shifted over the past century (see Vacelet et al. 2001), studies on recently-collected material of a broader geographical range in the Red Sea (e.g., Giles et al. 2015; Erpenbeck et al. 2016b; Voigt et al. 2017; van Soest and de Voogd 2018) are still scarce, which hampers the understanding of current distribution of species.

A thorough evaluation of publications and of the World Porifera Database (WPD, van Soest et al. 2018) revealed 261 valid sponge species (representing 114 genera) from the Red Sea (compiled in Table 6.1). New species of Red Sea sponges are still being described (Vacelet et al. 2001; Klautau and Valentine 2003; Helmy et al. 2004; Gugel et al. 2011; Voigt et al. 2017; van Soest and de Voogd 2018), and it can be expected that more research effort will further enhance the understanding of the biodiversity and endemism of the Red Sea sponge fauna and its relation to the biota of the adjacent regions of the Indian Ocean. The inclusion of historic material (preferably type material) in the molecular analyses in an integrative approach will greatly contribute to our understanding of biodiversity, distribution, endemism, and faunal changes of the Red Sea (see discussion in Erpenbeck et al. 2016a).

The Red Sea's recognized sponge biodiversity is mainly comprised of species from the classes Demospongiae (225 species) and Calcarea (32 species). Much less is known about Homoscleromorpha (2 species) and about the glass sponges (Hexactinellida), a generally more deep-water

Table 6.1 List of sponges reported from the Red Sea. "Species" indicates the current accepted name for the sponge species. "Citation" lists the oldest known record of the species in the Red Sea (to the best of our knowledge). Note that in some cases, the original description of the species was from another geographic location; in these cases, the citation for the original species description is included in parentheses. Because many of the sponges have had taxonomic revisions since the original records shown in the "Citation" column, the "Previous Name(s)" column indicates the name(s) used in the work cited. (For a full taxonomic history of each species / genus, see the World Porifera Database (van Soest et al. 2018)). Finally, notes are included regarding the distribution of each taxa in the "Distribution (WPD)" column. "Present" indicates that the WPD currently reflects that this species' distribution includes the Red Sea. Species listed as "Endemic" are shown in the WPD to only occur inside the Red Sea. In some cases, the publication listed in the "Citation" column has reported a species in the Red Sea although the species' distribution in the WPD does not include the Red Sea; these cases are indicated as "Unreported". This more likely reflects the ongoing work of the WPD editors and not an intentional omission. For some species, the WPD shows a distribution including the Red Sea but explicitly acknowledges that the distribution has not been reviewed by WPD editors ("Not Reviewed"). Finally, one species is indicated to occur in the Red Sea but WPD editors have flagged this as "Doubtful". Please note that Table 6.1 is available as an electronic file (Appendix) including additional taxonomic information for each species (i.e., Class and Order)

Species	Citation	Previous names	Distribution (WPD)
Class Demospongiae			
Acarnus bergquistae	Yosief et al. (1998a) (original van Soest, Hooper & Hiemstra 1991)		Unreported
Acarnus thielei	Lévi (1958)		Endemic
Acarnus wolffgangi	Keller (1889)		Present
Agelas marmarica	Lévi (1958)		Present
Agelas mauritiana	Lévi (1965) (original Carter 1879)		Present
Amphimedon chloros	Ilan et al. (2004)		Endemic
Amphimedon dinae	Helmy & van Soest (2005)		Endemic
Amphimedon hamadai	Helmy & van Soest (2005)		Endemic
Amphimedon jalae	Helmy & van Soest (2005)		Endemic
Amphimedon ochracea	Keller (1889)	Ceraochalina ochracea	Endemic
Antho (Jia) wunschorum	van Soest, Rützler & Sim (2016)		Endemic
Aplysilla lacunosa	Keller (1889)		Endemic
Aplysina reticulata	Burton (1926) (original Lendenfeld 1889)		Present
Arenosclera arabica	Keller (1889)	Arenochalina arabica	Present
Astrosclera willeyana	Karlinska-Batres & Wörheide (2015) (original Lister 1900)		Unreported
Axinella quercifolia	Keller (1889)	Antherochalina quercifolia, Querciclona quercifolia	Endemic
Axinyssa gravieri	Lévi (1965) (original Topsent 1906)	Pseudaxinyssa gravieri	Present
Batzella aurantiaca	Lévi (1958)	Prianos aurantiaca	Present
Biemna ehrenbergi	Keller (1889)	Acanthella ehrenbergi	Present
Biemna fortis	Fishelson (1971) (original Topsent 1897)		Present
Biemna trirhaphis	Lévi (1961) (Topsent 1879)		Present
Cacospongia ridleyi	Burton (1952)		Present
Callyspongia (Callyspongia) siphonella	Lévi (1965)	Siphonochalina siphonella	Endemic
Callyspongia (Callyspongia) tubulosa	Burton (1926) (original Esper 1797)	Siphonochalina tubulosa	Present
Callyspongia (Cladochalina) subarmigera	Burton (1959) (original Ridley 1884)	Callyspongia subarmigera	Unreported
Callyspongia (Euplacella) communis	Burton (1926) (original Carter 1881)	Siphonochalina communis	Present
Callyspongia (Euplacella) densa	Keller (1889)		Endemic
Callyspongia (Euplacella) paralia	Ilan et al. (2004)	Callyspongia paralia	Endemic
Callyspongia (Toxochalina) dendyi	Vine (1986) (original Burton 1931)		Not reviewed
Callyspongia calyx	Keller (1889)	Cacochalina calyx	Endemic

(continued)

Table 6.1 (continued)

Species	Citation	Previous names	Distribution (WPD)
Callyspongia clavata	Keller (1889)	*Crella cyathophora, Phylosiphonia clavata*	Endemic
Callyspongia conica	Keller (1889)	*Phylosiphonia conica*	Present
Callyspongia crassa	Keller (1889)	*Sclerochalina crassa*	Endemic
Callyspongia fistularis	Topsent (1892)	*Sclerochalina fistularis*	Endemic
Callyspongia implexa	Topsent (1892)	*Ceraochalina implexa*	Endemic
Callyspongia incrustans	Row (1911)	*Spinosella incrustans*	Endemic
Callyspongia maculata	Keller (1889)	*Cacochalina maculata*	Endemic
Callyspongia reticulata	Keller (1889)	*Siphonochalina reticulata*	Present
Callyspongia sinuosa	Topsent (1892)	*Sclerochalina sinuosa*	Endemic
Callyspongia spongionelloides	Fishelson (1971)		Endemic
Callyspongia vasseli	Keller (1889)	*Phylosiphonia vasseli*	Endemic
Carteriospongia foliascens	Lévi (1958) (original Pallas 1766)	*Phyllospongia foliascens*	Present
Chalinula saudiensis	Vacelet et al. (2001)		Endemic
Chelonaplysilla erecta	Row (1911)	*Megalopastas erectus*	Endemic
Chondrilla australiensis	Keller (1891) (original Carter 1873)	*Chondrilla globulifera*	Present
Chondrilla mixta	Lévi (1958) (original Schulze 1877)	*Chondrillastra mixta*	Present
Chondrilla nucula	El Bossery et al. (2017) (original Schmitt et al. 2012)		Unreported
Chondrilla sacciformis	Richter et al. (2001) (original Carter 1879)		Unreported
Chondrosia debilis	Lévi (1958) (original Thiele 1900)		Present
Cinachyrella albatridens	Lévi (1965) (original Lendenfeld 1907)	*Cinachyra alba tridens*	Present
Cinachyrella alloclada	Barnathan et al. (2003) (original Uliczka 1929)		Unreported
Cinachyrella eurystoma	Keller (1891)	*Cinachyra eurystoma*	Endemic
Cinachyrella ibis	Row (1911)	*Chrotella ibis*	Endemic
Cinachyrella kuekenthali	Barnathan et al. (2003) (original Uliczka 1929)		Unreported
Cinachyrella schulzei	Keller (1891)	*Cinachyra schulzei*	Present
Cinachyrella trochiformis	Keller (1891)	*Cinachyra trochiformis*	Endemic
Clathria (Clathria) arbuscula	Row (1911)	*Litaspongia arbuscula, Ophlitaspongia arbuscula*	Endemic
Clathria (Clathria) horrida	Row (1911)	*Clathria horrida, Ophlitaspongia horrida*	Endemic
Clathria (Clathria) maeandrina	Burton (1959) (original Ridley 1884)	*Clathria maeandrina*	Unreported
Clathria (Clathria) spongodes	Burton (1959) (original Dendy 1922)	*Clathria spongiosa*	Present
Clathria (Clathria) transiens	Burton (1959) (original Hallmann 1912)	*Clathria transiens*	Unreported
Clathria (Thalysias) abietina	Burton (1959) (Lamarck 1814)	*Clathria aculeata*	Present
Clathria (Thalysias) cactiformis	Hooper, Kelly & Kennedy (2000) (original Lamarck 1814)		Not reviewed
Clathria (Thalysias) fusterna	Hooper (1997)	*Clathria fusterna*	Present
Clathria (Thalysias) lambda	Lévi (1958)	*Leptoclathria lambda*	Endemic
Clathria (Thalysias) lendenfeldi	Hooper, Kelly & Kennedy (2000) (Ridley & Dendy 1886)		Not reviewed
Clathria (Thalysias) procera	Burton (1959) (original Ridley 1884)		Present

(continued)

Table 6.1 (continued)

Species	Citation	Previous names	Distribution (WPD)
Clathria (Thalysias) vulpina	Burton (1959) (original Lamarck 1814)		Present
Clathria granulata	Keller (1889)	Ceraochalina granulata	Endemic
Cliona orientalis	Lévi (1958) (original Thiele 1900)		Present
Crambe acuata	Lévi (1958)	Folitispa acuata	Present
Crella (Grayella) cyathophora	Lévi (1958) (original Carter 1869)	Grayella cyathophora	Present
Crella (Grayella) papillata	Lévi (1958)		Present
Dactylospongia elegans	Abdelmohsen et al. (2014a) (original Thiele 1899)		Unreported
Damiria simplex	Keller (1891)		Present
Darwinella gardineri	Lévi (1958) (original Topsent 1905)		Present
Dercitus (Halinastra) exostoticus	Keller (1891) (original Schmidt 1868)		Endemic
Diacarnus erythraeanus	Kelly-Borges & Vacelet (1995)		Present
Diplastrella gardineri	Lévi (1958) (original Topsent 1918)		Present
Discodermia stylifera	Keller (1891)		Endemic
Dragmacidon coccineum	Keller (1891)	Hymeniacidon coccinea, Pseudaxinella coccinea, Reniera coccinea, Stylissa coccinea,	Present
Dragmacidon durissimum	Burton (1959) (original Dendy 1905)	Axinella durissima	Present
Dysidea aedificanda	Row (1911)	Spongelia aedificanda	Endemic
Dysidea cinerea	Keller (1889)	Spongelia cinerea	Present
Echinoclathria digitiformis	Row (1911)	Ophlitaspongia digitiformis	Endemic
Echinoclathria gibbosa	Keller (1889)	Ceraochalina gibbosa, Xestospongia gibbosa	Endemic
Echinoclathria robusta	Keller (1889)	Halme robusta	Endemic
Echinodictyum flabelliforme	Keller (1889)	Acanthella flabelliformis	Endemic
Echinodictyum jousseaumi	Lévi (1958) (original Topsent 1892)		Present
Ecionemia arabica	Lévi (1958)	Hezekia arabica	Endemic
Ecionemia spinastra	Lévi (1958)		Endemic
Erylus lendenfeldi	Carmely et al. (1989) (original Sollas 1888)		Unreported
Erylus proximus	Lévi (1958) (original Dendy 1916)		Present
Eurypon calypsoi	Lévi (1958)		Present
Eurypon polyplumosum	Lévi (1958)		Endemic
Euryspongia lactea	Row (1911)		Present
Fascaplysinopsis reticulata	Helmy et al. (2004) (original Hentschel 1912)		Present
Fasciospongia cavernosa	Kashman et al. (1973) (original Schmidt 1862)		Unreported
Fasciospongia lordii	Lendenfeld (1889)	Stelospongia lordii	Present
Gelliodes incrustans	Lévi (1965) (original Dendy 1905)		Present
Geodia arabica	Topsent (1892) (original Carter 1869)		Present
Geodia jousseaumei	Topsent (1906)	Isops jousseaumei	Present
Geodia micropunctata	Row (1911)		Endemic
Guitarra indica	Burton (1959) (original Dendy 1916)	Guitarra fimbriata	Unreported
Halichondria (Halichondria) glabrata	Keller (1891)	Halichondria glabrata	Endemic
Halichondria (Halichondria) granulata	Keller (1891)	Halichondria granulata	Endemic
Halichondria (Halichondria) isthmica	Keller (1891) (original Keller 1883)	Amorphina isthmica	Endemic

(continued)

Table 6.1 (continued)

Species	Citation	Previous names	Distribution (WPD)
Halichondria (Halichondria) minuta	Keller (1891)	*Halichondria minuta*	Endemic
Haliclona (Gellius) bubastes	Row (1911)	*Halichondria bubastes*	Endemic
Haliclona (Gellius) flagellifera	Burton (1959) (original Ridley and Dendy 1886)	*Haliclona flagellifera*	Unreported
Haliclona (Gellius) toxia	Lévi (1958) (original Topsent 1897)	*Toxiclona toxius, Gellius toxius*	Present
Haliclona (Haliclona) violacea	Keller (1883)	*Lessepsia violacea*	Endemic
Haliclona (Reniera) tabernacula	Row (1911)	*Reniera tabernacula, Haliclona tabernacula*	Present
Haliclona decidua	Topsent (1906)	*Reniera decidua*	Present
Haliclona pigmentifera	Burton (1959) (original Dendy 1905)	*Adocia pigmentifera*	Present
Haliclona ramusculoides	Row (1911) (original Topsent 1893)	*Chalina minor*	Present
Haliclona spinosella	Row (1911)	*Reniera spinosella*	Endemic
Halisarca laxus	Lévi (1958) (original Lendenfeld 1889)	*Bajalus laxus*	Present
Hemimycale arabica	Ilan, Gugel & van Soest (2004)		Endemic
Higginsia arborea	Keller (1891)	*Allantella arborea, Trachytedania arborea*	Present
Higginsia higgini	Lévi (1958) (original Dendy 1922)		Present
Higginsia pumila	Keller (1889)	*Axinella pumila*	Endemic
Hyattella globosa	Lendenfeld (1889)		Endemic
Hyattella tubaria	Helmy et al. (2004) (Lendenfeld 1889)		Present
Hymedesmia (Hymedesmia) lancifera	Topsent (1906)	*Leptosia lancifera, Hymedesmia lancifera*	Present
Hymedesmia (Hymedesmia) rowi	Row (1911) (original van Soest 2017)	*Myxilla (Myxilla) tenuissima*	Endemic
Hymeniacidon calcifera	Row (1911)		Endemic
Hymeniacidon zosterae	Row (1911)		Endemic
Hyrtios communis	Row (1911) (original Carter 1885)	*Psammopemma commune*	Present
Hyrtios erectus	Keller (1889)	*Dysidea nigra, Heteronema erecta, Duriella nigra*	Present
Iotrochota baculifera	Lévi (1965) (original Ridley 1884)		Present
Ircinia atrovirens	Keller (1889)	*Hircinia atrovirens*	Endemic
Ircinia echinata	Keller (1889)	*Hircinia echinata*	Present
Ircinia ramosa	Keller (1889)	*Hircinia ramosa*	Present
Ircinia variabilis	Burton (1926) (original Schmidt 1862)	*Hircinia variabilis*	Unreported
Jaspis albescens	Row (1911)	*Coppatias albescens*	Endemic
Jaspis reptans	Lévi (1965) (original Dendy 1905)		Present
Jaspis sollasi	Burton & Rao (1932)	*Amphius sollasi*	Endemic
Jaspis virens	Lévi (1958)		Endemic
Lamellodysidea herbacea	Keller (1889)	*Carteriospongia cordifolia, Spongelia herbacea, Dysidea herbacea, Phyllospongia cordifolia, Spongelia delicatula*	Present
Levantiniella levantinensis	Tsurnamal (1969) (Vacelet, Bitar, Carteron, Zibrowius & Pérez 2007)	*Chrotella cavernosa*	Present
Lissodendoryx (Lissodendoryx) cratera	Row (1911)	*Myxilla cratera*	Endemic
Lissodendoryx (Waldoschmittia) schmidti	Lévi (1958) (original Ridley 1884)	*Damiriana schmidti*	Present
Lithoplocamia lithistoides	Burton (1959) (original Dendy 1922)		Present
Monanchora quadrangulata	Lévi (1958)	*Fasuberea quadrangulata*	Endemic

(continued)

Table 6.1 (continued)

Species	Citation	Previous names	Distribution (WPD)
Mycale (Aegogropila) sulevoidea	Burton (1959) (original Sollas 1902)	Mycale sulevoidea	Unreported
Mycale (Arenochalina) anomala	Burton (1952) (original Ridley & Dendy 1886)	Esperiopsis anomala, Parisociella anomala	Unreported
Mycale (Arenochalina) euplectellioides	Row (1911)	Esperella euplectellioides	Endemic
Mycale (Arenochalina) setosa	Keller (1889)	Gelliodes setosa	Endemic
Mycale (Carmia) erythraeana	Row (1911)	Esperella erythraeana	Endemic
Mycale (Carmia) fistulifera	Row (1911)	Esperella fistulifera	Endemic
Mycale (Carmia) suezza	Row (1911)	Esperella suezza	Endemic
Mycale (Mycale) dendyi	Row (1911)	Esperella dendyi	Endemic
Mycale (Mycale) grandis	Lévi (1958) (original Grey 1867)	Mycale grandis	Present
Myrmekioderma niveum	Row (1911)	Anacanthaea nivea	Endemic
Myrmekioderma tuberculatum	Keller (1891)	Halichondria tuberculatum	Endemic
Myxilla (Burtonanchora) gracilis	Lévi (1965)	Burtonanchora gracilis	Endemic
Negombata corticata	Carter (1879)		Present
Negombata magnifica	Keller (1889)	Latrunculia magnifica	Endemic
Neopetrosia contignata	Burton (1959) (original Thiele 1899)	Haliclona contignata	Present
Niphates furcata	Keller (1889)	Pachychalina furcata	Endemic
Niphates obtusispiculifera	Burton (1959) (original Dendy 1905)		Present
Niphates rowi	Ilan, Gugel & van Soest (2004)		Endemic
Oceanapia elastica	Keller (1891)	Reniera elastica	Present
Oceanapia incrustata	Burton (1959) (original Dendy 1922)		Present
Pachychalina alveopora	Topsent (1906)		Present
Paratetilla bacca	Row (1911) (original Selenka 1867)	Paratetilla eccentrica	Present
Petrosia (Petrosia) elephantotus	Ilan, Gugel & van Soest (2004)	Petrosia elephantotus	Endemic
Petrosia (Petrosia) nigricans	Burton (1959) (original Lindgren 1897)	Petrosia nigricans	Unreported
Phakellia palmata	Row (1911)		Endemic
Phakellia radiata	Burton (1959) (original Dendy 1916)		Present
Phorbas epizoaria	Lévi (1958)	Pronax epizoaria	Endemic
Phyllospongia lamellosa	Hassan et al. (2015) (original Esper (17940)		Unreported
Phyllospongia papyracea	Lévi (1958) (original Esper 1794)		Present
Pione mussae	Keller (1891)	Cliona mussae, Sapline mussae	Endemic
Pione vastifica	Ferrario et al. (2010) (original Hancock 1849)	paper says probably conspecific	Unreported
Psammoclema arenaceum	Lévi (1958)	Psammopemma arenaceum	Endemic
Psammoclema rubrum	Lévi (1958)	Psammopemma	Endemic
Pseudoceratina arabica	Keller (1889)	Psammaplysilla arabica	Present
Pseudoceratina purpurea	Rotem et al. (1983) (original Carter 1880)	Psammaplysilla purpurea	Unreported
Pseudosuberites andrewsi	Vine (1986) (original Kirkpatrick 1900)		Not reviewed
Ptilocaulis spiculifer	Rudi et al. (1999) (original Lamarck 1814)		Present
Rhabdastrella sterrastraea	Row (1911)	Diastra sterrastraea	Endemic

(continued)

Table 6.1 (continued)

Species	Citation	Previous names	Distribution (WPD)
Rhabderemia batatas	Ilan, Gugel & van Soest (2004)		Endemic
Rhabderemia indica	Burton (1959) (original Dendy 1905)		Present
Scalarispongia aqabaensis	Helmy, El Serehy, Mohamed & van Soest (2004)		Present
Spheciospongia inconstans	Lévi (1965) (original Dendy 1887)	*Spirastrella inconstans*	Present
Spheciospongia mastoidea	Keller (1891)	*Suberites mastoideus*	Endemic
Spheciospongia vagabunda var. *arabica*	Hooper and van Soest (2002) (original Topsent 1893)		Not reviewed
Spirastrella decumbens	Lévi (1958) (original Ridley 1884)		Present
Spirastrella pachyspira	Lévi (1958)		Present
Spongia (Spongia) arabica	Keller (1889)	*Spongia arabica, Spongia officinalis* var. *arabica*	Endemic
Spongia (Spongia) irregularis	Lévi (1965) (original Lendenfeld 1889)	*Spongia irregularis*	Present
Spongia (Spongia) lesleighae	Helmy, El Serehy, Mohamed & van Soest (2004)		Endemic
Spongia (Spongia) officinalis var. *exigua*	Lévi (1965) (original Schulze 1879)	*Spongia officinalis* f. *exigua*	Present
Spongia lacinulosa	Lamarck (1814)		Present
Stelletta parva	Row (1911)	*Pilochrota parva*	Endemic
Stelletta purpurea	Lévi (1958) (original Ridley 1884)	*Myriastra purpurea*	Present
Stelletta siemensi	Keller (1891)		Endemic
Stellettinopsis solida	Lévi (1965)		Present
Strongylacidon inaequale	Burton (1959) (original Hentschel 1911)	*Strongylacidon inaequalis*	Unreported
Stylissa carteri	Keller (1889) (original Dendy 1889)	*Acanthella aurantiaca, axinella carteri*	Present
Suberea mollis	Row (1911)	*Verongia mollis, Aplysina mollis*	Present
Suberea praetensa	Row (1911)	*Aplysina praetensa*	Present
Suberea purpureaflava	Gugel, Wagler & Brümmer (2011)		Endemic
Suberites clavatus	Keller (1891)		Endemic
Suberites kelleri	Keller (1891) (original Burton 1930)	*Suberites incrustans*	Present
Suberites tylobtusus	Lévi (1958)	*Suberites tylobtusa*	Endemic
Tedania (Tedania) anhelans	Burton (1959) (original Vio in Olivi 1792)	*Tedania nigrescens*	Unreported
Tedania (Tedania) assabensis	Keller (1891)	*Tedania assabensis, Tedania anhelans* var. *assabensis*	Present
Terpios lendenfeldi	Keller (1891)		Endemic
Terpios viridis	Keller (1891)		Endemic
Tethya japonica	Topsent (1906) (originial Sollas 1888)	*Donatia japonica*	Present
Tethya robusta	Burton (1926) (original Bowerbank 1873)	*Donatia robusta, Donatia arabica*	Present
Tethya seychellensis	Lévi (1958) (original Wright 1881)		Present
Tetilla diaenophora	Lévi (1958)		Endemic
Tetilla poculifera	Row (1911) (original Dendy 1905)		Present
Theonella conica	Lévi (1958) (original Kieschnick 1896)		Present
Theonella mirabilis	El Bossery et al. (2017) (original de Laubenfels 1954)		Unreported
Theonella swinhoei	Lévi (1958) (original Grey 1868)		Present
Timea intermedia	Lévi (1958)	*Timeopsis intermedia*	Endemic
Topsentia aqabaensis	Ilan, Gugel & van Soest (2004)	*Epipolasis aqabaensis*	Endemic
Topsentia halichondrioides	Burton (1926) (original Dendy 1905)	*Trachyopsis halichondrioides*	Present
Xestospongia ridleyi	Keller (1891)	*Reniera ridleyi*	Endemic

(continued)

Table 6.1 (continued)

Species	Citation	Previous names	Distribution (WPD)
Xestospongia testudinaria	Burton (1959) (Lamarck 1815)	*Petrosia testudinaria*	Present
Class Hexactinellida			
Neoaulocystis polae	Ijima (1927)	*Aulocystis polae*	Endemic
Tretocalyx polae	Schulze (1901)		Endemic
Class Homoscleromorpha			
Plakortis erythraena	Lévi (1958)		Endemic
Plakortis nigra	Lévi (1958) (original Levi 1953)		Present
Class Calcarea			
'*Arturia*' *adusta*	van Soest & de Voogd (2018) (original Wörheide & Hooper et al. 2000)	*Clathrina adusta*	Present. Genus affiliation to *Arturia* requires revision Voigt et al. (2017).
Arturia darwinii	Vine (1986) (original Haeckel 1870)	*Clathrina darwinii*	Not reviewed
Arturia suezinana	Row (1909) (original Klautau & Valentine 2003)	*Clathrina sueziana* Klautau, *Clathrina canariensis* var. *compacta*	Endemic
Arturia tenuipilosa	Burton 1952 (original Dendy 1905)	*Leucosolenia tenuipilosa*	Doubtful
Borojevia aff. *aspina*	Voigt et al. (2017)		Unreported
Borojevia voigti	van Soest & de Voogd (2018)		Endemic
Clathrina ceylonensis	Vine (1986) (original Dendy 1905)		Not reviewed
Clathrina maremeccae	van Soest & de Voogd (2018)		Endemic
Clathrina rotundata	Voigt et al. (2017)		Endemic
Clathrina rowi	Voigt et al. (2017)		Endemic
Clathrina sinusarabica	Klautau & Valentine (2003)		Endemic
Ernstia arabica	Voigt et al. (2017)		Endemic
Grantessa woerheidei	van Soest & de Voogd (2018)		Endemic
Grantilla quadriradiata	Row (1909)		Endemic
Kebira uteoides	Row (1909)		Endemic
Leucandra aspera	Row (1909)		Unreported
Leucandra bathybia	Lévi (1965) (original Haeckel 1869)	*Leuconia bathybia*	Present
Leucandra pulvinar	Haeckel (1872) (original Haeckel 1870)	*Mlea dohrnii* Maclay	Present
Leucandrilla intermedia	Row (1909)	*Leucilla intermedia*	Endemic
Leucetta chagosensis	Wörheide et al. (2008) (original Dendy 1913)		Present
Leucetta microraphis	Voigt et al. (2017) (original Haeckel 1872)		Unreported
Leucetta primigenia	Haeckel (1872)		Unreported
Leucetta pyriformis	van Soest & de Voogd (2018) (original Dendy 1913)		Present
Paraleucilla crosslandi	Row (1909)	*Leucilla crosslandi*	Endemic
Soleneiscus hamatus	Voigt et al. (2017)		Endemic
Sycettusa glabra	Row (1909)	*Grantessa glabra*	Endemic
Sycettusa hastifera	Row (1909)	*Grantessa hastifera*, *Grantilla hastifera*	Present
Sycettusa hirsutissima	van Soest & de Voogd (2018)		Endemic
Sycettusa stauridia	Row (1909) (original Haeckel 1872)	*Grantessa stauridia*	Present
Sycon ciliatum	Row (1909)	*Sycon coronatum*	Unreported
Sycon proboscideum	Haeckel (1872) (original Haeckel (1870)	*Syconella proboscideum*	Endemic 'species inquirenda van Soest & de Voogd (2018)'
Sycon raphanus	Haeckel (1872)		Unreported

affiliated class (2 species). The latter two classes likely have many more representatives in the Red Sea, but they have not yet been described.

6.2.1 Demosponge Diversity of the Red Sea

A large portion of the taxonomic work on Red Sea demosponges is based on monographs of Keller (1889, 1891), Row (1911), and Lévi (1958, 1961, 1965), subsequently complemented by other authors (e.g., Topsent 1892, 1906; Burton 1952, 1959; Kelly Borges and Vacelet 1995; Vacelet et al. 2001; Helmy et al. 2004; Ilan et al. 2004; Helmy and van Soest 2005; Gugel et al. 2011). The majority of these studies were almost entirely based on morphology. Studies on demosponges that use DNA sequencing were only recently employed to understand biodiversity patterns (e.g., Eid et al. 2011) and to apply integrative taxonomy methods (including phylogenetic analyses of DNA data, DNA barcodes and morphology).

Initial results to date on the largest molecular biodiversity survey on Red Sea demosponges (Erpenbeck et al. 2016b), summarizing the results of 1014 sponge specimens collected along the Saudi-Arabian coastline, revealed a dominance of dictyoceratid and haplosclerid operational taxonomic units (OTUs) collected from 0 m to approximately 30 m depth. Both orders constitute taxonomically challenging taxa, highlighting the need for more thorough research among those groups.

DNA sequence comparisons to other Indo-Pacific demosponge faunas indicated high endemism in the Red Sea with about 35% of the molecular sponge OTUs being shared with samples from other regions of the Indo-Pacific (Erpenbeck et al. 2016b). This study revealed several allegedly widespread Indo-Pacific species to be Red Sea endemics, such as the abundant keratose sponge *Hyrtios erectus* (Keller 1889). The Indo-Pacific "*Hyrtios erectus*" constitutes a species complex with *Hyrtios erectus* restricted to the Red Sea, and other, yet-unnamed species in other Indo-Pacific regions outside the Red Sea (Erpenbeck et al. 2017).

The results from *Hyrtios erectus* corroborated previous findings and hypotheses that the level of endemism among other marine invertebrates is underestimated (Klautau et al. 1999; Miloslavich et al. 2011): However, several Red Sea sponge species share DNA barcodes with Indonesian and other distant Indo-Pacific samples, such as *Spheciospongia vagabunda* and *Stylissa carteri* (Erpenbeck et al. 2017). *Stylissa carteri* (Fig. 6.2) has also been subject to the most extensive population genetic structure in the Red Sea at present. Giles et al. (2013a, 2015) analyzed microsatellite data of *S. carteri* samples collected from the Gulf of Aqaba to Socotra in the Arabian Sea and provided the first evidence for a latitudinal environmental gradient influencing sponge populations in the Red Sea. A gene flow barrier around the Farasan Islands, not fully explainable with the regional currents alone, separates the southern from the central and northern Red Sea *S. carteri* populations (Giles et al. 2015). In another demosponge species, *Astrosclera willeyana*, community structure and genetic diversity were explored across the species' range throughout the Indo-Pacific, with the Red Sea having its own haplotype (Wörheide 2006). Nevertheless, further studies in the Red Sea are clearly required to obtain a broader insight into Red Sea Sponge diversity and connectivity patterns.

6.2.2 Calcareous Sponge Diversity of the Red Sea

Most of the recognized species of calcareous sponges in the Red Sea were identified and described on material collected in the nineteenth and early twentieth century. Descriptions by Row (1909) are based on poorly preserved material, and often only one or very few specimens. These early reports include several species that were originally found in other oceans and climates but were considered to have a 'cosmopolitan' distribution. Newer studies have shown that some of these assumed 'cosmopolitan' species reported from the Red Sea constitute distinct species (e.g. *Clathrina sinusarabica* (Valentine and Klautau 2003), which was previously identified as the 'cosmopolitan' species *Clathrina primordialis*). Also, early reports of cold water species such as *Sycon 'coronatum'* (originally described from the North Sea) or *Sycon raphanus* (originally described from the Adriatic Sea) in the Red Sea have to be doubted. In particular, Haeckel (1872) sometimes provided relatively unspecific descriptions of species and did not specify type localities, making the identification of these species almost impossible (for example for *Leucetta primigenia*, Table 6.1). Newly collected material and the application of integrative taxonomic approaches revealed additional species (Voigt et al. 2017; van Soest and de Voogd 2018), as is the case in many other regions where such methods were applied to calcareous sponges (e.g., Imešek et al. 2014; Azevedo et al. 2015; Klautau et al. 2016).

At least 15 species of calcareous sponges are reported as Red Sea endemics (e.g., *Kebira uteoides* and *Clathrina sinusarabica* (Fig. 6.1)). However, because the calcareous sponge faunas of the adjacent Indian Ocean regions are also understudied, it is possible at least some of these typical Red Sea calcareous sponges will be found in adjacent regions in the future. Only a few Red Sea Calcarea appear to be more widespread in the Indo-Pacific, for example *Leucetta chagosensis* or *Leucetta microraphis*, but in both cases, it remains even unclear if the Red Sea specimens represent different (possibly cryptic) species (Wörheide et al. 2008; Voigt et al. 2017).

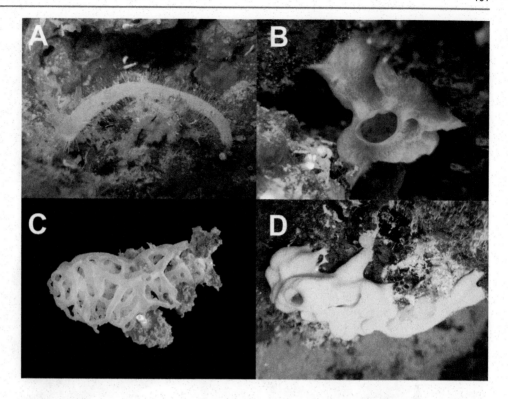

Fig. 6.1 Representative calcareous sponges (class Calcarea) of the Red Sea: (**a**) *Sycettusa hastifera* (Calcaronea, Heteropiidae); (**b**) *Kebira uteoides* (Calcaronea, Lelapiidae); (**c**) *Clathrina sinusarabica* (Calcinea, Clathrinidae); (**d**) *Leucetta chagosensis* (Calcinea, Leucettidae)

6.3 Publications on Red Sea Sponge Biology

A total of 236 publications including sponges of the Red Sea (including a handful of taxonomic papers from larger regions that only have 1 or 2 species in the Red Sea (but a first Red Sea record of a species)) were reviewed. These publications date from 1814 to early 2018 and cover 5 broad research categories: bioactive compounds, ecology, microbiology, molecular biology, and taxonomy. Red Sea sponge research increased rapidly in the last two decades with 167 of the 236 papers published since 2000. Only 51 of the pre-2000 publications were not among the reports of early taxonomy exploration. Of the 236 publications, 122 were related to bioactive compounds, 47 were ecology-focused, 27 addressed microbial aspects, 33 were taxonomy-oriented, and 7 explored molecular biology of Red Sea sponges. In publications not solely focused on taxonomy, a total of 62 sponge species were identified (i.e., at the species level and not only at a genus or higher level). Among these 62 sponge species, 29 species were noted in more than one publication and the other 33 species each appeared in only one publication.

6.3.1 Bioactive Compounds of Red Sea Sponges

A large portion of the reviewed publications describe studies of secondary metabolites of Red Sea sponges. Marine sponges are a potentially valuable source of novel natural compounds because of the diversity of secondary biological compounds they produce, most of which are not found in terrestrial organisms. Sponges are the most prolific source of new organic structures with bioactive properties; sponge-derived compounds exceed those recovered from other marine organisms by far (see Blunt et al. 2007). From the compounds extracted from Red Sea sponges to date, many have been shown to possess anti-cancer, anti-microbial, anti-inflammatory, anti-malarial, and anti-viral properties, along with other specificities as seen in Table 6.2. A multitude of studies suggests a further development of these structures towards medical applications (Faulkner 2000; Newman and Cragg 2004; Sipkema et al. 2005; Mehbub et al. 2014), such as new antibiotics for the ongoing race against antibiotic resistance. With relatively high species diversity and possibly high endemism in the Red Sea, there is a reasonable expectation that Red Sea sponges could yet yield a large number of novel compounds. Table 6.2 shows the sponges studied in the bioactive molecule publications and the associated source references.

6.3.2 Ecology of Red Sea Sponges

Although there are only 47 ecological publications on Red Sea sponges, they span a wide range of topics. Seven publications (not all exclusively focused on ecology) addressed the reproductive biology of several species (Ilan and Loya 1988, 1990, 1995; Ilan and Vacelet 1993; Meroz and Ilan

Fig. 6.2 Some common demosponges (Class Demospongiae) of the Red Sea. (**a**) *Xestospongia testudinaria* (Haplosclerida, Petrosiidae), a large massive sponge; (**b**) *Hyrtios* cf. *erectus* (Dictyoceratida, Thorectidae) with erect and branching growth form; (**c**) *Chalinula*? sp. (Haplosclerida, Chalinidae), an encrusting sponge; (**d**) *Stylissa carteri* (Scopalinida, Scopalinidae); (**e**) *Pione* cf. sp. (Clionaida, Clionaidae), a bioeroding sponge; (**f**) *Carteriospongia* sp. (Dictyoceratida, Thorectidae), a foliose sponge; (**g**) *Ircinia echinata* (Dictyoceratida, Iciniidae), a common, massive sponge; (**h**) *Crella* (*Grayella*) *cyathophora* (Poecilosclerida, Crellidae); (**i**) *Cinachyrella* sp. (Tetractinellida, Tetillidae)

1995b; Ilan et al. 2004; Oren et al. 2005). Seven publications examined feeding ecology, such as diet composition and feeding rates, including uptake of DOC, DOM, POM, viruses, and plankton (Yahel et al. 1998, 2003, 2005; Hadas et al. 2006, 2009; Genin et al. 2009). Rix et al. (2016) specifically studied the uptake of coral mucus (DOM) by encrusting sponges, which converted the DOM into detritus and thus brought it back into the food web. Rix et al. (2017) further considered DOM uptake in sponges by looking at the uptake rates of DOM from different sources (coral and algae). Rix et al. (2018) showed the transfer of organic matter from corals to sponges whose detritus was taken up by detritivores using stable isotopes. Several papers explored the symbiotic relationship between sponges and other associated macrofauna such as cirripedia, scyphozoans, mesostigmatid mites, polychaetes, barnacles, and many others (Kolbasov 1990; Meroz and Ilan 1995a; Ramadan 1997; Magnino et al. 1999; Ilan et al. 1999; Kandler 2015). A variety of other topics are addressed in some papers, such as dispersal out of the Red Sea (Tsurnamal 1969), publications identifying specific habitats and environmental conditions of sponges (Fishelson 1966; Fishelson 1971; Ilan and Vacelet 1993; Ilan and Abelson 1995; Reitner et al. 1996; Steindler et al. 2001), aspects of life history (Meroz and Ilan 1995b), the role of sponges in nitrogen cycling and primary production (Rix et al. 2015), various physical and chemical defenses employed by sponges (Burns and Ilan 2003, Burns et al. 2003, Ilan and Loya 1995), uptake of chemical defense by a spongivorous nudibranch (Mebs 1985), competition among benthic fauna (Rinkevich et al. 1993), bioerosion (Zundelevich et al. 2007), metabolism, O_2 dynamics inside the sponge, and photosynthetic responses to dim light

Table 6.2 Sponge species that have been the subject of secondary metabolite research

Species	References	Category of study
Aaptos aaptos	Rudi and Kashman (1993)	NP
Acarnus bergquistae	Yosief et al. (1998a)	NP
Amphimedon chloros	Ajabnoor et al. (1991), Kelman et al. (2009)	Blood glucose levels, AM
Amphimedon viridis	Kelman et al. (2001)	AM
Biemna ehrenbergi	Kelman et al. (2009), Youssef et al. (2015a)	AM, NP and AM
Biemna fortis	Delseth et al. (1979)	NP
Callyspongia crassa	Ibrahim et al. (2017)	(AC, AI, AM)
Callyspongia (Callyspongia) siphonella	Shmueli et al. (1981), Carmely and Kashman (1986), Kashman et al. (2001), Jain et al. (2007a), Jain et al. (2007b), Jain et al. (2009), Kelman et al. (2009), Abraham et al. (2010), Angawi et al. (2014), Foudah et al. (2014), Al-Massarani et al. (2015), Amina et al. (2016), Ibrahim et al. (2017), Ahmed A et al. (2018)	NP, NP, NP, reversal of cancer chemotherapy resistance with new NP, AC, AC multidrug resistance with new NP, AM, AC, AC with new NP, AC (new NP), AC and AM and Anti-viral), Binding properties to BSA, NP (AC and AM), new NP and inhibitationn of RANKL indused osteoclastogensis
Callyspongia aff. *implexa*	Abdelmohsen et al. (2010), Abdelmohsen et al. (2015), Elsayed et al. (2017)	AM, AM and new NP, new NP (AM and antitrypanosomal)
Callyspongia fistularis	Youssef et al. (2003a)	NP
Callyspongia spp.	Youssef et al. (2003b), Youssef et al. (2000), Abdelwahed et al. (2014), Shaala et al. (2016)	NP and AC, NP, Extract from fungus, NP and AC
Chalinula saudiensis	Al-Sofyani et al. (2011)	NP
Cinachyrella alloclada,	Barnathan et al. (2003)	NP
Cinachyrella kuekenthali	Barnathan et al. (2003)	NP
Clathria sp.	Rudi et al. (2001)	NP and anti-HIV RT
Crella (Grayella) cyathophora	El-Damhougy et al. (2017)	(AC, AM, AI)
Diacarnus erythraeanus	El Sayed et al. (2001), Lefranc et al. (2013), Youssef et al. (2001), Youssef (2004)	(Antimalarial, Antiviral, and Antitoxoplasmosis with new NP), AC with new NP, NP, NP and AC
Dragmacidon coccineum	Abou-Hussein et al. (2014)	NP and AI
Dysidea herbacea	Carmely et al. (1990)	NP
Dysidea sp.	Gebreyesusa et al. (1988)	NP
Echinoclathria gibbosa	Mohamed et al. (2014b)	(AC, AM, AI, antipyretic, and hepatoprotective activities)
Echinoclathria sp.	Abdelhameed et al. (2017)	New NP (AC and AI)
Erylus sp. (possibly *lendenfeldi*)	Goobes et al. (1996)	NP
Erylus lendenfeldi	Carmely et al. (1989), Fouad et al. (2004), Sandler et al. (2005)	Anti-tumor and AF (new NP), NP and AM and AF, NP and cytotoxicity against a yeast strain (Δrad50)
Fasciospongia cavernosa	Kashman et al. (1973)	NP
Haliclona sp.	Al-Massarani et al. (2016)	NP and AC
Hemimycale arabica	Mudit et al. (2009), Youssef et al. (2015b), Ahmed H et al. (2018)	NP and AC, NP (AM and AC), AC
Hippospongia sp.	Guo et al. (1997), Guo and Trivellone (2000)	NP
Hyrtios erectus	Youssef et al. (2002), Kashman and Rudi (1977), Youssef (2005), Youssef et al. (2005), Sauleau et al. (2006), Ashour et al. (2007), Abdelmohsen et al. (2010), Alarif et al. (2016), Elhady et al. (2016a, b), Sameh et al. (2016), Walied et al. (2016), El-Gendy et al. (2017), Hawas et al. (2018), Alahdal et al. (2018), Abd El Moneam et al. (2018)	NP and AC, NP, NP and AC, AM with new NP, NP and anti-venom, AC and AM and new NP, AM, NP and AC, NP and AC, NP and AC, AC with new NP, AC with new NP, AC and hepatitis inhibition from endophytic fungi AC with new NP, AC with (Anti Helicobacter and Antitubercular, new NP and liver toxicity
Hyrtios spp.	Youssef et al. (2004), Youssef et al. (2013), Shady et al. (2017)	NP and AM, NP and (AM, free radical scavenging and AC), NP and (AI, anti-pyretic, analgesic activities)
Lamellodysidea herbacea	Kashman and Zviely (1979), Sauleau and Bourguet-Kondracki (2005), Sauleau et al. (2005)	NP, NP and AF, NP
Latrunculia corticata	Řezanka and Dembitsky (2003)	NP and antifeeding (chemical defense)
Leucetta cf. *chagosensis*	Dunbar et al. (2000)	AF and Nitric Oxide Synthase Inhibitory with new NP
Mycale (Arenochalina) euplectellioides	Mohamed et al. (2014a), Gamal et al. (2014), Abdelhameed et al.(2016)	AI and hepato-protective, NP with (AM, AI, hepato-protective, AC), NP with anti-choline esterase activity
Negombata corticata	Ahmed et al. (2008)	Anti-epileptic with new NP

(continued)

Table 6.2 (continued)

Species	References	Category of study
Negombata magnifica	Neeman et al. (1975), Kashman et al. (1980), Spector et al. (1983), Mebs (1985), Gillor et al. (2000), Vilozny et al. (2004), Abdelmohsen et al. (2010), El-Damhougy et al. (2017), Ahmed H et al. (2018)	NP, NP, disrupt microfilament organization in cultured cells, NP and chemical defense, Immunolocalization, NP, AM, AC, AC
Niphates rowi	Gesner et al. (2005)	NP
Niphates sp.	Talpir et al. (1992)	NP
Petrosia sp.	Abdel-Lateff et al. (2014)	NP and AC
Phyllospongia lamellosa	Hassan et al. (2015)	NP and AC and AM
Prianos sp. (could be *Batzella aurantiaca*)	Kashman and Rotem (1979), Sokoloff et al. (1982)	NP, AM and AF
Pseudoceratina arabica	Badr et al. (2008), Shaala et al. (2015b), Shaala et al. (2012)	NP and parasympatholytic effects, NP and AC, NP and AC
Pseudoceratina purpurea	Rotem et al. (1983)	NP
Ptilocaulis spiculifer	Rudi et al. (1998), Rudi et al. (1999)	NP, NP
Raspailia sp.	Yosief et al. (1998b), Yosief et al. (2000)	NP, AC with NP
Siphonochalina sp.	Rotem and Kashman (1979)	NP
Spheciospongia vagabunda var. *arabica*	Abdelmohsen et al. (2010), Eltamany et al. (2014a), Eltamany et al. (2014b), Abdelmohsen et al. (2014b), Eltamany et al. (2015)	AM, AM with new NP, AC with new NP, NP and antiparasitic (from sponge associated bacterium), NP and AC
Stylissa carteri	Mancini et al. (1997), O'Rourke et al. (2016), Hamed et al. (2018)	NP, Anti-HIV, AC
Suberea mollis	Abou-Shoer et al. (2008), Shaala et al. (2011), Shaala et al. (2012), Abbas et al. (2014)	NP and AM and anti-oxidant, NP (AM, anti-oxidant, AC), NP and AC, hepatoprotective
Suberea spp.	Shaala et al. (2015a), Shaala and Almohammadi (2017)	NP and AC, new NP (AC and AM)
Theonella mirabilis	Abou-Hussein and Youssef (2016)	NP and AC
Theonella swinhoei	Youssef and Mooberry (2006), Tabares et al. (2012), Youssef et al. (2014)	NP and AC, NP, NP and AF and AC
Toxiclona toxius	Isaacs and Kashman (1992)	NP
Xestospongia testudinaria	El-Shitany et al.(2015), El-Gamal et al. (2016)	NP (AI, antioxidant, and immunomodulatory), NP and AC
Reviews	Kashman et al. (1982), Kashman et al. (1989), Kalinin et al. (2012), El-Ezz et al. (2017)	
Other	Shaaban et al. (2012), Abdelmohsen et al. (2014a), Afifi and Khabour (2017)	Inhibits oxidative stress, New Actinomycetes, AM
Undescribed sp.	Guo et al. (1996)	NP

The broad category of each study is indicated: Natural products (NP); Anti-microbial (AM); Anti-cancer (AC), Anti-fungal (AF), Anti-inflammatory (AI), etc. Where multiple studies used a given species, the categories are listed respectively for each study separated by a comma. See References section for full details of each study

(Hadas et al. 2008; Lavy et al. 2016; Beer and Ilan 1998), heavy metal accumulation (Pan et al. 2011), elemental composition of some sponge species (Mayzel et al. 2014), the discovery of chitin in skeletons of non-verongiid demosponges (Ehrlich et al. 2018; Żółtowska-Aksamitowska et al. 2018), sea ranching (Hadas et al. 2005), and effects of water movement on zonation (Sara et al. 1979). There were several publications that we placed in the ecology category which were not actually specifically focused on sponges but instead more generally assessed benthic fauna with only peripheral attention to sponges (e.g., Hoeksema et al. 2016). Many of these publications are discussed further with respect to reported distribution and density of Red Sea sponges.

Beyond biodiversity studies, there is a distinct lack of data about the abundance and coverage of sponges in the Red Sea. Benthic surveys in the Red Sea typically report very low values. For example, Benayahu and Loya (1981) found that sponges constitute about 1% of reef cover and that their space utilization is negligible. Surveys that quantify sponges on Red Sea reefs are very rare and do not exist for most parts of the Red Sea. Some publications do offer some examples of abundance of the community or at least of the study species. Meroz and Ilan (1995b) used belt transects (10 m × 0.6 m) to survey the number and percent coverage of the sponge *Mycale fistulifera* at 3, 6, 10, and 20 m depths at 2 reefs at the end of winter and summer. The number of colonies generally decreased with depth. The highest occurrence of colonies was found on one of the reefs with an average of 9 colonies at 3 m depth, but the overall density was approximately 3.7 individuals per transect at one of the sites. The highest percent coverage was ~ 33 cm^2 / m^2 at 6 m depths on one reef (see Figure 2 in Meroz and Ilan (1995b)). Yahel

(1998) used 0.25 m² quadrats and quantified sponges in four taxon categories: *Mycale fistulifera*, *Cliona* sp., unrecognized blue sponge, and "other sponges"; these had average densities of 0.12, 0.38, 0.91, and 0.31 individuals/quadrat, respectively, and % occurrences (i.e., presence/absence from 65 quadrats) of 9.2, 24.6, 50.8, and 18.5%, respectively. Perkol-Finkel and Benayahu (2005) followed stages of benthic community development on a purpose-planned artificial reef. They primarily focused on corals and compared the developing community to a nearby natural reef. After 10 years, the artificial reef was dominated by the sponge *Crella cyatophora*, which contributed 35% (± SE 13.38) of the living cover on the artificial reef compared to 0.08% (± 0.18 SE) on the natural reef. In another example of a temporal study, Rix et al. (2015) noted that visible sponge cover was constant throughout the year, averaging 1.2 ± 0.9% and that the non-cryptic sponge community was dominated by the abundant encrusting sponge *Mycale fistulifera*, which accounted for 65% of the visible sponge cover at 10 m depth. More recently, Ellis et al. (2017) conducted cross-shelf benthic surveys using point-count assessment of 1m² photo quadrats. They found a higher sponge abundance on inshore reefs (up to 5% cover).

It is important to note that standard (visual) benthic survey approaches may not fully reveal the abundance of sponges on a reef. Many species are endolithic or otherwise reside inside the reef matrix, thus masking their potential ecological importance. Richter et al. (2001) developed endoscopic techniques to explore the extensive crevices common in the physical framework of Red Sea reefs. Their approach revealed a large internal surface (2.5–7.4 m² per projected m² of reef); sponges dominated in the posterior sections of these crevices, constituting 51–73% of the coelobite cover. This highlights that sponges are more abundant than they seem because they are often living out of sight of most standard visual surveys. Pearman et al. (2016) compared various benthic diversity assessments including standard visual reef surveys, photo analysis of Autonomous Reef Monitoring Structures (ARMS) plates, and metabarcoding fauna collected from ARMS. Visual surveys found ~1% of benthic cover to be sponges, while the photo analysis of the ARMS plates had up to about 25% cover, and metabarcoding revealed slightly under 30% of the reads as sponges. Again, these results demonstrate the inadequacy of standard visual surveys for quantifying sponge abundance.

Compared with other important reef organisms, sponges have been greatly neglected in quantitative studies. The principal reasons for this are taxonomic problems, due to great variability in shape and size, and difficulties in quantification, partly because most sponge biomass is not readily visible using standard survey techniques. In the Red Sea, sponges are sometimes left out of or not reported in reef surveys because of their relatively low abundance (Benayahu and Loya 1981; Furby et al. 2013). Even in the few Red Sea studies that did attempt to include sponges, point-intercept or line-intercept transect methods were used, but these methods only provide insight to the relative rate of occurrence of individual sponge colonies (or even just morphologies) with limited further quantitative application (e.g., Roberts et al. 2016). In some cases, even when grouped into one single "sponge" category, the cover measured by these methods is typically <1% (Khalil et al. 2017). To more fully assess the ecological role of sponges, more detailed measurements, such as surface area or biomass, are needed. Methods using photographic quadrats are potentially useful for sponge surveys because percent cover can be easily derived from the photographs, and morphometric measurements can be used to calculate volume or size for more uniformly-shaped sponge species. When conducting surveys *in situ*, the use of quadrats may help to focus a researcher and enable intense investigations in crevices and the space in-between corals to search for sponges that are out of sight (and thus typically overlooked in belt transects or intercept-based methods). While there is reasonable knowledge about the diversity of sponges occurring in the Red Sea (see Table 6.1), data from a wider variety of depths and geographic locations would be helpful. Most of the published surveys conducted today to date are at a single depth (usually about 10 m) with the majority in the Gulf of Aqaba (Berumen et al. 2013).

6.3.3 Microbiology of Red Sea Sponges

Microbiology-related studies are a major area of research for sponges. Sponges can have a large proportion of their biomass comprised of bacteria and it is believed that many of the bioactive molecules from sponges are synthesized by bacterial inhabitants (Taylor et al. 2007). The 27 studies that focused on the microbial communities of the sponges from the Red Sea do not seem to have applied a systematic approach within this collective body of work, and they represent a wide range of topics that are not directly addressing questions that are specific to the Red Sea. The publications included, for example, aspects of nitrogen fixation (Wilkinson and Fay 1979), better bacterial culturing methods (Lavy et al. 2014; Keren et al. 2016), and bacterial tolerance to heavy metals (Keren et al. 2015, 2017). Several studies investigated the composition of the bacterial communities in sponges (Radwan et al. 2010; Lee et al. 2011; Schmitt et al. 2012, Karlinska-Batres and Wörheide et al. 2015), with some researchers employing phylogenetic approaches (Steindler et al. 2005). There was one phylogenetic analysis and biological evaluation of marine fungi isolated from the Red Sea *Hyrtios erectus* (El-Gendy et al. 2017). The bacterial phylum Actinobacter was sometimes specifically targeted as it includes Actinomycetes (particularly sought after for bioactive molecules) (Bergman et al. 2011; Abdelmohsen et al. 2014a, b; Kämpfer et al. 2015; Elsayed et al. 2018). One study found a novel lineage of *Marinobacter* (Lee et al. 2012) while another found a new N-Acyl homoserine lactone synthase in an uncultured symbiont (Britstein et al. 2016). Sponge species are commonly assigned either to a group that contains a high abundance of microbes (high microbial abundance (HMA)), or a group with relatively low abundances of microbes (low microbial abundance (LMA)) (Gloeckner et al. 2014). Giles et al. (2013b) specifically investigated the bacterial community of three LMA sponges. Beyond characterizations of the microbial communities, Gao et al. (2015) examined the changes in microbial communities found in healthy tissues compared to disease-like tissues; the healthy tissue hosted mostly Proteobacteria, Cyanobacteria, and Bacteroidetes, while there was a shift in the disease-like tissues that was enriched with a novel clade affiliated with the phylum Verrucomicrobia. Gao et al. (2014a) also showed shifts in bacterial communities between healthy and abnormal tissue. Oren et al. (2005) found evidence of vertical transmission of cyanobacteria from adults to larvae. Others have looked at the microbial meta-genomes and attempted to determine the functional role of the bacteria contained within the sponges (Gao et al. 2014b; Bayer et al. 2014; Moitinho-Silva et al. 2014a, b). One study found quorum sensing signal production by sponge associated bacteria (Yahia et al. 2017).

6.4 Potential Future Research Directions

Even though there has been increased interest in sponge research in the region recently, many questions remain to be answered and the Red Sea remains a poorly-understood system in many aspects. A large effort is still needed to fully document the sponge species present in the Red Sea. This knowledge gap has important consequences beyond simply cataloging biodiversity. Our present understanding of sponge diversity in the region is insufficient to recognize, for example, if there are changes to the community caused by anthropogenic disturbances, such as rapid coastal development or changes due to coral bleaching events. Although there are many gaps and important information missing, insight may be gained from reviewing the early natural history expeditions and the valuable historical records they provide. Future sponge biodiversity surveys could explicitly check to see if the same species are still present in the areas where they were originally recorded more than 100 years ago (although the fragmentary nature of the earliest data may present some challenges in such a comparison). The application of modern molecular techniques to accompany traditional morphological identifications is important ongoing work that will provide more clarity and will help to resolve some taxonomic uncertainties.

We have noted the need for a better understanding of the biodiversity, distribution, and the coverage of sponges so that a proper baseline knowledge of the Red Sea sponge community is available. Even though previous works have covered a wide range of topics, most of the topics have been addressed by only a small number of studies and there remain important topics that have yet to be investigated at all. For example, we were not able to find any studies addressing whether sponge population sizes or community composition in the Red Sea are controlled by top-down or bottom-up effects. As far as we are aware, no publication systematically examined predation on sponges in the Red Sea, nor did we find any publications discussing food or nutrient limitations impacting sponges. This type of information could provide some insight to more general questions about Red Sea sponge ecology. Are sponges in the Red Sea found more readily in crevices in shallower water because they need to be hidden from predators or is it simply because there is too much competition for space with more efficient colonizers?

The Red Sea may yet hold important insights for the ecology of coral reefs under predicted climate change scenarios (Voolstra et al. 2016). With further understanding of the biogeography and biology of the regional sponge fauna, it may be possible to ask questions about how sponges elsewhere will cope with expected environmental changes. For example, do endemic Red Sea sponges possess (or express) genetic traits or adaptations that may be latent (or unexpressed) traits in their Indian Ocean ancestors? Such traits could for the basis for the rapid emergence of heat tolerant phenotypes (e.g., Dixon et al. 2015). Very little is currently known about the biogeography or evolutionary history of Red Sea sponges, but work in other taxa suggests that the Red Sea has the potential to export biodiversity to the wider Indo-Pacific (Bowen et al. 2013; Berumen et al. 2017). There is clearly much work left to do to enhance our understanding of Red Sea sponge communities, how they interact with other organisms in the ecosystems, and the potential role they will play in Indo-Pacfic reefs under future climate scenarios.

Appendix

Electronic version of the list of sponges reported from the Red Sea. "Species" indicates the current accepted name for the sponge species. Taxonomic classification ("Class" and "Order") are provided for each species. "Citation" lists the oldest known record of the species in the Red Sea (to the best of our knowledge). Note that in some cases, the original description of the species was from another geographic location; in these cases, the citation for the original species description is included in parentheses. Because many of the sponges have had taxonomic revisions since the original records shown in the "Citation" column, the "Previous Name(s)" column indicates the name(s) used in the work cited. (For a full taxonomic history of each species / genus, see the World Porifera Database (van Soest et al. 2018)). Finally, notes are included regarding the distribution of each taxa in the "Distribution (WPD)" column. "Present" indicates that the WPD currently reflects that this species' distribution includes the Red Sea. Species listed as "Endemic" are shown in the WPD to only occur inside the Red Sea. In some cases, the publication listed in the "Citation" column has reported a species in the Red Sea although the species' distribution in the WPD does not include the Red Sea; these cases are indicated as "Unreported". This more likely reflects the ongoing work of the WPD editors and not an intentional omission. For some species, the WPD shows a distribution including the Red Sea but explicitly acknowledges that the distribution has not been reviewed by WPD editors ("Not Reviewed"). Finally, one species is indicated to occur in the Red Sea but WPD editors have flagged this as "Doubtful".

Class	Order	Species	Citation	Previous names	Distribution (WPD)
Calcarea	Clathrinida	'Arturia' adusta	van Soest & de Voogd (2018) (original Wörheide & Hooper 2000)	Clathrina adusta	Present. Genus affiliation to Arturia requires revision (Voigt et al. 2017).
Calcarea	Clathrinida	Arturia darwinii	Vine (1986) (original Haeckel 1870)	Clathrina darwinii	Not reviewed
Calcarea	Clathrinida	Arturia sueziana	Row (1909) (original Klautau & Valentine 2003)	Clathrina sueziana Klautau, Clathrina canariensis var. compacta	Endemic
Calcarea	Clathrinida	Arturia tenuipilosa	Burton (1952) (original Dendy 1905)	Leucosolenia tenuipilosa	Doubtful
Calcarea	Clathrinida	Borojevia aff. aspina	Voigt et al. (2017)		Unreported
Calcarea	Clathrinida	Borojevia voigti	van Soest & de Voogd (2018)		Endemic
Calcarea	Clathrinida	Clathrina ceylonensis	Vine (1986) (original Dendy 1905)		Not reviewed
Calcarea	Clathrinida	Clathrina maremeccae	van Soest & de Voogd (2018)		Endemic
Calcarea	Clathrinida	Clathrina rotundata	Voigt et al. (2017)		Endemic
Calcarea	Clathrinida	Clathrina rowi	Voigt et al. (2017)		Endemic
Calcarea	Clathrinida	Clathrina sinusarabica	Klautau & Valentine (2003)		Endemic
Calcarea	Clathrinida	Ernstia arabica	Voigt et al. (2017)		Endemic
Calcarea	Clathrinida	Leucetta chagosensis	Wörheide et al. (2008) (original Dendy 1913)		Present
Calcarea	Clathrinida	Leucetta microraphis	Voigt et al. (2017) (original Haeckel 1872)	Leucetta primigenia var. microraphis	Unreported
Calcarea	Clathrinida	Leucetta primigenia	Haeckel (1872)		Unreported
Calcarea	Clathrinida	Leucetta pyriformis	van Soest & de Voogd (2018) (original Dendy 1913)		Present
Calcarea	Clathrinida	Soleneiscus hamatus	Voigt et al. (2017)		Endemic
Calcarea	Leucosolenida	Grantessa woerheidei	van Soest & de Voogd (2018)		Endemic
Calcarea	Leucosolenida	Grantilla quadriradiata	Row (1909)		Endemic
Calcarea	Leucosolenida	Kebira uteoides	Row (1909)		Endemic
Calcarea	Leucosolenida	Leucandra aspera	Row (1909)		Unreported
Calcarea	Leucosolenida	Leucandra bathybia	Lévi (1965) (original Haeckel 1869)	Leuconia bathybia	Present
Calcarea	Leucosolenida	Leucandra pulvinar	Haeckel (1872) (original Haeckel 1870)	Mlea dohrnii Maclay	Present
Calcarea	Leucosolenida	Leucandrilla intermedia	Row (1909)	Leucilla intermedia	Endemic

(continued)

Class	Order	Species	Citation	Previous names	Distribution (WPD)
Calcarea	Leucosolenida	*Paraleucilla crosslandi*	Row (1909)	*Leucilla crosslandi*	Endemic
Calcarea	Leucosolenida	*Sycettusa glabra*	Row (1909)	*Grantessa glabra*	Endemic
Calcarea	Leucosolenida	*Sycettusa hastifera*	Row (1909)	*Grantessa hastifera, Grantilla hastifera*	Present
Calcarea	Leucosolenida	*Sycettusa hirsutissima*	van Soest & de Voogd (2018)		Endemic
Calcarea	Leucosolenida	*Sycettusa stauridia*	Row (1909) (original Haeckel 1872)	*Grantessa stauridia*	Present
Calcarea	Leucosolenida	*Sycon ciliatum*	Row (1909)	*Sycon coronatum*	Unreported
Calcarea	Leucosolenida	*Sycon proboscideum*	Haeckel (1872) (original Haeckel 1870)	*Syconella proboscideum*	Endemic 'species inquirenda van Soest & de Voogd 2018'
Calcarea	Leucosolenida	*Sycon raphanus*	Haeckel (1872)		Unreported
Demospongiae	Agelasida	*Agelas marmarica*	Lévi (1958)		Present
Demospongiae	Agelasida	*Agelas mauritiana*	Lévi (1965) (original Carter 1883)		Present
Demospongiae	Agelasida	*Astrosclera willeyana*	Karlinska-Batres & Wörheide (2015) (original Lister 1900)		Unreported
Demospongiae	Axinellida	*Axinella quercifolia*	Keller (1889)	*Antherochalina quercifolia, Querciclona quercifolia*	Endemic
Demospongiae	Axinellida	*Dragmacidon coccineum*	Keller (1891)	*Hymeniacidon coccinea, Pseudaxinella coccinea, Reniera coccinea, Stylissa coccinea,*	Present
Demospongiae	Axinellida	*Dragmacidon durissimum*	Burton (1959) (original Dendy 1905)	*Axinella durissima*	Present
Demospongiae	Axinellida	*Echinodictyum flabelliforme*	Keller (1889)	*Acanthella flabelliformis*	Endemic
Demospongiae	Axinellida	*Echinodictyum jousseaumi*	Lévi (1958) (original Topsent 1892)		Present
Demospongiae	Axinellida	*Eurypon calypsoi*	Lévi (1958)		Present
Demospongiae	Axinellida	*Eurypon polyplumosum*	Lévi (1958)		Endemic
Demospongiae	Axinellida	*Higginsia arborea*	Keller (1891)	*Allantella arborea, Trachytedania arborea*	Present
Demospongiae	Axinellida	*Higginsia higgini*	Lévi (1958) (original Dendy 1922)		Present
Demospongiae	Axinellida	*Higginsia pumila*	Keller (1889)	*Axinella pumila*	Endemic
Demospongiae	Axinellida	*Lithoplocamia lithistoides*	Burton (1959) (original Dendy 1922)		Present
Demospongiae	Axinellida	*Myrmekioderma niveum*	Row (1911)	*Anacanthaea nivea*	Endemic
Demospongiae	Axinellida	*Myrmekioderma tuberculatum*	Keller (1891)	*Halichondria tuberculatum*	Endemic
Demospongiae	Axinellida	*Phakellia palmata*	Row (1911)		Endemic
Demospongiae	Axinellida	*Phakellia radiata*	Burton (1959) (original Dendy 1916)		Present
Demospongiae	Axinellida	*Ptilocaulis spiculifer*	Rudi et al. (1999) (original Lamarck 1814)		Present
Demospongiae	Biemnida	*Biemna ehrenbergi*	Keller (1889)	*Acanthella ehrenbergi*	Present
Demospongiae	Biemnida	*Biemna fortis*	Fishelson (1971) (original Topsent 1897)		Present
Demospongiae	Biemnida	*Biemna trirhaphis*	Lévi (1961) (Topsent 1879)		Present
Demospongiae	Biemnida	*Rhabderemia batatas*	Ilan, Gugel & van Soest (2004)		Endemic
Demospongiae	Biemnida	*Rhabderemia indica*	Burton (1959) (original Dendy 1905)		Present

(continued)

Class	Order	Species	Citation	Previous names	Distribution (WPD)
Demospongiae	Chondrillida	*Chondrilla australiensis*	Keller (1891) (original Carter 1873)	*Chondrilla globulifera*	Present
Demospongiae	Chondrillida	*Chondrilla mixta*	Lévi (1958) (original Schulze 1877)	*Chondrillastra mixta*	Present
Demospongiae	Chondrillida	*Chondrilla nucula*	El Bossery et al. (2017) (original Schmidt 1862)		Unreported
Demospongiae	Chondrillida	*Chondrilla sacciformis*	Richter et al. (2001) (original Carter 1879)		Unreported
Demospongiae	Chondrillida	*Halisarca laxus*	Lévi (1958) (original Lendenfeld 1889)	*Bajalus laxus*	Present
Demospongiae	Chondrosiida	*Chondrosia debilis*	Lévi (1958) (original Thiele 1900)		Present
Demospongiae	Clionaida	*Cliona orientalis*	Lévi (1958) (original Thiele 1900)		Present
Demospongiae	Clionaida	*Diplastrella gardineri*	Lévi (1958) (original Topsent 1918)		Present
Demospongiae	Clionaida	*Pione mussae*	Keller (1891)	*Cliona mussae, Sapline mussae*	Endemic
Demospongiae	Clionaida	*Pione vastifica*	Ferrario et al. (2010) (original Hancock 1849)	paper says probably conspecific	Unreported
Demospongiae	Clionaida	*Spheciospongia inconstans*	Lévi (1965) (original Dendy 1887)	*Spirastrella inconstans*	Present
Demospongiae	Clionaida	*Spheciospongia mastoidea*	Keller (1891)	*Suberites mastoideus*	Endemic
Demospongiae	Clionaida	*Spheciospongia vagabunda var. arabica*	Hooper and van Soest (2002) (original Topsent 1893)		Not reviewed
Demospongiae	Clionaida	*Spirastrella decumbens*	Lévi (1958) (original Ridley 1884)		Present
Demospongiae	Clionaida	*Spirastrella pachyspira*	Lévi (1958)		Present
Demospongiae	Dendroceratida	*Aplysilla lacunosa*	Keller (1889)		Endemic
Demospongiae	Dendroceratida	*Chelonaplysilla erecta*	Row (1911)	*Megalopastas erectus*	Endemic
Demospongiae	Dendroceratida	*Darwinella gardineri*	Lévi (1958) (original Topsent 1905)		Present
Demospongiae	Dictyoceratida	*Cacospongia ridleyi*	Burton (1952)		Present
Demospongiae	Dictyoceratida	*Carteriospongia foliascens*	Lévi (1958) (original Pallas 1766)	*Phyllospongia foliascens*	Present
Demospongiae	Dictyoceratida	*Dactylospongia elegans*	Abdelmohsen et al. (2014a) (original Thiele 1899)		Unreported
Demospongiae	Dictyoceratida	*Dysidea aedificanda*	Row (1911)	*Spongelia aedificanda*	Endemic
Demospongiae	Dictyoceratida	*Dysidea cinerea*	Keller (1889)	*Spongelia cinerea*	Present
Demospongiae	Dictyoceratida	*Euryspongia lactea*	Row (1911)		Present
Demospongiae	Dictyoceratida	*Fascaplysinopsis reticulata*	Helmy et al. (2004) (original Hentschel 1912)		Present
Demospongiae	Dictyoceratida	*Fasciospongia cavernosa*	Kashman et al. (1973) (original Schmidt 1862)		Unreported
Demospongiae	Dictyoceratida	*Fasciospongia lordii*	Lendenfeld (1889)	*Stelospongia lordii*	Present
Demospongiae	Dictyoceratida	*Hyattella globosa*	Lendenfeld (1889)		Endemic
Demospongiae	Dictyoceratida	*Hyattella tubaria*	Helmy et al. (2004) (Lendenfeld 1889)		Present
Demospongiae	Dictyoceratida	*Hyrtios communis*	Row (1911) (original Carter 1885)	*Psammopemma commune*	Present
Demospongiae	Dictyoceratida	*Hyrtios erectus*	Keller (1889)	*Dysidea nigra, Heteronema erecta, Duriella nigra*	Present
Demospongiae	Dictyoceratida	*Ircinia atrovirens*	Keller (1889)	*Hircinia atrovirens*	Endemic
Demospongiae	Dictyoceratida	*Ircinia echinata*	Keller (1889)	*Hircinia echinata*	Present
Demospongiae	Dictyoceratida	*Ircinia ramosa*	Keller (1889)	*Hircinia ramosa*	Present

(continued)

Class	Order	Species	Citation	Previous names	Distribution (WPD)
Demospongiae	Dictyoceratida	*Ircinia variabilis*	Burton (1926) (original Schmidt 1862)	*Hircinia variabilis*	Unreported
Demospongiae	Dictyoceratida	*Lamellodysidea herbacea*	Keller (1889)	*Carteriospongia cordifolia, Spongelia herbacea, Dysidea herbacea, Phyllospongia cordifolia, Spongelia delicatula*	Present
Demospongiae	Dictyoceratida	*Phyllospongia lamellosa*	Hassan et al. (2015) (original Esper 1794)		Unreported
Demospongiae	Dictyoceratida	*Phyllospongia papyracea*	Lévi (1958) (original Esper 1794)		Present
Demospongiae	Dictyoceratida	*Scalarispongia aqabaensis*	Helmy, El Serehy, Mohamed & van Soest (2004)		Present
Demospongiae	Dictyoceratida	*Spongia (Spongia) arabica*	Keller (1889)	*Spongia arabica, Spongia officinalis* var. *arabica*	Endemic
Demospongiae	Dictyoceratida	*Spongia (Spongia) irregularis*	Lévi (1965) (original Lendenfeld 1889)	*Spongia irregularis*	Present
Demospongiae	Dictyoceratida	*Spongia (Spongia) lesleighae*	Helmy, El Serehy, Mohamed & van Soest (2004)		Endemic
Demospongiae	Dictyoceratida	*Spongia (Spongia) officinalis* var. *exigua*	Lévi (1965) (original Schulze 1879)	*Spongia officinalis* f. *exigua*	Present
Demospongiae	Dictyoceratida	*Spongia lacinulosa*	Lamarck (1814)		Present
Demospongiae	Haplosclerida	*Amphimedon chloros*	Ilan, Gugel & van Soest (2004)		Endemic
Demospongiae	Haplosclerida	*Amphimedon dinae*	Helmy & van Soest (2005)		Endemic
Demospongiae	Haplosclerida	*Amphimedon hamadai*	Helmy & van Soest (2005)		Endemic
Demospongiae	Haplosclerida	*Amphimedon jalae*	Helmy & van Soest (2005)		Endemic
Demospongiae	Haplosclerida	*Amphimedon ochracea*	Keller (1889)	*Ceraochalina ochracea*	Endemic
Demospongiae	Haplosclerida	*Arenosclera arabica*	Keller (1889)	*Arenochalina arabica*	Present
Demospongiae	Haplosclerida	*Callyspongia (Callyspongia) siphonella*	Lévi (1965)	*Siphonochalina siphonella*	Endemic
Demospongiae	Haplosclerida	*Callyspongia (Callyspongia) tubulosa*	Burton (1926) (original Esper 1797)	*Siphonochalina tubulosa*	Present
Demospongiae	Haplosclerida	*Callyspongia (Cladochalina) subarmigera*	Burton (1959) (original Ridley 1884)	*Callyspongia subarmigera*	Unreported
Demospongiae	Haplosclerida	*Callyspongia (Euplacella) communis*	Burton (1926) (original Carter 1881)	*Siphonochalina communis*	Present
Demospongiae	Haplosclerida	*Callyspongia (Euplacella) densa*	Keller (1889)		Endemic
Demospongiae	Haplosclerida	*Callyspongia (Euplacella) paralia*	Ilan, Gugel & van Soest (2004)	*Callyspongia paralia*	Endemic
Demospongiae	Haplosclerida	*Callyspongia (Toxochalina) dendyi*	Vine (1986) (original Burton 1931)		Not reviewed
Demospongiae	Haplosclerida	*Callyspongia calyx*	Keller (1889)	*Cacochalina calyx*	Endemic
Demospongiae	Haplosclerida	*Callyspongia clavata*	Keller (1889)	*Crella cyathophora, Phylosiphonia clavata*	Endemic
Demospongiae	Haplosclerida	*Callyspongia conica*	Keller (1889)	*Phylosiphonia conica*	Present
Demospongiae	Haplosclerida	*Callyspongia crassa*	Keller (1889)	*Sclerochalina crassa*	Endemic
Demospongiae	Haplosclerida	*Callyspongia fistularis*	Topsent (1892)	*Sclerochalina fistularis*	Endemic
Demospongiae	Haplosclerida	*Callyspongia implexa*	Topsent (1892)	*Ceraochalina implexa*	Endemic
Demospongiae	Haplosclerida	*Callyspongia incrustans*	Row (1911)	*Spinosella incrustans*	Endemic

(continued)

Class	Order	Species	Citation	Previous names	Distribution (WPD)
Demospongiae	Haplosclerida	*Callyspongia maculata*	Keller (1889)	*Cacochalina maculata*	Endemic
Demospongiae	Haplosclerida	*Callyspongia reticulata*	Keller (1889)	*Siphonochalina reticulata*	Present
Demospongiae	Haplosclerida	*Callyspongia sinuosa*	Topsent (1892)	*Sclerochalina sinuosa*	Endemic
Demospongiae	Haplosclerida	*Callyspongia spongionelloides*	Fishelson (1971)		Endemic
Demospongiae	Haplosclerida	*Callyspongia vasseli*	Keller (1889)	*Phylosiphonia vasseli*	Endemic
Demospongiae	Haplosclerida	*Chalinula saudiensis*	Vacelet et al. (2001)		Endemic
Demospongiae	Haplosclerida	*Gelliodes incrustans*	Lévi (1965) (original Dendy 1905)		Present
Demospongiae	Haplosclerida	*Haliclona (Gellius) bubastes*	Row (1911)	*Halichondria bubastes*	Endemic
Demospongiae	Haplosclerida	*Haliclona (Gellius) flagellifera*	Burton (1959) (original Ridley and Dendy 1886)	*Haliclona flagellifera*	Unreported
Demospongiae	Haplosclerida	*Haliclona (Gellius) toxia*	Lévi (1958) (original Topsent 1897)	*Toxiclona toxius, Gellius toxius*	Present
Demospongiae	Haplosclerida	*Haliclona (Haliclona) violacea*	Keller (1883)	*Lessepsia violacea*	Endemic
Demospongiae	Haplosclerida	*Haliclona (Reniera) tabernacula*	Row (1911)	*Reniera tabernacula, Haliclona tabernacula*	Present
Demospongiae	Haplosclerida	*Haliclona decidua*	Topsent (1906)	*Reniera decidua*	Present
Demospongiae	Haplosclerida	*Haliclona pigmentifera*	Burton (1959) (original Dendy (1905)	*Adocia pigmentifera*	Present
Demospongiae	Haplosclerida	*Haliclona ramusculoides*	Row (1911) (original Topsent 1893)	*Chalina minor*	Present
Demospongiae	Haplosclerida	*Haliclona spinosella*	Row (1911)	*Reniera spinosella*	Endemic
Demospongiae	Haplosclerida	*Neopetrosia contignata*	Burton (1959) (original Thiele 1899)	*Haliclona contignata*	Present
Demospongiae	Haplosclerida	*Niphates furcata*	Keller (1889)	*Pachychalina furcata*	Endemic
Demospongiae	Haplosclerida	*Niphates obtusispiculifera*	Burton (1959) (original Dendy 1905)		Present
Demospongiae	Haplosclerida	*Niphates rowi*	Ilan, Gugel & van Soest (2004)		Endemic
Demospongiae	Haplosclerida	*Oceanapia elastica*	Keller (1891)	*Reniera elastica*	Present
Demospongiae	Haplosclerida	*Oceanapia incrustata*	Burton (1959) (original Dendy 1922)		Present
Demospongiae	Haplosclerida	*Pachychalina alveopora*	Topsent (1906)		Present
Demospongiae	Haplosclerida	*Petrosia (Petrosia) elephantotus*	Ilan, Gugel & van Soest (2004)	*Petrosia elephantotus*	Endemic
Demospongiae	Haplosclerida	*Petrosia (Petrosia) nigricans*	Burton (1959) (original Lindgren 1897)	*Petrosia nigricans*	Unreported
Demospongiae	Haplosclerida	*Xestospongia ridleyi*	Keller (1891)	*Reniera ridleyi*	Endemic
Demospongiae	Haplosclerida	*Xestospongia testudinaria*	Burton (1959) (Lamarck 1815)	*Petrosia testudinaria*	Present
Demospongiae	Poecilosclerida	*Acarnus bergquistae*	Yosief et al. (1998a) (original van Soest, Hooper & Hiemstra 1991)		Unreported
Demospongiae	Poecilosclerida	*Acarnus thielei*	Lévi (1958)		Endemic
Demospongiae	Poecilosclerida	*Acarnus wolffgangi*	Keller (1889)		Present
Demospongiae	Poecilosclerida	*Antho (Jia) wunschorum*	van Soest, Rützler & Sim (2016)		Endemic
Demospongiae	Poecilosclerida	*Batzella aurantiaca*	Lévi (1958)	*Prianos aurantiaca*	Present
Demospongiae	Poecilosclerida	*Clathria (Clathria) arbuscula*	Row (1911)	*Litaspongia arbuscula, Ophlitaspongia arbuscula*	Endemic
Demospongiae	Poecilosclerida	*Clathria (Clathria) horrida*	Row (1911)	*Clathria horrida, Ophlitaspongia horrida*	Endemic
Demospongiae	Poecilosclerida	*Clathria (Clathria) maeandrina*	Burton (1959) (original Ridley 1884)	*Clathria maeandrina*	Unreported

(continued)

Class	Order	Species	Citation	Previous names	Distribution (WPD)
Demospongiae	Poecilosclerida	Clathria (Clathria) spongodes	Burton (1959) (original Dendy 1922)	Clathria spongiosa	Present
Demospongiae	Poecilosclerida	Clathria (Clathria) transiens	Burton (1959) (original Hallmann 1912)	Clathria transiens	Unreported
Demospongiae	Poecilosclerida	Clathria (Thalysias) abietina	Burton (1959) (Lamarck 1814)	Clathria aculeata	Present
Demospongiae	Poecilosclerida	Clathria (Thalysias) cactiformis	Hooper, Kelly & Kennedy (2000) (original Lamarck 1814)		Not reviewed
Demospongiae	Poecilosclerida	Clathria (Thalysias) fusterna	Hooper (1997)	Clathria fusterna	Present
Demospongiae	Poecilosclerida	Clathria (Thalysias) lambda	Lévi (1958)	Leptoclathria lambda	Endemic
Demospongiae	Poecilosclerida	Clathria (Thalysias) lendenfeldi	Hooper, Kelly & Kennedy (2000) (Ridley & Dendy 1886)		Not reviewed
Demospongiae	Poecilosclerida	Clathria (Thalysias) procera	Burton (1959) (original Ridley 1884)		Present
Demospongiae	Poecilosclerida	Clathria (Thalysias) vulpina	Burton (1959) (original Lamarck 1814)		Present
Demospongiae	Poecilosclerida	Clathria granulata	Keller (1889)	Ceraochalina granulata	Endemic
Demospongiae	Poecilosclerida	Crambe acuata	Lévi (1958)	Folitispa acuata	Present
Demospongiae	Poecilosclerida	Crella (Grayella) cyathophora	Lévi (1958) (original Carter 1869)	Grayella cyathophora	Present
Demospongiae	Poecilosclerida	Crella (Grayella) papillata	Lévi (1958)		Present
Demospongiae	Poecilosclerida	Damiria simplex	Keller (1891)		Present
Demospongiae	Poecilosclerida	Diacarnus erythraeanus	Kelly-Borges & Vacelet (1995)		Present
Demospongiae	Poecilosclerida	Echinoclathria digitiformis	Row (1911)	Ophlitaspongia digitiformis	Endemic
Demospongiae	Poecilosclerida	Echinoclathria gibbosa	Keller (1889)	Ceraochalina gibbosa, Xestospongia gibbosa	Endemic
Demospongiae	Poecilosclerida	Echinoclathria robusta	Keller (1889)	Halme robusta	Endemic
Demospongiae	Poecilosclerida	Guitarra indica	Burton (1959) (original Dendy 1916)	Guitarra fimbriata	Unreported
Demospongiae	Poecilosclerida	Hemimycale arabica	Ilan, Gugel & van Soest (2004)		Endemic
Demospongiae	Poecilosclerida	Hymedesmia (Hymedesmia) lancifera	Topsent (1906)	Leptosia lancifera, Hymedesmia lancifera	Present
Demospongiae	Poecilosclerida	Hymedesmia (Hymedesmia) rowi	Row (1911) (original van Soest 2017)	Myxilla (Myxilla) tenuissima	Endemic
Demospongiae	Poecilosclerida	Iotrochota baculifera	Lévi (1965) (original Ridley 1884)		Present
Demospongiae	Poecilosclerida	Lissodendoryx (Lissodendoryx) cratera	Row (1911)	Myxilla cratera	Endemic
Demospongiae	Poecilosclerida	Lissodendoryx (Waldoschmittia) schmidti	Lévi (1958) (original Ridley 1884)	Damiriana schmidti	Present
Demospongiae	Poecilosclerida	Monanchora quadrangulata	Lévi (1958)	Fasuberea quadrangulata	Endemic
Demospongiae	Poecilosclerida	Mycale (Aegogropila) sulevoidea	Burton (1959) (original Sollas 1902)	Mycale sulevoidea	Unreported
Demospongiae	Poecilosclerida	Mycale (Arenochalina) anomala	Burton (1952) (original Ridley & Dendy 1886)	Esperiopsis anomala, Parisociella anomala	Unreported
Demospongiae	Poecilosclerida	Mycale (Arenochalina) euplectellioides	Row (1911)	Esperella euplectellioides	Endemic
Demospongiae	Poecilosclerida	Mycale (Arenochalina) setosa	Keller (1889)	Gelliodes setosa	Endemic

(continued)

Class	Order	Species	Citation	Previous names	Distribution (WPD)
Demospongiae	Poecilosclerida	Mycale (Carmia) erythraeana	Row (1911)	Esperella erythraeana	Endemic
Demospongiae	Poecilosclerida	Mycale (Carmia) fistulifera	Row (1911)	Esperella fistulifera	Endemic
Demospongiae	Poecilosclerida	Mycale (Carmia) suezza	Row (1911)	Esperella suezza	Endemic
Demospongiae	Poecilosclerida	Mycale (Mycale) dendyi	Row (1911)	Esperella dendyi	Endemic
Demospongiae	Poecilosclerida	Mycale (Mycale) grandis	Lévi (1958) (original Grey 1867)	Mycale grandis	Present
Demospongiae	Poecilosclerida	Myxilla (Burtonanchora) gracilis	Lévi (1965)	Burtonanchora gracilis	Endemic
Demospongiae	Poecilosclerida	Negombata corticata	Carter (1879)		Present
Demospongiae	Poecilosclerida	Negombata magnifica	Keller (1889)	Latrunculia magnifica	Endemic
Demospongiae	Poecilosclerida	Phorbas epizoaria	Lévi (1958)	Pronax epizoaria	Endemic
Demospongiae	Poecilosclerida	Psammoclema arenaceum	Lévi (1958)	Psammopemma arenaceum	Endemic
Demospongiae	Poecilosclerida	Psammoclema rubrum	Lévi (1958)	Psammopemma	Endemic
Demospongiae	Poecilosclerida	Strongylacidon inaequale	Burton (1959) (original Hentschel 1911)	Strongylacidon inaequalis	Unreported
Demospongiae	Poecilosclerida	Tedania (Tedania) anhelans	Burton (1959) (original Vio in Olivi 1792)	Tedania nigrescens	Unreported
Demospongiae	Poecilosclerida	Tedania (Tedania) assabensis	Keller (1891)	Tedania assabensis, Tedania anhelans var. assabensis	Present
Demospongiae	Scopalinida	Stylissa carteri	Keller (1889) (original Dendy 1889)	Acanthella aurantiaca, axinella carteri	Present
Demospongiae	Suberitida	Axinyssa gravieri	Lévi (1965) (original Topsent 1906)	Pseudaxinyssa gravieri	Present
Demospongiae	Suberitida	Halichondria (Halichondria) glabrata	Keller (1891)	Halichondria glabrata	Endemic
Demospongiae	Suberitida	Halichondria (Halichondria) granulata	Keller (1891)	Halichondria granulata	Endemic
Demospongiae	Suberitida	Halichondria (Halichondria) isthmica	Keller (1891) (original Keller 1883)	Amorphina isthmica	Endemic
Demospongiae	Suberitida	Halichondria (Halichondria) minuta	Keller (1891)	Halichondria minuta	Endemic
Demospongiae	Suberitida	Hymeniacidon calcifera	Row (1911)		Endemic
Demospongiae	Suberitida	Hymeniacidon zosterae	Row (1911)		Endemic
Demospongiae	Suberitida	Pseudosuberites andrewsi	Vine (1986) (original Kirkpatrick 1900)		Not reviewed
Demospongiae	Suberitida	Suberites clavatus	Keller (1891)		Endemic
Demospongiae	Suberitida	Suberites kelleri	Keller (1891) (original Burton 1930)	Suberites incrustans	Present
Demospongiae	Suberitida	Suberites tylobtusus	Lévi (1958)	Suberites tylobtusa	Endemic
Demospongiae	Suberitida	Terpios lendenfeldi	Keller (1891)		Endemic
Demospongiae	Suberitida	Terpios viridis	Keller (1891)		Endemic
Demospongiae	Suberitida	Topsentia aqabaensis	Ilan, Gugel & van Soest (2004)	Epipolasis aqabaensis	Endemic
Demospongiae	Suberitida	Topsentia halichondrioides	Burton (1926) (original Dendy 1905)	Trachyopsis halichondrioides	Present
Demospongiae	Tethyida	Tethya japonica	Topsent (1906) (originial Sollas 1888)	Donatia japonica	Present
Demospongiae	Tethyida	Tethya robusta	Burton (1926) (original Bowerbank 1873)	Donatia robusta, Donatia arabica	Present

(continued)

Class	Order	Species	Citation	Previous names	Distribution (WPD)
Demospongiae	Tethyida	*Tethya seychellensis*	Lévi (1958) (original Wright 1881)		Present
Demospongiae	Tethyida	*Timea intermedia*	Lévi (1958)	*Timeopsis intermedia*	Endemic
Demospongiae	Tetractinellida	*Cinachyrella albatridens*	Lévi (1965) (original Lendenfeld 1907)	*Cinachyra alba tridens*	Present
Demospongiae	Tetractinellida	*Cinachyrella alloclada*	Barnathan et al. (2003) (original Uliczka 1929)		Unreported
Demospongiae	Tetractinellida	*Cinachyrella eurystoma*	Keller (1891)	*Cinachyra eurystoma*	Endemic
Demospongiae	Tetractinellida	*Cinachyrella ibis*	Row (1911)	*Chrotella ibis*	Endemic
Demospongiae	Tetractinellida	*Cinachyrella kuekenthali*	Barnathan et al. (2003) (original Uliczka (1929)		Unreported
Demospongiae	Tetractinellida	*Cinachyrella schulzei*	Keller (1891)	*Cinachyra schulzei*	Present
Demospongiae	Tetractinellida	*Cinachyrella trochiformis*	Keller (1891)	*Cinachyra trochiformis*	Endemic
Demospongiae	Tetractinellida	*Dercitus (Halinastra) exostoticus*	Keller (1891) (original Schmidt (1868)		Endemic
Demospongiae	Tetractinellida	*Discodermia stylifera*	Keller (1891)		Endemic
Demospongiae	Tetractinellida	*Ecionemia arabica*	Lévi (1958)	*Hezekia arabica*	Endemic
Demospongiae	Tetractinellida	*Ecionemia spinastra*	Lévi (1958)		Endemic
Demospongiae	Tetractinellida	*Erylus lendenfeldi*	Carmely et al. (1989) (original Sollas 1888)		Unreported
Demospongiae	Tetractinellida	*Erylus proximus*	Lévi (1958) (original Dendy 1916)		Present
Demospongiae	Tetractinellida	*Geodia arabica*	Topsent (1892) (original Carter 1869)		Present
Demospongiae	Tetractinellida	*Geodia jousseaumei*	Topsent (1906)	*Isops jousseaumei*	Present
Demospongiae	Tetractinellida	*Geodia micropunctata*	Row (1911)		Endemic
Demospongiae	Tetractinellida	*Jaspis albescens*	Row (1911)	*Coppatias albescens*	Endemic
Demospongiae	Tetractinellida	*Jaspis reptans*	Lévi (1965) (original Dendy 1905)		Present
Demospongiae	Tetractinellida	*Jaspis sollasi*	Burton & Rao (1932)	*Amphius sollasi*	Endemic
Demospongiae	Tetractinellida	*Jaspis virens*	Lévi (1958)		Endemic
Demospongiae	Tetractinellida	*Levantiniella levantinensis*	Tsurnamal (1969) (Vacelet, Bitar, Carteron, Zibrowius & Pérez 2007)	*Chrotella cavernosa*	Present
Demospongiae	Tetractinellida	*Paratetilla bacca*	Row (1911) (original Selenka 1867)	*Paratetilla eccentrica*	Present
Demospongiae	Tetractinellida	*Rhabdastrella sterrastraea*	Row (1911)	*Diastra sterrastraea*	Endemic
Demospongiae	Tetractinellida	*Stelletta parva*	Row (1911)	*Pilochrota parva*	Endemic
Demospongiae	Tetractinellida	*Stelletta purpurea*	Lévi (1958) (original Ridley 1884)	*Myriastra purpurea*	Present
Demospongiae	Tetractinellida	*Stelletta siemensi*	Keller (1891)		Endemic
Demospongiae	Tetractinellida	*Stellettinopsis solida*	Lévi (1965)		Present
Demospongiae	Tetractinellida	*Tetilla diaenophora*	Lévi (1958)		Endemic
Demospongiae	Tetractinellida	*Tetilla poculifera*	Row (1911) (original Dendy 1905)		Present
Demospongiae	Tetractinellida	*Theonella conica*	Lévi (1958) (original Kieschnick 1896)		Present
Demospongiae	Tetractinellida	*Theonella mirabilis*	El Bossery et al. (2017) (original de Laubenfels 1954)		Unreported
Demospongiae	Tetractinellida	*Theonella swinhoei*	Lévi (1958) (original Grey 1868)		Present
Demospongiae	Verongiida	*Aplysina reticulata*	Burton (1926) (original Lendenfeld 1889)		Present
Demospongiae	Verongiida	*Pseudoceratina arabica*	Keller (1889)	*Psammaplysilla arabica*	Present

(continued)

Class	Order	Species	Citation	Previous names	Distribution (WPD)
Demospongiae	Verongiida	*Pseudoceratina purpurea*	Rotem et al. (1983) (original Carter 1880)	*Psammaplysilla purpurea*	Unreported
Demospongiae	Verongiida	*Suberea mollis*	Row (1911)	*Verongia mollis*, *Aplysina mollis*	Present
Demospongiae	Verongiida	*Suberea praetensa*	Row (1911)	*Aplysina praetensa*	Present
Demospongiae	Verongiida	*Suberea purpureaflava*	Gugel, Wagler & Brümmer (2011)		Endemic
Hexactinellida	Lychniscosida	*Neoaulocystis polae*	Ijima (1927)	*Aulocystis polae*	Endemic
Hexactinellida	Sceptrulophora	*Tretocalyx polae*	Schulze (1901)		Endemic
Homoscleromorpha	Homosclerophorida	*Plakortis erythraena*	Lévi (1958)		Endemic
Homoscleromorpha	Homosclerophorida	*Plakortis nigra*	Lévi (1958) (original Lévi 1953)		Present

References

Abbas AT, El-Shitany NA, Shaala LA, Ali SS, Azhar EI, Abdel-Dayem UA, Youssef DT (2014) Red Sea *Suberea mollis* sponge extract protects against CCl4-induced acute liver injury in rats via an antioxidant mechanism. Evid Based Complement Alternat Med 2014:745606

Abd El Moneam NM, Shreadah MA, El-Assar SA, de Voogd NJ, Nabil-Adam A (2018) Hepatoprotective effect of Red Sea sponge extract against the toxicity of a real-life mixture of persistent organic pollutants. Biotechnol Biotechnol Equip 32:734–743

Abdelhameed R, Elgawish MS, Mira A, Ibrahim AK, Ahmed SA, Shimizu K, Yamada K (2016) Anti-choline esterase activity of ceramides from the Red Sea marine sponge *Mycale euplectellioides*. RSC Adv 6:20422–20430

Abdelhameed RF, Ibrahim AK, Temraz TA, Yamada K, Ahmed SA (2017) Chemical and biological investigation of the red sea sponge *Echinoclathria* species. J Pharm Sci Res 9:1324

Abdel-Lateff A, Alarif WM, Asfour HZ, Ayyad SEN, Khedr A, Badria FA, Al-lihaibi SS (2014) Cytotoxic effects of three new metabolites from Red Sea marine sponge, *Petrosia sp*. Environ Toxicol Pharmacol 37:928–935

Abdelmohsen UR, Pimentel-Elardo SM, Hanora A, Radwan M, Abou-El-Ela SH, Ahmed S, Hentschel U (2010) Isolation, phylogenetic analysis and anti-infective activity screening of marine sponge-associated Actinomycetes. Mar Drugs 8:399–412

Abdelmohsen UR, Yang C, Horn H, Hajjar D, Ravasi T, Hentschel U (2014a) Actinomycetes from Red Sea sponges: sources for chemical and phylogenetic diversity. Mar Drugs 12:2771–2789

Abdelmohsen UR, Cheng C, Viegelmann C, Zhang T, Grkovic T, Ahmed S, Quinn RJ, Hentschel U, Edrada-Ebel R (2014b) Dereplication strategies for targeted isolation of new antitrypanosomal actinosporins A and B from a marine sponge associated-*Actinokineospora sp*. EG49. Mar Drugs 12:1220–1244

Abdelmohsen UR, Cheng C, Reimer A, Kozjak-Pavlovic V, Ibrahim AK, Rudel T, Hentschel U, Edrada-Ebel R, Ahmed SA (2015) Antichlamydial sterol from the Red Sea sponge *Callyspongia aff. implexa*. Planta Med 81:382–387

Abdelwahed NA, Ahmed EF, El-Gammal EW, Hawas UW (2014) Application of statistical design for the optimization of dextranase production by a novel fungus isolated from Red Sea sponge. 3 Biotech 4:533–544

Abou-Hussein DR, Youssef DT (2016) Mirabolides A and B; new cytotoxic glycerides from the Red Sea sponge *Theonella mirabilis*. Mar Drugs 14:155

Abou-Hussein DR, Badr JM, Youssef DT (2014) Dragmacidoside: A new nucleoside from the Red Sea sponge *Dragmacidon coccinea*. Nat Prod Res 28:1134–1141

Abou-Shoer MI, Shaala LA, Youssef DT, Badr JM, Habib AAM (2008) Bioactive brominated metabolites from the Red Sea sponge *Suberea mollis*. J Nat Prod 71:1464–1467

Abraham I, Jain S, Wu CP, Khanfar MA, Kuang Y, Dai CL, Shi Z, Chen X, Fu L, Ambudkar SV, El Sayed K (2010) Marine sponge-derived sipholane triterpenoids reverse P-glycoprotein (ABCB1)-mediated multidrug resistance in cancer cells. Biochem Pharmacol 80:1497–1506

Aerts LAM (1998) Sponge/coral interactions in Caribbean reefs: analysis of overgrowth patterns in relation to species identity and cover. Mar Ecol Prog Ser 175:241–249

Afifi R, Khabour OF (2017) Antibacterial activity of the Saudi Red Sea sponges against Gram-positive pathogens. J King Saud Univ Sci. https://doi.org/10.1016/j.jksus.2017.08.009

Ahmed SA, Khalifa SI, Hamann MT (2008) Antiepileptic ceramides from the Red Sea sponge *Negombata corticata*. J Nat Prod 71:513–515

Ahmed A, El-Desoky AH, Al-hammady MA, Elshamy AI, Hegazy MEF, Kato H, Tsukamoto S (2018) New inhibitors of RANKL-induced Osteoclastogenesis from the marine sponge *Siphonochalina siphonella*. Fitoterapia 128:43–49

Ahmed HH, Rady HM, Kotob SE (2018) Evidences for the antitumor potentiality of *Hemimycale arabica* and *Negombata magnifica* mesohyls in hepatocellular carcinoma rat model. Med Chem Res:1–11

Ajabnoor MAM, Tilmisany AK, Taha AM, Antonius A (1991) Effect of red sea sponge extracts on blood glucose levels in normal mice. J Ethnopharmacol 33:103–106

Alahdal AM, Asfour HZ, Ahmed SA, Noor AO, Al-Abd AM, Elfaky MA, Elhady SS (2018) Anti-Helicobacter, antitubercular and cytotoxic activities of Scalaranes from the Red Sea sponge *Hyrtios erectus*. Molecules 23:978

Alarif WM, Al-Lihaibi SS, Ghandourah MA, Orif MI, Basaif SA, Ayyad SEN (2016) Cytotoxic scalarane-type sesterterpenes from the Saudi Red Sea sponge *Hyrtios erectus*. J Asian Nat Prod Res 18:611–617

Al-Massarani SM, El-Gamal AA, Al-Said MS, Al-Lihaibi SS, Basoudan OA (2015) In vitro cytotoxic, antibacterial and antiviral activities of triterpenes from the Red Sea sponge, *Siphonochalina siphonella*. Trop J Pharm Res 14:33–40

Al-Massarani SM, El-Gamal AA, Al-Said MS, Abdel-Kader MS, Ashour AE, Kumar A, Abdel-Mageed WM, Al-Rehaily AJ, Ghabbour HA, Fun HK (2016) Studies on the Red Sea sponge

Haliclona sp. for its chemical and cytotoxic properties. Pharmacogn Mag 12:114

Al-Sofyani A, Al-Farawati RK, ElMaradny AA, Niaz GR (2011) Long-chain aliphatic wax esters isolated from the sponge *Chalinula saudensis* (Demospongia) along the Jeddah coast of the Red Sea. Braz J Oceanogr 59:1–6

Amina M, Ali MS, Al-Musayeib NM, Al-Lohedan HA (2016) Biophysical characterization of the interaction of bovine serum albumin with anticancer sipholane triterpenoid from the Red Sea sponge. J Mol Liq 220:931–938

Angawi RF, Saqer E, Abdel-Lateff A, Badria FA, Ayyad SEN (2014) Cytotoxic neviotane triterpene-type from the Red Sea sponge *Siphonochalina siphonella*. Pharmacogn Mag 10:334

Ashour MA, Elkhayat ES, Ebel R, Edrada R, Proksch P (2007) Indole alkaloid from the Red Sea sponge *Hyrtios erectus*. ARKIVOC (15):225–231

Badr JM, Shaala LA, Abou-Shoer MI, Tawfik MK, Habib AAM (2008) Bioactive brominated metabolites from the Red Sea sponge *Pseudoceratina arabica*. J Nat Prod 71:1472–1474

Barnathan G, Genin E, Velosaotsy NE, Kornprobst JM, Al-Lihaibi S, Al-Sofyani A, Nongonierma R (2003) Phospholipid fatty acids and sterols of two *Cinachyrella* sponges from the Saudi Arabian Red Sea: comparison with *Cinachyrella* species from other origins. Comp Biochem Physiol B Biochem Mol Biol 135:297–308

Bayer K, Moitinho-Silva L, Brümmer F, Cannistraci CV, Ravasi T, Hentschel U (2014) GeoChip-based insights into the microbial functional gene repertoire of marine sponges (high microbial abundance, low microbial abundance) and seawater. FEMS Microbiol Ecol 90:832–843

Beer S, Ilan M (1998) In situ measurements of photosynthetic irradiance responses of two Red Sea sponges growing under dim light conditions. Mar Biol 131:613–617

Bell JJ (2008) The functional roles of marine sponges. Estuar Coast Shelf Sci 79:341–353

Benayahu Y, Loya Y (1981) Competition for space among coral-reef sessile organisms at Eilat, Red Sea. Bull Mar Sci 31:514–522

Bergman O, Haber M, Mayzel B, Anderson MA, Shpigel M, Hill RT, Ilan M (2011) Marine-based cultivation of *Diacarnus* sponges and the bacterial community composition of wild and maricultured sponges and their larvae. Mar Biotechnol 13:1169–1182

Berumen ML, Hoey AS, Bass WH, Bouwmeester J, Catania D, Cochran JE, Khalil MT, Miyake S, Mughal MR, Spät JL, Saenz-Agudelo P (2013) The status of coral reef ecology research in the Red Sea. Coral Reefs 32:737–748

Berumen ML, DiBattista JD, Rocha LA (2017) Introduction to virtual issue on Red Sea and Western Indian Ocean biogeography. J Biogeogr 44:1923–1926

Blunt JW, Copp BR, Hu WP, Munro MH, Northcote PT, Prinsep MR (2007) Marine natural products. Nat Prod Rep 24:31–86

Bowen BW, Rocha LA, Toonen RJ, Karl SA (2013) The origins of tropical marine biodiversity. Trends Ecol Evol 28:359–366

Britstein M, Devescovi G, Handley KM, Malik A, Haber M, Saurav K, Teta R, Costantino V, Burgsdorf I, Gilbert JA, Sher N (2016) A new N-Acyl homoserine lactone synthase in an uncultured symbiont of the Red Sea sponge *Theonella swinhoei*. Appl Environ Microbiol 82:1274–1285

Burns E, Ilan M (2003) Comparison of anti-predatory defenses of Red Sea and Caribbean sponges. II. physical defense. Mar Ecol Prog Ser 252:115–123

Burns E, Ifrach I, Carmeli S, Pawlik JR, Ilan M (2003) Comparison of anti-predatory defenses of Red Sea and Caribbean sponges. I. chemical defense. Mar Ecol Prog Ser 252:105–114

Burton M (1926) Sponges. Zoological results of the Suez Canal expedition. Trans Zool Soc Lond 22:71–83

Burton M (1952) The 'Manihine' expedition to the Gulf of Aqaba 1948–1949 – Sponges. Bull Br Mus Nat Hist Zool 1(8):163–174

Burton M (1959) Sponges. In: Scientific Reports. John Murray Expedition 1933–34. Br Mus Nat Hist Lond 10(5):151–281

Burton M, Rao HS (1932) Report on the shallow-water marine sponges in the collection of the Indian museum. Part I. Rec Indian Mus 34:299–358

Carmely S, Kashman Y (1986) Neviotine-A, a new triterpene from the Red Sea sponge *Siphonochalina siphonella*. J Org Chem 51:784–788

Carmely S, Roll M, Loya Y, Kashman Y (1989) The structure of eryloside A, a new antitumor and antifungal 4-methylated steroidal glycoside from the sponge *Erylus lendenfeldi*. J Nat Prod 52:167–170

Carmely S, Gebreyesus T, Kashman Y, Skelton BW, White AH, Yosief T (1990) Dysidamide, a novel metabolite from a Red Sea sponge *Dysidea herbacea*. Aust J Chem 43:1881–1888

Carter HJ (1879) Contributions to our knowledge of the Spongida. Ann Mag Nat Hist 5:284–304, 343–360, pls XXV-XXVII

Dayton PK, Kim S, Jarrell SC, Oliver JS, Hammerstrom K, Fisher JL, O'Connor K, Barber JS, Robilliard G, Barry J, Thurber AR, Conlan K (2013) Recruitment, growth and mortality of an Antarctic hexactinellid sponge, *Anoxycalyx joubini*. PLoS One 8:e56939

De Goeij JM, Van Oevelen D, Vermeij MJ, Osinga R, Middelburg JJ, de Goeij AF, Admiraal W (2013) Surviving in a marine desert: the sponge loop retains resources within coral reefs. Science 342:108–110

Delseth C, Kashman Y, Djerassi C (1979) Ergosta-5, 7, 9 (11), 22-tetraen-3β-ol and its 24ξ-Ethyl homolog, two new marine sterols from the Red Sea sponge *Biemna fortis*. Helv Chim Acta 62:2037–2045

DiBattista JD, Roberts MB, Bouwmeester J, Bowen BW, Coker DJ, Lozano-Cortés DF, Howard Choat J, Gaither MR, Hobbs JPA, Khalil MT, Kochzius M (2016) A review of contemporary patterns of endemism for shallow water reef fauna in the Red Sea. J Biogeogr 43:423–439

Dixon GB, Davies SW, Aglyamova GV, Meyer E, Bay LK, Matz MV (2015) Genomic determinants of coral heat tolerance across latitudes. Science 348:1460–1462

Duffy JE (1992) Host use patterns and demography in a guild of tropical sponge-dwelling shrimps. Mar Ecol Prog Ser 90:127–138

Dunbar DC, Rimoldi JM, Clark AM, Kelly M, Hamann MT (2000) Anti-cryptococcal and nitric oxide synthase inhibitory imidazole alkaloids from the calcareous sponge *Leucetta cf chagosensis*. Tetrahedron 56:8795–8798

Edwards FJ (1987) Climate and oceanography. In: Edwards AJ, Head S (eds) Key environments: Red Sea. Pergamon Press, Oxford, pp 45–68

Ehrlich H, Shaala LA, Youssef DT, Żółtowska-Aksamitowska S, Tsurkan M, Galli R, Meissner H, Wysokowski M, Petrenko I, Tabachnick KR, Ivanenko VN (2018) Discovery of chitin in skeletons of non-verongiid Red Sea demosponges. PLoS One 13:e0195803

Eid ES, Abo-Elmatty DM, Hanora A, Mesbah NM, Abou-El-Ela SH (2011) Molecular and protein characterization of two species of the latrunculin-producing sponge *Negombata* from the Red Sea. J Pharm Biomed Anal 56:911–915

El Bossery AM, Shoukr F, El Komy MM, Rady HM, El-Arab MALE (2017) Sponges from Elphinstone Reef, Northern Red Sea, Egypt. Egypt J Exp Biol Zool 13:79–89

El Sayed KA, Hamann MT, Hashish NE, Shier WT, Kelly M, Khan AA (2001) Antimalarial, antiviral, and antitoxoplasmosis norsesterterpene peroxide acids from the Red Sea sponge *Diacarnus erythraeanus*. J Nat Prod 64:522–524

El-Damhougy KA, El-Naggar HA, Ibrahim HA, Bashar MA, Senna FMA (2017) Biological activities of some marine sponge extracts from Aqaba Gulf, Red Sea, Egypt. Int J Fish Aquat Stud 5:652–659

El-Ezz RA, Ibrahim A, Habib E, Kamel H, Afifi M, Hassanean H, Ahmed S (2017) Review of natural products from marine organisms in the Red Sea. Int J Pharm Sci Res 8:940

El-Gamal AA, Al-Massarani SM, Shaala LA, Alahdald AM, Al-Said MS, Ashour AE, Kumar A, Abdel-Kader MS, Abdel-Mageed WM, Youssef DT (2016) Cytotoxic compounds from the Saudi Red Sea sponge *Xestospongia testudinaria*. Mar Drugs 14:82

El-Gendy MMAA, Yahya SM, Hamed AR, Soltan MM, El-Bondkly AMA (2017) Phylogenetic analysis and biological evaluation of marine endophytic fungi derived from Red Sea Sponge *Hyrtios erectus*. Appl Biochem Biotechnol, first online. https://doi.org/10.1007/s12010-017-2679-x.

Elhady SS, El-Halawany AM, Alahdal AM, Hassanean HA, Ahmed SA (2016a) A new bioactive metabolite isolated from the Red Sea marine sponge *Hyrtios erectus*. Molecules 21:82

Elhady SS, Al-Abd AM, El-Halawany AM, Alahdal AM, Hassanean HA, Ahmed SA (2016b) Antiproliferative scalarane-based metabolites from the Red Sea sponge *Hyrtios erectus*. Mar Drugs 14:130

Ellis J, Anlauf H, Kürten S, Lozano-Cortés D, Alsaffar Z, Cúrdia J, Jones B, Carvalho S (2017) Cross shelf benthic biodiversity patterns in the southern Red Sea. Sci Rep 7

Elsayed Y, Refaat J, Abdelmohsen UR, Ahmed S, Fouad MA (2017) Rhodozepinone, a new antitrypanosomal azepino-diindole alkaloid from the marine sponge-derived bacterium *Rhodococcus sp. UA13*. Med Chem Res 26:2751–2760

Elsayed Y, Refaat J, Abdelmohsen UR, Othman EM, Stopper H, Fouad MA (2018) Metabolomic profiling and biological investigation of the marine sponge-derived bacterium *Rhodococcus sp. UA13*. Phytochem Anal. https://doi.org/10.1002/pca.2765

El-Shitany NA, Shaala LA, Abbas AT, Abdel-dayem UA, Azhar EI, Ali SS, van Soest RW, Youssef DT (2015) Evaluation of the anti-inflammatory, antioxidant and immunomodulatory effects of the organic extract of the Red Sea marine sponge *Xestospongia testudinaria* against carrageenan induced rat paw inflammation. PLoS One 10:e0138917

Eltamany EE, Radwan MM, Ibrahim AK, ElSohly M, Hassanean HA, Ahmed SA (2014a) Antitumor metabolites from the Red Sea sponge *Spheciospongia vagabunda*. Planta Med 80:5

Eltamany EE, Abdelmohsen UR, Ibrahim AK, Hassanean HA, Hentschel U, Ahmed SA (2014b) New antibacterial xanthone from the marine sponge-derived *Micrococcus sp.* EG45. Bioorg Med Chem Lett 24:4939–4942

Eltamany EE, Ibrahim AK, Radwan MM, ElSohly MA, Hassanean HA, Ahmed SA (2015) Cytotoxic ceramides from the Red Sea sponge *Spheciospongia vagabunda*. Med Chem Res 24:3467–3473

Erpenbeck D, Ekins M, Enghuber N, Hooper JN, Lehnert H, Poliseno A, Schuster A, Setiawan E, de Voogd NJ, Wörheide G, van Soest RW (2016a) Nothing in (sponge) biology makes sense–except when based on holotypes. J Mar Biol Assoc UK 96:305–311

Erpenbeck D, Voigt O, Al-Aidaroos AM, Berumen ML, Büttner G, Catania D, Guirguis AN, Paulay G, Schätzle S, Wörheide G (2016b) Molecular biodiversity of Red Sea demosponges. Mar Pollut Bull 105:507–514

Erpenbeck D, Aryasari R, Benning S, Debitus C, Kaltenbacher E, Al-Aidaroos AM, Schupp P, Hall K, Hooper JNA, Voigt O, de Voogd NJ, Wörheide G (2017) Diversity of two widespread Indo-Pacific demosponge species revisited. Mar Biodivers 47:1035–1043

Erwin PM, Thacker RW (2007) Incidence and identity of photosynthetic symbionts in Caribbean coral reef sponge assemblages. J Mar Biol Assoc UK 87:1683–1692

Faulkner DJ (2000) Marine pharmacology. Antonie Van Leeuwenhoek 77:135–145

Ferrario F, Calcinai B, Erpenbeck D, Galli P, Wörheide G (2010) Two *Pione* species (Hadromerida, Clionaidae) from the Red Sea: a taxonomical challenge. Org Divers Evol 10:275–285

Fishelson L (1966) *Spirastrella inconstans* Dendy (Porifera) as an ecological niche in the littoral zone of the Dahlak Archipelago (Eritrea). Bull Sea Fish Res Stat Isr 41:17–25

Fishelson L (1971) Ecology and distribution of the benthic fauna in the shallow waters of the Red Sea. Mar Biol 10:113–133

Fouad M, Al-Trabeen K, Badran M, Wray V, Edrada R, Proksch P, Ebel R (2004) New steroidal saponins from the sponge *Erylus lendenfeldi*. ARKIVOC 2004(13):17–27

Foudah AI, Sallam AA, Akl MR, El Sayed KA (2014) Optimization, pharmacophore modeling and 3D-QSAR studies of sipholanes as breast cancer migration and proliferation inhibitors. Eur J Med Chem 73:310–324

Furby KA, Bouwmeester J, Berumen ML (2013) Susceptibility of central Red Sea corals during a major bleaching event. Coral Reefs 32:505–513

Gao ZM, Wang Y, Lee OO, Tian RM, Wong YH, Bougouffa S, Batang Z, Al-Suwailem A, Lafi FF, Bajic VB, Qian PY (2014a) Pyrosequencing reveals the microbial communities in the Red Sea sponge *Carteriospongia foliascens* and their impressive shifts in abnormal tissues. Microb Ecol 68:621–632

Gao ZM, Wang Y, Tian RM, Wong YH, Batang ZB, Al-Suwailem AM, Bajic VB, Qian PY (2014b) Symbiotic adaptation drives genome streamlining of the cyanobacterial sponge symbiont "*Candidatus Synechococcus spongiarum*". MBio 5:e00079–e00014

Gao ZM, Wang Y, Tian RM, Lee OO, Wong YH, Batang ZB, Al-Suwailem A, Lafi FF, Bajic VB, Qian PY (2015) Pyrosequencing revealed shifts of prokaryotic communities between healthy and disease-like tissues of the Red Sea sponge *Crella cyathophora*. PeerJ 3:e890

Gebreyesusa T, Yosief T, Carmely S, Kashmanb Y (1988) Dysidamide, a novel hexachloro-metabolite from a Red Sea sponge *Dysidea sp.* Tetrahedron Lett 29:3863–3864

Genin A, Monismith SG, Reidenbach MA, Yahel G, Koseff JR (2009) Intense benthic grazing of phytoplankton in a coral reef. Limnol Oceanogr 54:938–951

Gesner S, Cohen N, Ilan M, Yarden O, Carmeli S (2005) Pandangolide 1a, a metabolite of the sponge-associated fungus *Cladosporium sp.*, and the absolute stereochemistry of pandangolide 1 and iso-cladospolide B. J Nat Prod 68:1350–1353

Gibson PJ (2011) Ecosystem impacts of carbon and nitrogen cycling by coral reef sponges. PhD thesis, UNC – Chapel Hill, 161

Giles EC, Saenz-Agudelo P, Berumen ML, Ravasi T (2013a) Novel polymorphic microsatellite markers developed for a common reef sponge, *Stylissa carteri*. Mar Biodivers 43:237–241

Giles EC, Kamke J, Moitinho-Silva L, Taylor MW, Hentschel U, Ravasi T, Schmitt S (2013b) Bacterial community profiles in low microbial abundance sponges. FEMS Microbiol Ecol 83:232–241

Giles EC, Saenz-Agudelo P, Hussey NE, Ravasi T, Berumen ML (2015) Exploring seascape genetics and kinship in the reef sponge *Stylissa carteri* in the Red Sea. Ecol Evol 5:2487–2502

Gillor O, Carmeli S, Rahamim Y, Fishelson Z, Ilan M (2000) Immunolocalization of the toxin latrunculin B within the Red Sea sponge *Negombata magnifica* (Demospongiae, Latrunculiidae). Mar Biotechnol 2:213–223

Gloeckner V, Wehrl M, Moitinho-Silva L, Gernert C, Schupp P, Pawlik JR, Lindquist NL, Erpenbeck D, Wörheide G, Hentschel U (2014) The HMA-LMA dichotomy revisited: an electron microscopical survey of 56 sponge species. Biol Bull 227:78–88

Goobes R, Rudi A, Kashman Y, Ilan M, Loya Y (1996) Three new glycolipids from a Red Sea sponge of the genus *Erylus*. Tetrahedron 52:7921–7928

Gugel J, Wagler M, Brümmer F (2011) Porifera, one new species *Suberea purpureaflava* n. sp. (Demospongiae, Verongida, Aplysinellidae) from northern Red Sea coral reefs, with short descriptions of Red Sea Verongida and known *Suberea* species. Zootaxa 2994:60–68

Guo YW, Trivellone E (2000) New hurghamids from a Red Sea sponge of the genus *Hippospongia*. J Asian Nat Prod Res 2:251–256

Guo Y, Gavagnin M, Mollo E, Trivellone E, Cimino G, Hamdy NA, Fakhr I, Pansini M (1996) A new norsesterterpene peroxide from a Red Sea sponge. Nat Prod Lett 9:105–112

Guo Y, Gavagnin M, Mollo E, Cimino G, Hamdy NA, Fakhr I, Pansini M (1997) Hurghamides AD, new N-acyl-2-methylene-β-alanine methyl esters from Red Sea *Hippospongia sp.* Nat Prod Lett 10:143–150

Hadas E, Shpigel M, Ilan M (2005) Sea ranching of the marine sponge *Negombata magnifica* (Demospongiae, Latrunculiidae) as a first step for latrunculin B mass production. Aquaculture 244:159–169

Hadas E, Marie D, Shpigel M, Ilan M (2006) Virus predation by sponges is a new nutrient-flow pathway in coral reef food webs. Limnol Oceanogr 51:1548–1550

Hadas E, Ilan M, Shpigel M (2008) Oxygen consumption by a coral reef sponge. J Exp Biol 211:2185–2190

Hadas E, Shpigel M, Ilan M (2009) Particulate organic matter as a food source for a coral reef sponge. J Exp Biol 212:3643–3650

Haeckel E (1870) XVIII. Prodromus of a system of the calcareous sponges. J Nat Hist 5:176–191

Haeckel E (1872) Die Kalkschwämme. Eine Monographie in zwei Bänden Text und einem Atlas mit 60 Tafeln Abbildungen. G. Reimer, Berlin. (1:1–484) 2:1-418 (3:pls 1–60)

Hamed AN, Schmitz R, Bergermann A, Totzke F, Kubbutat M, Müller WE, Youssef DT, Bishr MM, Kamel MS, Edrada-Ebel R, Wätjen W (2018) Bioactive pyrrole alkaloids isolated from the Red Sea: marine sponge *Stylissa carteri*. Zeitschrift für Naturforschung C 73:199–210

Hassan MH, Rateb ME, Hetta M, Abdelaziz TA, Sleim MA, Jaspars M, Mohammed R (2015) Scalarane sesterterpenes from the Egyptian Red Sea sponge *Phyllospongia lamellosa*. Tetrahedron 71:577–583

Hawas UW, Abou El-Kassem LT, Abdelfattah MS, Elmallah MI, Eid MAG, Monier M, Marimuthu N (2018) Cytotoxic activity of alkyl benzoate and fatty acids from the red sea sponge *Hyrtios erectus*. Nat Prod Res 32:1369–1374

Helmy T, van Soest RW (2005) *Amphimedon* species (Porifera: Niphatidae) from the Gulf of Aqaba, Northern Red Sea: filling the gaps in the distribution of a common pantropical genus. Zootaxa 859(1):18

Helmy T, Mohamed SZ, van Soest RW (2004) Description and classification of dictyoceratid sponges from the Northern Red Sea. Beaufortia 54:81–91

Henkel T, Pawlik JR (2005) Habitat use by sponge-dwelling brittlestars. Mar Biol 146:301–313

Hoeksema BW, Ten Hove HA, Berumen ML (2016) Christmas tree worms evade smothering by a coral-killing sponge in the Red Sea. Mar Biodivers 46:15–16

Hooper JN (1997) Revision of Microcionidae (Porifera: Poecilosclerida: Demospongiae), with description of Australian species. Oceanogr Lit Rev 3:247

Hooper JN, van Soest RW (2002) Systema Porifera. A guide to the classification of sponges. In: Systema Porifera. Springer, US, pp 1–7

Hooper JN, Kelly M, Kennedy A (2000) A new *Clathria* (Porifera: Demospongiae: Microcionidae) from the Western Indian Ocean. Mem Queensland Mus 45:427–444

Ibrahim HA, El-Naggar HA, El-Damhougy KA, Bashar MA, Senna FMA (2017) *Callyspongia crassa* and *C. siphonella* (Porifera, Callyspongiidae) as a potential source for medical bioactive substances, Aqaba Gulf, Red Sea, Egypt. J Basic Appl Zool 78:7

Ilan M (1995) Reproductive biology, taxonomy, and aspects of chemical ecology of Latrunculiidae (Porifera). Biol Bull 188:306–312

Ilan M, Abelson A (1995) The life of a sponge in a sandy lagoon. Biol Bull 189:363–369

Ilan M, Loya Y (1988) Reproduction and settlement of the coral reef sponge *Niphates sp.* (Red Sea). In: Proc 6th Internat Coral Reef Symp, Townsville Australia, vol 2:745–749

Ilan M, Loya Y (1990) Sexual reproduction and settlement of the coral reef sponge *Chalinula sp.* from the Red Sea. Mar Biol 105:25–31

Ilan M, Vacelet J (1993) *Kebira uteoides* (Porifera, Calcarea) a recent "pharetronid" sponge from coral reefs. Ophelia 38:107–116

Ilan M, Loya Y, Kolbasov GA, Brickner I (1999) Sponge-inhabiting barnacles on Red Sea coral reefs. Mar Biol 133:709–716

Ilan M, Gugel J, van Soest R (2004) Taxonomy, reproduction and ecology of new and known Red Sea sponges. Sarsia.: North Atlantic Marine Science 89:388–410

Imešek M, Pleše B, Pfannkuchen M, Godrijan J, Pfannkuchen DM, Klautau M, Ćetković H (2014) Integrative taxonomy of four *Clathrina* species of the Adriatic Sea, with the first formal description of *Clathrina rubra* Sarà, 1958. Org Divers Evol 14:21–29

Isaacs S, Kashman Y (1992) Shaagrockol B and C; two hexaprenylhydroquinone disulfates from the Red Sea sponge *Toxiclona toxius*. Tetrahedron Lett 33:2227–2230

Jain S, Shirode A, Yacoub S, Barbo A, Sylvester PW, Huntimer E, Halaweish F, El Sayed KA (2007a) Biocatalysis of the anticancer sipholane triterpenoids. Planta Med 73:591–596

Jain S, Laphookhieo S, Shi Z, Fu LW, Akiyama SI, Chen ZS, Youssef DT, van Soest RW, El Sayed KA (2007b) Reversal of P-glycoprotein-mediated multidrug resistance by sipholane triterpenoids. J Nat Prod 70:928–931

Jain S, Abraham I, Carvalho P, Kuang YH, Shaala LA, Youssef DT, Avery MA, Chen ZS, El Sayed KA (2009) Sipholane triterpenoids: chemistry, reversal of ABCB1/P-glycoprotein-mediated multidrug resistance, and pharmacophore modeling. J Nat Prod 72:1291–1298

Kalinin VI, Ivanchina NV, Krasokhin VB, Makarieva TN, Stonik VA (2012) Glycosides from marine sponges (Porifera, Demospongiae): structures, taxonomical distribution, biological activities and biological roles. Mar Drugs 10:1671–1710

Kämpfer P, Glaeser SP, Busse HJ, Abdelmohsen UR, Ahmed S, Hentschel U (2015) Actinokineospora *spheciospongiae sp.* nov., isolated from the marine sponge *Spheciospongia vagabunda*. Int J Syst Evol Microbiol 65:879–884

Kandler N (2015) Biodiversity *of Macrofauna Associated with Sponges across Ecological Gradients in the Central Red Sea*, Master's Thesis, King Abdullah University of Science and Technology, Thuwal Saudi Arabia.

Karlińska-Batres K, Wörheide G (2015) Spatial variability of microbial communities of the coralline demosponge *Astrosclera willeyana* across the Indo-Pacific. Aquat Microb Ecol 74:143–156

Kashman Y, Rotem M (1979) Muqubilin, a new C24-isoprenoid from a marine sponge. Tetrahedron Lett 20:1707–1708

Kashman Y, Rudi A (1977) The 13C-NMR spectrum and stereochemistry of heteronemin. Tetrahedron 33:2997–2998

Kashman Y, Zviely M (1979) New alkylated scalarins from the sponge *Dysidea herbacea*. Tetrahedron Lett 20:3879–3882

Kashman Y, Fishelson L, Ne'eman I (1973) N-Acyl-2-methylene-β-alanine methyl esters from the sponge *Fasciospongia cavernosa*. Tetrahedron 29:3655–3657

Kashman Y, Groweiss A, Shmueli U (1980) Latrunculin, a new 2-thiazolidinone macrolide from the marine sponge *Latrunculia magnifica*. Tetrahedron Lett 21:3629–3632

Kashman Y, Groweiss A, Carmely S, Kinamoni Z, Czarkie D, Rotem M (1982) Recent research in marine natural products from the Red Sea. Pure Appl Chem 54:1995–2010

Kashman Y, Carmely S, Blasberger D, Hirsch S, Green D (1989) Marine natural products: new results from Red Sea invertebrates. Pure Appl Chem 61:517–520

Kashman Y, Yosief T, Carmeli S (2001) New triterpenoids from the Red Sea sponge *Siphonochalina siphonella*. J Nat Prod 64:175–180

Keller C (1883) Die Fauna im Suez-Kanal und die Diffusion der mediterranen und erythräischen Thierwelt: eine thiergeographische Untersuchung, vol 28. Allgemeine schweizerische Gesellschaft für die gesammten Naturwissenschaften, Zürich, pp 1–39

Keller C (1889) Die Spongienfauna des rothen Meeres (I. Hälfte). Z wiss Zool 48:311–405. pls XX-XXV

Keller C (1891) Die Spongienfauna des Rothen Meeres (II. Hälfte). Z Wiss Zool 52:294–368. pls XVI-XX

Kelly-Borges M, Vacelet J (1995) A revision of *Diacamus* Burton and *Negombata* de Laubenfels (Demospongiae: Latrunculiidae) with descriptions of new species from the west central Pacific and the Red Sea. Mem Queensland Mus 38:477–504

Kelman D, Kashman Y, Rosenberg E, Ilan M, Ifrach I, Loya Y (2001) Antimicrobial activity of the reef sponge *Amphimedon viridis* from the Red Sea: evidence for selective toxicity. Aquat Microb Ecol 24:9–16

Kelman D, Kashman Y, Hill RT, Rosenberg E, Loya Y (2009) Chemical warfare in the sea: the search for antibiotics from Red Sea corals and sponges. Pure Appl Chem 81:1113–1121

Keren R, Lavy A, Mayzel B, Ilan M (2015) Culturable associated-bacteria of the sponge *Theonella swinhoei* show tolerance to high arsenic concentrations. Front Microbiol 6:154

Keren R, Lavy A, Ilan M (2016) Increasing the richness of culturable arsenic-tolerant bacteria from *Theonella swinhoei* by addition of sponge skeleton to the growth medium. Microb Ecol 71:873–886

Keren R, Mayzel B, Lavy A, Polishchuk I, Levy D, Fakra SC, Pokroy B, Ilan M (2017) Sponge-associated bacteria mineralize arsenic and barium on intracellular vesicles. Nat Commun 8:14393

Khalil MT, Bouwmeester J, Berumen ML (2017) Spatial variation in coral reef fish and benthic communities in the central Saudi Arabian Red Sea. PeerJ 5:e3410

Klautau M, Valentine C (2003) Revision of the genus *Clathrina* (Porifera, Calcarea). Zool J Linnean Soc 139:1–62

Klautau M, Russo CA, Lazoski C, Boury-Esnault N, Thorpe JP, Solé-Cava AM (1999) Does cosmopolitanism result from overconservative systematics? a case study using the marine sponge *Chondrilla nucula*. Evolution 53:1414–1422

Klautau M, Imešek M, Azevedo F, Pleše B, Nikolić V, Ćetković H (2016) Adriatic calcarean sponges (Porifera, Calcarea), with the description of six new species and a richness analysis. Eur J Taxon (178):1–52

Kolbasov GA (1990) *Acasta-pertusa* sp-n (Cirripedia, Thoracica) from the Red Sea. Zool Zhurnal 69:142–145

Lamarck JBPA (1814) Sur les polypiers empâtés. In: Annales du Muséum d'histoire Naturelle, vol 20, Paris, pp 294–312

Lavy A, Keren R, Haber M, Schwartz I, Ilan M (2014) Implementing sponge physiological and genomic information to enhance the diversity of its culturable associated bacteria. FEMS Microbiol Ecol 87:486–502

Lavy A, Keren R, Yahel G, Ilan M (2016) Intermittent hypoxia and prolonged suboxia measured in situ in a marine sponge. Front Mar Sci 3:263

Lee OO, Wang Y, Yang J, Lafi FF, Al-Suwailem A, Qian PY (2011) Pyrosequencing reveals highly diverse and species-specific microbial communities in sponges from the Red Sea. ISME J 5:650

Lee OO, Lai PY, Wu HX, Zhou XJ, Miao L, Wang H, Qian PY (2012) *Marinobacter xestospongiae* sp. nov., isolated from the marine sponge *Xestospongia testudinaria* collected from the Red Sea. Int J Syst Evol Microbiol 62:1980–1985

Lefranc F, Nuzzo G, Hamdy NA, Fakhr I, Moreno Y, Banuls L, Van Goietsenoven G, Villani G, Mathieu V, van Soest R, Kiss R, Ciavatta ML (2013) In vitro pharmacological and toxicological effects of norterpene peroxides isolated from the Red Sea sponge *Diacarnus erythraeanus* on normal and cancer cells. J Nat Prod 76:1541–1547

Lendenfeld R (1889) A monograph of the horny sponges. Royal Society by Trübner and Co, London

Lévi C (1958) Spongiaires de Mer Rouge, recueillis par la Calypso (1951–1952). Annales de l'Institut océanographique, Monaco 34:3–46

Lévi C (1961) Résultats scientifiques des Campagnes de la 'Calypso'. Campagne 1954 dans l'Océan Indien (suite). 2. Les spongiaires de l'Ile Aldabra. Annales de l'Institut océanographique 39:1–32

Lévi C (1965) Spongiaires récoltés par l'Expedition israelienne dans le sud de la Mer Rouge en 1962. Sea Fish Res Station Haifa Bull 39:3–27 (Israel South Red Sea Exped. 1962 Rep. 13)

Loh TL, McMurray SE, Henkel TP, Vicente J, Pawlik JR (2015) Indirect effects of overfishing on Caribbean reefs: sponges overgrow reef-building corals. PeerJ 3:e901. https://doi.org/10.7717/peerj.901

Magnino G, Sarà A, Lancioni T, Gaino E (1999) Endobionts of the coral reef sponge *Theonella swinhoei* (Porifera, Demospongiae). Invertebr Biol:213–220

Mancini I, Guella G, Pietra F, Amade P (1997) Hanishenols AB, novel linear or methyl-branched glycerol enol ethers of the axinellid sponge *Acanthella carteri* (= *Acanthella aurantiaca*) from the Hanish Islands, southern Red Sea. Tetrahedron 53:2625–2628

Mayzel B, Aizenberg J, Ilan M (2014) The elemental composition of demospongiae from the Red Sea, Gulf of Aqaba. PLoS One 9:e95775

McClintock JB, Amsler CD, Baker BJ, van Soest RWM (2005) Ecology of Antarctic marine sponges: an overview. Integr Comp Biol 45:359–368

Mebs D (1985) Chemical defense of a dorid nudibranch, *Glossodoris quadricolor*, from the Red Sea. J Chem Ecol 11:713–716

Mehbub MF, Lei J, Franco C, Zhang W (2014) Marine sponge derived natural products between 2001 and 2010: trends and opportunities for discovery of bioactives. Mar Drugs 12:4539–4577

Meroz E, Ilan M (1995a) Life history characteristics of a coral reef sponge. Mar Biol 124:443–451

Meroz E, Ilan M (1995b) Cohabitation of a coral reef sponge and a colonial scyphozoan. Mar Biol 124:453–459

Miloslavich P, Klein E, Díaz JM, Hernández CE, Bigatti G, Campos L, Artigas F, Castillo J, Penchaszadeh PE, Neill PE, Carranza A (2011) Marine biodiversity in the Atlantic and Pacific coasts of South America: knowledge and gaps. PLoS One 6:e14631

Mohamed GA, Abd-Elrazek AE, Hassanean HA, Alahdal AM, Almohammadi A, Youssef DT (2014a) New fatty acids from the Red Sea sponge *Mycale euplectellioides*. Nat Prod Res 28:1082–1090

Mohamed GA, Abd-Elrazek AE, Hassanean HA, Youssef DT, van Soest R (2014b) New compounds from the Red Sea marine sponge *Echinoclathria gibbosa*. Phytochem Lett 9:51–58

Moitinho-Silva L, Bayer K, Cannistraci CV, Giles EC, Ryu T, Seridi L, Ravasi T, Hentschel U (2014a) Specificity and transcriptional activity of microbiota associated with low and high microbial abundance sponges from the Red Sea. Mol Ecol 23:1348–1363

Moitinho-Silva L, Seridi L, Ryu T, Voolstra CR, Ravasi T, Hentschel U (2014b) Revealing microbial functional activities in the Red Sea sponge *Stylissa carteri* by metatranscriptomics. Environ Microbiol 16:3683–3698

Mudit M, Khanfar M, Muralidharan A, Thomas S, Shah GV, van Soest RW, El Sayed KA (2009) Discovery, design, and synthesis of anti-metastatic lead phenylmethylene hydantoins inspired by marine natural products. Bioorg Med Chem 17:1731–1738

Neeman I, Fishelson L, Kashman Y (1975) Isolation of a new toxin from the sponge *Latrunculia magnifica* in the Gulf of Aquaba (Red Sea). Mar Biol 30:293–296

Newman DJ, Cragg GM (2004) Marine natural products and related compounds in clinical and advanced preclinical trials. J Nat Prod 67:1216–1238

O'Rourke A, Kremb S, Bader TM, Helfer M, Schmitt-Kopplin P, Gerwick WH, Brack-Werner R, Voolstra CR (2016) Alkaloids from the sponge *Stylissa carteri* present prospective scaffolds for the inhibition of human immunodeficiency Virus 1 (HIV-1). Mar Drugs 14:28

Oren M, Steindler L, Ilan M (2005) Transmission, plasticity and the molecular identification of cyanobacterial symbionts in the Red Sea sponge *Diacarnus erythraenus*. Mar Biol 148:35–41

Ormond RFG, Edwards AJ (1987) Red Sea fishes. In: Red Sea, pp 251–287

Pan K, Lee OO, Qian PY, Wang WX (2011) Sponges and sediments as monitoring tools of metal contamination in the eastern coast of the Red Sea, Saudi Arabia. Mar Pollut Bull 62:1140–1146

Pawlik JR (1983) A sponge-eating worm from Bermuda: *Branchiosyllis oculata* (Polychaeta, Syllidae). PSZNI Mar Ecol 4:65–79

Pawlik JR, Loh TL (2016) Biogeographical homogeneity of caribbean coral reef benthos. J Biogeogr 44:950–952. https://doi.org/10.1111/jbi.12858

Pearman JK, Anlauf H, Irigoien X, Carvalho S (2016) Please mind the gap–visual census and cryptic biodiversity assessment at central Red Sea coral reefs. Mar Environ Res 118:20–30

Perkol-Finkel S, Benayahu Y (2005) Recruitment of benthic organisms onto a planned artificial reef: shifts in community structure one decade post-deployment. Mar Environ Res 59:79–99

Peters KJ, Amsler CD, McClintock JB, van Soest RWM, Baker BJ (2009) Palatability and chemical defenses of sponges from the western Antarctic Peninsula. Mar Ecol Prog Ser 385:77–85

Peterson BJ, Chester CM, Jochem FJ, Fourqurean JW (2006) Potential role of sponge communities in controlling phytoplankton blooms in Florida Bay. Mar Ecol Prog Ser 328:93–103

Pile AJ, Patterson MR, Savarese M, Chernykh VI, Fialkov VA (1997) Trophic effects of sponge feeding within Lake Baikal's littoral zone. 2. Sponge abundance, diet, feeding efficiency, and carbon flux. Limnol Oceanogr 42:178–184

Radwan M, Hanora A, Zan J, Mohamed NM, Abo-Elmatty DM, Abou-El-Ela SH, Hill RT (2010) Bacterial community analyses of two Red Sea sponges. Mar Biotechnol 12:350–360

Ramadan SA (1997) Two new species of mesostigmatid mites (*Acari*) associated with sponges from the Red Sea, Egypt. Assiut Vet Med J 38:191–204

Reiswig HM (1974) Water transport, respiration and energetics of three tropical marine sponges. J Exp Mar Biol Ecol 14:231–249

Reitner J, Wörheide G, Thiel V, Gautret P (1996) Reef caves and cryptic habitats of Indo-Pacific reefs—distribution patterns of coralline sponges and microbialites. Global and Regional Controls on Biogenic Sedimentation. I. Reef Evolution: Göttinger Arbeiten zur Geologie und Paläontologie, 2:91–100.

Řezanka T, Dembitsky VM (2003) Ten-membered substituted cyclic 2-oxecanone (Decalactone) derivatives from *Latrunculia corticata*, a Red Sea sponge. Eur J Org Chem 2003:2144–2152

Richter C, Wunsch M, Rasheed M, Kotter I, Badran MI (2001) Endoscopic exploration of Red Sea coral reefs reveals dense populations of cavity-dwelling sponges. Nature 413:726

Rinkevich B, Shashar N, Liberman T (1993) Nontransitive xenogeneic interactions between four common Red Sea sessile invertebrates. In: Proceedings of the Seventh International Coral Reef Symposium, vol 2, pp 833–839

Rix L, Bednarz VN, Cardini U, van Hoytema N, Al-Horani FA, Wild C, Naumann MS (2015) Seasonality in dinitrogen fixation and primary productivity by coral reef framework substrates from the Northern Red Sea. Mar Ecol Prog Ser 533:79–92

Rix L, De Goeij JM, Mueller CE, Struck U, Middelburg JJ, Van Duyl FC, Al-Horani FA, Wild C, Naumann MS, Van Oevelen D (2016) Coral mucus fuels the sponge loop in warm-and cold-water coral reef ecosystems. Sci Rep 6:18715

Rix L, Goeij JM, Oevelen D, Struck U, Al-Horani FA, Wild C, Naumann MS (2017) Differential recycling of coral and algal dissolved organic matter via the sponge loop. Funct Ecol 31:778–789

Rix L, de Goeij JM, van Oevelen D, Struck U, Al-Horani FA, Wild C, Naumann MS (2018) Reef sponges facilitate the transfer of coral-derived organic matter to their associated fauna via the sponge loop. Mar Ecol Prog Ser 589:85–96

Roberts MB, Jones GP, McCormick MI, Munday PL, Neale S, Thorrold S, Robitzch VS, Berumen ML (2016) Homogeneity of coral reef communities across 8 degrees of latitude in the Saudi Arabian Red Sea. Mar Pollut Bull 105:558–565

Rotem M, Kashman Y (1979) New polyacetylenes from the sponge *Siphonochalina sp.* Tetrahedron Lett 20:3193–3196

Rotem M, Carmely S, Kashman Y, Loya Y (1983) Two new antibiotics from the red sea sponge *Psammaplysilla purpurea*: total 13C-NMR line assignment of psammaplysins A and B and aerothionin. Tetrahedron 39:667–676

Row RW (1909) Reports on the marine biology of the Sudanese Red Sea.–XIII. Report on the sponges, collected by Mr. Cyril Crossland in 1904-5.—Part I. Calcarea. Zool J Linnean Soc 31:182–214

Row RW (1911) Reports on the marine biology of the Sudanese Red Sea.—XIX. Report on the sponges collected by Mr. Cyril Crossland in 1904-5. Part II. Non-Calcarea. Zool J Linnean Soc 31:287–400

Rudi A, Kashman Y (1993) Aaptosine-a new cytotoxic 5, 8-diazabenz [cd] azulene alkaloid from the Red Sea sponge *Aaptos aaptos*. Tetrahedron Lett 34:4683–4684

Rudi A, Yosief T, Schleyer M, Kashman Y (1999) Several new isoprenoids from two marine sponges of the family Axinellidae. Tetrahedron 55:5555–5566

Rudi A, Yosief T, Loya S, Hizi A, Schleyer M, Kashman Y (2001) Clathsterol, a novel anti-HIV-1 RT sulfated sterol from the sponge *Clathria* species. J Nat Prod 64:1451–1453

Rützler K, Rieger G (1973) Sponge burrowing: fine structure of *Cliona lampa* penetrating calcareous substrata. Mar Biol 21:144–162

Ryu T, Seridi L, Moitinho-Silva L, Oates M, Liew YJ, Mavromatis C, Wang X, Haywood A, Lafi FF, Kupresanin M, Sougrat R (2016) Hologenome analysis of two marine sponges with different microbiomes. BMC Genomics 17:158

Sandler JS, Forsburg SL, Faulkner DJ (2005) Bioactive steroidal glycosides from the marine sponge *Erylus lendenfeldi*. Tetrahedron 61:1199–1206

Sara M, Pansini M, Pronzato R (1979) Zonation of photophilous sponges related to water movement in reef biotopes of Obhor Creek (Red Sea). Sponge Biology, Colloques Internationaux du Centre National de la Recherche Scientifique 291:271–282

Sauleau P, Bourguet-Kondracki ML (2005) Novel polyhydroxysterols from the Red Sea marine sponge *Lamellodysidea herbacea*. Steroids 70:954–959

Sauleau P, Retailleau P, Vacelet J, Bourguet-Kondracki ML (2005) New polychlorinated pyrrolidinones from the Red Sea marine sponge *Lamellodysidea herbacea*. Tetrahedron 61:955–963

Sauleau P, Martin MT, Dau METH, Youssef DT, Bourguet-Kondracki ML (2006) Hyrtiazepine, an azepino-indole-type alkaloid from the Red Sea marine sponge *Hyrtios erectus*⊥. J Nat Prod 69:1676–1679

Savarese M, Patterson MR, Chernykh VI, Fialkov VA (1997) Trophic effects of sponge feeding within Lake Baikal's littoral zone. 1. In situ pumping rates. Limnol Oceanogr 42:171–178

Schmitt S, Tsai P, Bell J, Fromont J, Ilan M, Lindquist N, Perez T, Rodrigo A, Schupp PJ, Vacelet J, Webster N (2012) Assessing the complex sponge microbiota: core, variable and species-specific bacterial communities in marine sponges. ISME J 6:564

Schulze FE (1901) Berichte der Commission für oceanographische Forschungen. Zoologische Ergebnisse XVI Hexactinelliden des Rothen Meeres Denkschriften der Kaiserlichen Akademie der Wissenschaften Mathematisch-Naturwissenschaftliche Classe 69:311–324

Shaaban M, Abd-Alla HI, Hassan AZ, Aly HF, Ghani MA (2012) Chemical characterization, antioxidant and inhibitory effects of some marine sponges against carbohydrate metabolizing enzymes. Org Med Chem Lett 2:30

Shaala LA, Almohammadi A (2017) Biologically active compounds from the red sea sponge *Suberea sp.* Pak J Pharm Sci:30

Shaala LA, Bamane FH, Badr JM, Youssef DT (2011) Brominated arginine-derived alkaloids from the Red Sea sponge *Suberea mollis*. J Nat Prod 74:1517–1520

Shaala LA, Youssef DT, Sulaiman M, Behery FA, Foudah AI, Sayed KAE (2012) Subereamolline A as a potent breast cancer migration, invasion and proliferation inhibitor and bioactive dibrominated alkaloids from the Red Sea sponge *Pseudoceratina arabica*. Mar Drugs 10:2492–2508

Shaala LA, Youssef DT, Badr JM, Sulaiman M, Khedr A (2015a) Bioactive secondary metabolites from the Red Sea marine Verongid sponge *Suberea* species. Mar Drugs 13:1621–1631

Shaala LA, Youssef DT, Badr JM, Sulaiman M, Khedr A, El Sayed KA (2015b) Bioactive alkaloids from the Red Sea marine Verongid sponge *Pseudoceratina arabica*. Tetrahedron 71:7837–7841

Shaala LA, Youssef DT, Ibrahim SR, Mohamed GA (2016) Callyptide A, a new cytotoxic peptide from the Red Sea marine sponge *Callyspongia* species. Nat Prod Res 30:2783–2790

Shady NH, Abdelmohsen UR, Safwat A, Fouad M, Kamel MS (2017) Phytochemical and biological investigation of the Red Sea marine sponge *Hyrtios sp*. J Pharmacogn Phytochem 6:241

Shmueli U, Carmely S, Groweiss A, Kashman Y (1981) Sipholenol and sipholenone, two new triterpenes from the marine sponge *siphonochalina siphonella* (Lévi). Tetrahedron Lett 22:709–712

Simister R, Taylor MW, Tsai P, Webster N (2012) Sponge-microbe associations survive high nutrients and temperatures. PLoS One 7:e52220

Sipkema D, Franssen MC, Osinga R, Tramper J, Wijffels RH (2005) Marine sponges as pharmacy. Mar Biotechnol 7:142

Sokoloff S, Halevy S, Usieli V, Colorni A, Sarel S (1982) Prianicin A and B, nor-sesterterpenoid peroxide antibiotics from Red Sea sponges. Experientia 38:337–338

Southwell MW, Weisz JB, Martens CS, Lindquist N (2008a) In situ fluxes of dissolved inorganic nitrogen from the sponge community on Conch Reef, Key Largo, Florida. Limnol Oceanogr 53:986–996

Southwell MW, Popp BN, Martens CS (2008b) Nitrification controls on fluxes and isotopic composition of nitrate from Florida Keys sponges. Mar Chem 108:96–108

Spaet JL, Thorrold SR, Berumen ML (2012) A review of elasmobranch research in the Red Sea. J Fish Biol 80:952–965

Spector I, Shochet NR, Kashman Y, Groweiss A (1983) Latrunculins: novel marine toxins that disrupt microfilament organization in cultured cells. Science 219:493–495

Stehli FG, Wells JW (1971) Diversity and age patterns in hermatypic corals. Syst Zool 20:115–126

Steindler L, Beer S, Peretzman-Shemer A, Nyberg C, Ilan M (2001) Photoadaptation of zooxanthellae in the sponge *Cliona vastifica* from the Red Sea, as measured in situ. Mar Biol 138:511–515

Steindler L, Huchon D, Avni A, Ilan M (2005) 16S rRNA phylogeny of sponge-associated cyanobacteria. Appl Environ Microbiol 71:4127–4131

Suchanek TH, Carpenter RC, Witman JD, Harvell CD (1985) Sponges as important space competitors in deep Caribbean coral reef communities. In: Reaka ML (ed) The ecology of deep and shallow coral reefs, symposia series for undersea research. NOAA/NURP, Rockville, pp 55–59

Tabares P, Degel B, Schaschke N, Hentschel U, Schirmeister T (2012) Identification of the protease inhibitor miraziridine A in the Red sea sponge *Theonella swinhoei*. Pharm Res 4:63

Talpir R, Rudi A, Ilan M, Kashman Y (1992) Niphatoxin A and B; two new ichthyo-and cytotoxic tripyridine alkaloids from a marine sponge. Tetrahedron Lett 33:3033–3034

Targett NM, Schmahl GP (1984) Chemical ecology and distribution of sponges in the Salt River Canyon, St. Croix, U.S.V.I. NOAA Technical Memorandum OAR NURP-1, Rockville

Taylor MW, Radax R, Steger D, Wagner M (2007) Sponge-associated microorganisms: evolution, ecology, and biotechnological potential. Microbiol Mol Biol Rev 71:295–347

Thomas T, Moitinho-Silva L, Lurgi M, Björk JR, Easson C, Astudillo-García C, Olson JB, Erwin PM, López-Legentil S, Luter H, Chaves-Fonnegra A (2016) Diversity, structure and convergent evolution of the global sponge microbiome. Nat Commun 7

Topsent E (1892) Éponges de la Mer Rouge. Mémoires de la Société Zoologique de France 5:21–29. pl. I

Topsent E (1906) Éponges recueillies par M. Ch. Gravier dans la Mer Rouge. Bulletin du Muséum National d'Histoire Naturelle 12:557–570

Tsurnamal M (1969) Sponges of Red Sea origin on the Mediterranean coast of Israel. Isr J Zool 18:149–155

Vacelet J, Al Sofyani A, Al Lihaibi S, Kornprobst JM (2001) A new haplosclerid sponge species from the Red Sea. J Mar Biol Assoc U K 81:943–948

van Soest RWM, Beglinger EJ (2008) Tetractinellid and hadromerid sponges of the Sultanate of Oman. Zoologische Mededelingen 82:749–790

van Soest RW, de Voogd NJ (2018) Calcareous sponges of the Western Indian Ocean and Red Sea. Zootaxa 4426(1):160

van Soest RW, Boury-Esnault N, Vacelet J, Dohrmann M, Erpenbeck D, de Voogd NJ, Santodomingo N, Vanhoorne B, Kelly M, Hooper JN (2012) Global diversity of sponges (Porifera). PLoS One 7:e35105

van Soest RWM, Boury-Esnault N, Hooper JNA, Rützler K, de Voogd NJ, Alvarez de Glasby B, Hajdu E, Pisera AB, Manconi R, Schoenberg C, Klautau M, Picton B, Kelly M, Vacelet J, Dohrmann M, Díaz MC, Cárdenas P, Carballo JL, Rios Lopez P (2018) World Porifera database. Accessed 6 Dec 2018 http://www.marinespecies.org/porifera.

Vaughan GO, Burt JA (2016) The changing dynamics of coral reef science in Arabia. Mar Pollut Bull 105:441–458

Vilozny B, Amagata T, Mooberry SL, Crews P (2004) A new dimension to the biosynthetic products isolated from the sponge *Negombata magnifica*. J Nat Prod 67:1055–1057

Voigt O, Erpenbeck D, González-Pech RA, Al-Aidaroos AM, Berumen ML, Wörheide G (2017) Calcinea of the Red Sea: providing a DNA barcode inventory with description of four new species. Mar Biodivers:1–26

Voolstra CR, Miller DJ, Ragan MA, Hoffmann A, Hoegh-Guldberg O, Bourne D, Ball E, Ying H, Foret S, Takahashi S, Weynberg KD (2015) The ReFuGe 2020 consortium—using "omics" approaches to explore the adaptability and resilience of coral holobionts to environmental change. Front Mar Sci 2:68

Westinga E, Hoetjes PC (1981) The intrasponge fauna of *Speciospongia vesparia* (Porifera, Demospongiae) at Curaçao and Bonaire. Mar Biol 62:139–150

Wilkinson CR, Fay P (1979) Nitrogen fixation in coral reef sponges with symbiotic cyanobacteria. Nature 279:527–529

Wörheide G (2006) Low variation in partial Cytochrome Oxidase Subunit I (COI) mitochondrial sequences in the coralline demosponge *Astrosclera willeyana* across the Indo-Pacific. Mar Biol 148:907–912

Wörheide G, Epp LS, Macis L (2008) Deep genetic divergences among Indo-Pacific populations of the coral reef sponge *Leucetta chagosensis* (Leucettidae): founder effects, vicariance, or both? BMC Evol Biol 8:24

Wulff JL (1984) Sponge-mediated coral reef growth and rejuvenation. Coral Reefs 3:157–164

Yahel G, Post AF, Fabricius K, Marie D, Vaulot D, Genin A (1998) Phytoplankton distribution and grazing near coral reefs. Limnol Oceanogr 43:551–563

Yahel G, Sharp JH, Marie D, Häse C, Genin A (2003) In situ feeding and element removal in the symbiont-bearing sponge *Theonella swinhoei*: bulk DOC is the major source for carbon. Limnol Oceanogr 48:141–149

Yahel G, Marie D, Genin A (2005) InEx—a direct in situ method to measure filtration rates, nutrition, and metabolism of active suspension feeders. Limnol Oceanogr Methods 3:46–58

Yahia R, Hanora A, Fahmy N, Aly KA (2017) Quorum sensing signal production by sponge-associated bacteria isolated from the Red Sea, Egypt. Afr J Biotechnol 16:1688–1698

Yosief T, Rudi A, Wolde-ab Y, Kashman Y (1998a) Two new C22 1, 2-dioxane polyketides from the marine sponge *Acarnus cf. bergquistae*. J Nat Prod 61:491–493

Yosief T, Rudi A, Stein Z, Goldberg I, Gravalos GM, Schleyer M, Kashman Y (1998b) Asmarines AC; three novel cytotoxic metabolites from the marine sponge *Raspailia sp.* Tetrahedron Lett 39:3323–3326

Yosief T, Rudi A, Kashman Y (2000) Asmarines A–F, novel cytotoxic compounds from the marine sponge *Raspailia* species. J Nat Prod 63:299–304

Youssef DT (2004) Tasnemoxides A–C, new cytotoxic cyclic norsesterterpene peroxides from the Red Sea sponge *Diacarnus erythraenus*. J Nat Prod 67:112–114

Youssef DT (2005) Hyrtioerectines A–C, cytotoxic alkaloids from the Red Sea sponge *Hyrtios erectus*. J Nat Prod 68:1416–1419

Youssef DT, Mooberry SL (2006) Hurghadolide A and swinholide I, potent actin-microfilament disrupters from the Red Sea sponge *Theonella swinhoei*. J Nat Prod 69:154–157

Youssef DT, Yoshida WY, Kelly M, Scheuer PJ (2000) Polyacetylenes from a Red Sea sponge *Callyspongia* species. J Nat Prod 63:1406–1410

Youssef DT, Yoshida WY, Kelly M, Scheuer PJ (2001) Cytotoxic cyclic norterpene peroxides from a Red Sea sponge *Diacarnus erythraenus*. J Nat Prod 64:1332–1335

Youssef DT, Yamaki RK, Kelly M, Scheuer PJ (2002) Salmahyrtisol A, a novel cytotoxic sesterterpene from the Red Sea sponge *Hyrtios erecta*. J Nat Prod 65:2–6

Youssef DT, van Soest RW, Fusetani N (2003a) Callyspongenols A–C, new cytotoxic C22-polyacetylenic alcohols from a Red Sea sponge, *Callyspongia* species. J Nat Prod 66:679–681

Youssef DT, van Soest RW, Fusetani N (2003b) Callyspongamide A, a new cytotoxic polyacetylenic amide from the Red Sea sponge *Callyspongia fistularis*. J Nat Prod 66:861–862

Youssef DT, Singab ANB, van Soest RW, Fusetani N (2004) Hyrtiosenolides A and B, two new sesquiterpene γ-methoxybutenolides and a new sterol from a Red Sea sponge *Hyrtios* species. J Nat Prod 67:1736–1739

Youssef DT, Shaala LA, Emara S (2005) Antimycobacterial scalaranebased sesterterpenes from the Red Sea sponge *Hyrtios erecta*. J Nat Prod 68:1782–1784

Youssef DT, Shaala LA, Asfour HZ (2013) Bioactive compounds from the Red Sea marine sponge *Hyrtios* species. Mar Drugs 11:1061–1070

Youssef DT, Shaala LA, Mohamed GA, Badr JM, Bamanie FH, Ibrahim SR (2014) Theonellamide G, a potent antifungal and cytotoxic bicyclic glycopeptide from the Red Sea marine sponge *Theonella swinhoei*. Mar Drugs 12:1911–1923

Youssef DT, Badr JM, Shaala LA, Mohamed GA, Bamanie FH (2015a) Ehrenasterol and biemnic acid; new bioactive compounds from the Red Sea sponge *Biemna ehrenbergi*. Phytochem Lett 12:296–301

Youssef DT, Shaala LA, Alshali KZ (2015b) Bioactive hydantoin alkaloids from the Red Sea marine sponge *Hemimycale arabica*. Mar Drugs 13:6609–6619

Żółtowska-Aksamitowska S, Shaala LA, Youssef DT, Elhady SS, Tsurkan MV, Petrenko I, Wysokowski M, Tabachnick K, Meissner H, Ivanenko VN, Bechmann N (2018) First report on chitin in a non-verongiid marine demosponge: the *Mycale euplectellioides* case. Mar Drugs 16:68

Zundelevich A, Lazar B, Ilan M (2007) Chemical versus mechanical bioerosion of coral reefs by boring sponges-lessons from *Pione cf. vastifica*. J Exp Biol 210:91–96

Corals of the Red Sea

Michael L. Berumen, Roberto Arrigoni, Jessica Bouwmeester, Tullia I. Terraneo, and Francesca Benzoni

Abstract

The biodiversity of Red Sea corals captured the attention of some of the earliest European natural historians. Many of the first descriptions of tropical reef corals were based on Red Sea material. Modern approaches to resolve the notorious challenges of coral taxonomy have only recently been applied to Red Sea taxa. This chapter reviews current knowledge of the distributions of coral species in the Arabian region, including assessments of endemism. We also review new species described (or resurrected) since the last major assessment (in 2002). Where sufficient data is available, we highlight within-region distribution patterns. The Red Sea has the highest levels of endemism among all regions of the Indian Ocean. Analysis of the similarity of species composition among the Arabian subregions shows that the Red Sea and Socotra Island are the most speciose, but also have distinct community compositions. The regional diversity of Red Sea corals is likely influenced by the unique environmental gradients of the Arabian region. Despite evolving in testing conditions, Red Sea corals have been impacted by global climate change. Recent thermal bleaching events in the Red Sea highlight the pressures and challenges to future recovery.

M. L. Berumen (✉) · R. Arrigoni · T. I. Terraneo
Red Sea Research Center, Division of Biological and Environmental Science and Engineering, King Abdullah University of Science and Technology, Thuwal, Saudi Arabia
e-mail: michael.berumen@kaust.edu.sa

J. Bouwmeester
Red Sea Research Center, Division of Biological and Environmental Science and Engineering, King Abdullah University of Science and Technology, Thuwal, Saudi Arabia

Smithsonian Conservation Biology Institute, Front Royal, VA, USA

F. Benzoni
Department of Biotechnologies and Biosciences, University of Milano – Bicocca, Milan, Italy

UMR ENTROPIE (IRD, Université de La Réunion, CNRS), Laboratoire d'excellence-CORAIL, Centre IRD de Nouméa, Noumea Cedex, New Caledonia, France

Keywords

Biodiversity · Climate change · Coral · Coral bleaching · Endemism · Genetic connectivity · Indo-Pacific · Taxonomy

7.1 Diversity and Patterns within Red Sea

A reliable taxonomic framework is fundamental for the fields of biology, ecology, palaeontology, and biogeography (Wheeler 2004). In order to quantify the diversity of a given group, informative characters enabling the detection of species boundaries are necessary. In certain metazoans, such as scleractinian corals, these can be particularly difficult to detect due to the high plasticity of the morphological features traditionally used to address their taxonomy and systematics (Todd 2008). In the last decade, considerable advances have been made in the current understanding of scleractinian coral evolution and systematics, and, as a consequence, their taxonomy has undergone some radical changes. An approach integrating a constantly growing body of genetic, reproductive, and ecological data combined with novel morphological characters has led to descriptions of new taxa and to the discovery of unexpected phylogenetic relationships and cryptic lineages (Kitahara et al. 2016). Advances have been made for most genera belonging to major reef building families, including, e.g., the Mussidae, Lobophylliidae, Merulinidae, Coscinaraeidae, Psammocoridae, and Pocilloporidae (Benzoni et al. 2010, 2012a; Budd et al. 2012; Keshavmurthy et al. 2013; Huang et al. 2014a, b, 2016; Schmidt-Roach et al. 2014; Arrigoni et al. 2015, 2016a, b; Gélin et al. 2017). However, the process is far from complete, especially concerning deep water and azooxanthellate taxa (Kitahara et al. 2016). To date, there are still unresolved groups which are either yet to be investigated or to be formally revised: these are currently considered *incertae sedis*. Among the reef-dwelling taxa, diverse, widespread, and ecologically important genera such as *Acropora*, *Montipora*, and *Porites* await reassessment through an integrated morpho-molecular approach. For example, in the genus *Acropora*, some morphological characters show high variability among locations and under different environmental conditions, resulting in mostly unclear species boundaries. The high plasticity within the genus is reflected in the discrepancies in numbers of *Acropora* species reported throughout the globe among authors (e.g., Wallace 1999; Veron 2000; Wallace et al. 2012). Nevertheless, from the 161 valid *Acropora* species listed in the World Register of Marine Species (Hoeksema 2014), species boundaries were only successfully identified in a couple of regions and for a handful of species (e.g., Van Oppen et al. 2000; Wolstenholme et al. 2003). A revision of the genus has not yet been attempted in the Red Sea and, as is the case for other problematic groups, still awaits the identification of genetic markers informative at the species-level or genetic markers that would at least reliably separate the genus into smaller, more manageable taxonomic groups.

Ultimately, the ongoing revolution in coral systematics is leading to the reconsideration of biogeography patterns for corals. In such a rapidly changing taxonomic framework, the regional quantification of coral diversity needs to be considered as a process in constant evolution. Nonetheless, one of the objectives of this chapter is to provide an up-to-date overview of the biogeography of Red Sea corals. Despite the known imperfections of taxonomy in corals, they still provide one of the most complete datasets among non-vertebrate taxa for this type of analysis in the region. Since the last global compilation of reef-dwelling scleractinian corals was published (Veron 2000), a number of coral taxa have been described or resurrected throughout the Indo-Pacific region (e.g., Ditlev 2003; Kitahara et al. 2010; Wallace et al. 2011) and are known from the Red Sea (Table 7.1, Fig. 7.1, Appendix 1).

For corals and several other marine organisms, a distinct Arabian region has been recognized based on the rate of endemism and on the great diversity of habitat and environment types concentrated therein compared to other coastal sites in the Indian Ocean (Sheppard and Sheppard 1991; Sheppard et al. 1992; Obura 2012; DiBattista et al. 2016a). There is agreement that within the Arabian region the Red Sea harbors the highest diversity of scleractinian corals and their highest rate of endemism (Sheppard et al. 1992; Veron 1995; Hughes et al. 2002; Obura 2012; DiBattista et al. 2016a). However, to put it in Head's (1987) words "it is rather difficult to compare the coral faunas of different regions, because the intensity of sampling is so variable, and workers differ in their taxonomic interpretations". The latest available list of coral species for the Red Sea and the rest of the Arabian region was published by DiBattista et al. (2016a) in a review of the contemporary patterns of endemism for shallow water reef fauna in the region. The list, including zooxanthellate and azooxanthellate corals, aggregated records from reference collection-based papers (Scheer and Pillai 1983; Sheppard and Sheppard 1991; Claereboudt 2006; Riegl et al. 2012) and checklists (Head 1980; DeVantier et al. 2004, 2010; Turak et al. 2007), adopting the World Register of Marine Species nomenclature [available from http://www.marinespecies.org]. Geographic records based on existing collections have the advantage of being re-assessable in further studies as new methods and informative characters become available, however checklists based on *in situ* identifications regretfully do not (Rocha et al. 2014). Therefore, some of the species records included in the list could not be verified. For example, *Pseudosiderastrea tayamai* is included in checklists for the northern and central Red Sea as well as for Socotra (DeVantier et al. 2004, 2010). Previous records of this coral based on reference collections and illustrations of the examined material (Figs 6a-b in Sheppard and Sheppard 1991; Claereboudt 2006) were revised, revealing that the specimens actually belonged to *Anomastraea irregularis*. This species is macro-morphologically similar to *P. tayamai in situ* and is typically present throughout the Arabian region. As a complete reassessment of the previous Red Sea coral records goes beyond the scopes of this chapter, the central Red Sea and at Socotra records of *P. tayamai* are conservatively maintained herein but remain unverifiable. Unfortunately, current records of deep-water and azooxanthellate coral taxa result from a much smaller sampling effort in the region. Therefore, our knowledge and understanding of the distribution patterns and biogeography of these corals is likely incomplete.

Table 7.1 List of species described or resurrected since Veron (2000) and recorded in the Red Sea (see also Appendix 1)

Species	Taxonomic authority	References
Craterastrea levis[a, b]	Head 1983	Benzoni et al. (2012a)
Cyphastrea kausti[b]	Bouwmeester and Benzoni 2015	Bouwmeester et al. (2015)
Cyphastrea magna[b]	Benzoni and Arrigoni 2017	Arrigoni et al. (2017)
Echinophyllia bulbosa[b]	Arrigoni, Benzoni and Berumen 2016	Arrigoni et al. (2016c)
Goniopora tantillus	(Claereboudt and Al-Amri 2004)	Claereboudt and Al-Amri (2004)
Micromussa indiana	Benzoni and Arrigoni 2016	Arrigoni et al. (2016b)
Pachyseris inattesa[b]	Benzoni and Terraneo 2014	Terraneo et al. (2014)
Porites fontanesii	Benzoni and Stefani 2012	Benzoni and Stefani (2012)
Psammocora albopicta	Benzoni 2006	Benzoni (2006)
Sclerophyllia margariticola[a, b]	Klunzinger 1879	Arrigoni et al. (2015)

[a]Genus type locality in the Red Sea
[b]Species type locality in the Red Sea

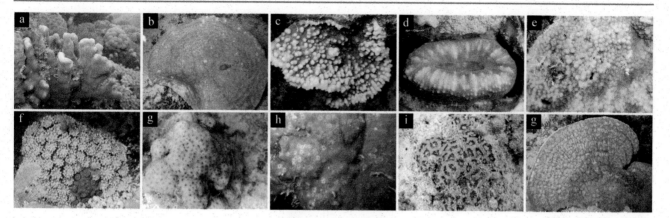

Fig. 7.1 Red Sea scleractinian coral species described or resurrected since Veron (2000) (see also Table 7.1). (**a**) *Porites fontanesii* Benzoni and Stefani 2012; (**b**) *Pachyseris inattesa* Terraneo and Benzoni, 2014; (**c**) *Echinophyllia bulbosa* Arrigoni, Benzoni, and Berumen, 2016; (**d**) *Sclerophyllia margariticola* Klunzinger, 1879; (**e**) *Cyphastrea kausti* Bouwmeester and Benzoni, 2014; (**f**) *Goniopora tantillus* (Claereboudt and Al Amri, 2004); (**g**) *Cyphastrea magna* Benzoni and Arrigoni, 2017; (**h**) *Psammocora albopicta* Benzoni 2006; (**i**) *Micromussa indiana* Benzoni and Arrigoni, 2016; (**j**) *Craterastrea levis* Head, 1983

In Appendix 1, we report the known distribution of the 401 species of zooxanthellate (91.5% of the 401 species) and azooxanthellate (8.5%) corals currently known to occur in the Arabian region. The list of DiBattista et al. (2016a) was modified with the inclusion of records from recently-published papers (Bouwmeester et al. 2015; Arrigoni et al. 2016a, b, c, 2017; Terraneo et al. 2016) and additional references (Veron 2002; Pichon et al. 2010; Benzoni et al. 2010, 2011, 2012b-c; Stefani et al. 2011; Arrigoni et al. 2014). Several additional distribution records came from the study of the coral reference collection from the King Abdullah University of Science and Technology (KAUST) biodiversity surveys (2013–2017) and are presented here for the first time. Recently, an assessment of the genetic diversity of the Agariciidae in the Red Sea has revealed that the molecular boundaries among 14 of the 20 examined species remain unclear (Terraneo et al. 2017) and ongoing studies of various genera indicate that the Red Sea coral diversity is to be further updated soon.

Based on the currently available information, the Red Sea harbors 94 genera and 359 species of scleractinian corals (91.6% of which are zooxanthellate). (Table 7.2, Appendix 1). This coral fauna includes Red Sea endemics (23 spp.), Arabian region endemics, and Indo-Pacific species with different distributions in the Arabian region, several of which are disjunct. In total, 48.5% of the Red Sea corals are found in at least one of the other partitions of the Arabian region (Fig. 7.2a–e, i–l). Among those, some corals, such as *Pavona cactus* and *Plesiastrea versipora*, are actually recorded from throughout the Arabian region and the Indo-Pacific (Fig. 7.2a). Others, including *Galaxea fascicularis* (Fig. 7.2b) and *Gardineroseris planulata*, have a similar distribution but are absent from the Arabian Gulf, likely due to the extreme environmental conditions found there (Sheppard et al. 1997; Riegl et al. 2012). Some Indo-Pacific corals present in the

Table 7.2 Total number of species, endemic species, and endemism for the regions considered in this study. The corresponding Marine Ecoregions of the World (MEOWs) defined by Spalding et al. (2007) are given in parentheses. List modified from DiBattista et al. (2016a)

Region (MEOW)	Total number of species	Number of endemics	% endemism
North & Central Red Sea (87)	314	14	4.5
South Red Sea (88)	263	0	0.0
Consensus Red Sea	359	23	6.4
Arabian Gulf (90)	66	0	0.0
Gulf of Oman (91)	126	0	0.0
Socotra Archipelago (89 *partim*)	233	0	0.0
Arabian Sea and Gulf of Aden (89 *partim*, 92)	124	0	0.0
Arabian Region (all of the above)	401	45	11.2

Red Sea are also found in the Gulf of Aden and in Socotra, but are not found in the Arabian Sea or the Arabian Gulf. This is the case for species such as *Cycloseris wellsi* (Fig. 7.2d) and some entire genera, including *Leptoria, Paramontastrea, Fungia, Herpolitha,* and *Pleuractis.* Thus, within the Arabian region, a remarkable barrier seems to be present for some taxa, perhaps due to the upwelling on the coast of the Arabian Sea. The seasonal upwelling brings cold and nutrient-rich waters that may limit coral distributions and generally inhibit reef-building processes (Sheppard and Sheppard 1991; Sheppard et al. 1997; Sheppard et al. 2000).

Very recent species descriptions (i.e., 2016–2018) utilizing integrative systematics approaches have increased the level of known endemism in Red Sea corals from 5.5% (DiBattista et al. 2016a) to 6.4% (Table 7.2). This figure is interestingly similar to the 6.1% previously reported by Sheppard et al. (1992), particularly when considering the increase in observation records and species descriptions in

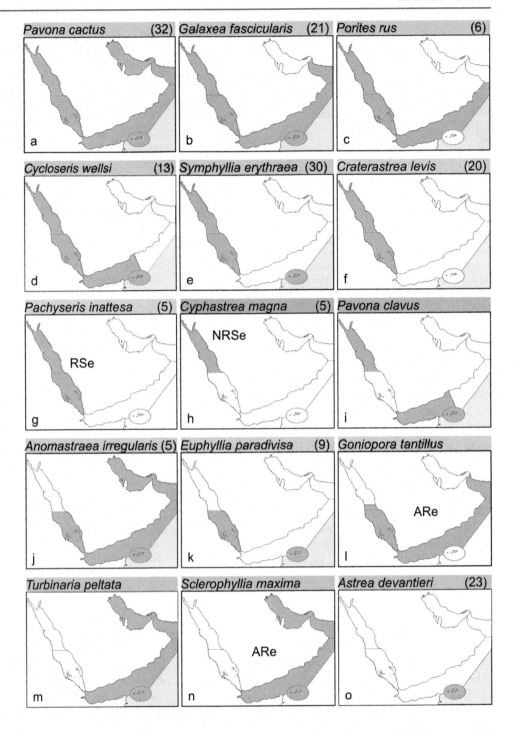

Fig. 7.2 Distribution patterns of scleractinian corals around the Arabian Peninsula. Each panel shows a representative species with presence confirmed in at least one of six subdivisions of Arabian waters (i.e., north-central Red Sea, southern Red Sea, Gulf of Aden, Socotra Island, Arabian Sea, Gulf of Oman, and Arabian Gulf), indicated by the dark grey shading. Light grey shading in the bottom right of a panel indicates presence of the species in other regions of the Indian Ocean. The total number of species known to exhibit the same distribution pattern is given in parentheses. *RSe* Red Sea endemic, *NRSe* North Red Sea endemic, *ARe* Arabian Region endemic

the last 25 years of Red Sea coral studies (Table 7.3). Among the Red Sea endemics, some species are found throughout the Red Sea, such as *Pachyseris inattesa* (Fig. 7.2g) and *Sclerophyllia margariticola*. Some species are currently recorded from the north-central Red Sea but not from the south, such as *Acropora squarrosa* and *Cyphastrea magna* (Fig. 7.2h). Interestingly, 29% of the coral species occurring in the Red Sea present a disjunct distribution because they are absent from the rest of the Arabian region but are recorded from the Indian Ocean (e.g. *Craterastrea levis*, Fig. 7.2f) or from the wider Indo-Pacific (e.g. *Diploastrea heliopora*). Another 27% of the Red Sea corals have a similar disjunct distribution and are found in the Socotra archipelago but nowhere else in the Arabian region (e.g. *Symphyllia erythraea*, Fig. 7.2e). In total, more than half of the Red Sea corals have a disjunct distribution as they do not occur in the Gulf of Aden and/or the Arabian Sea. Similar distributions have been observed for other reef organisms and the possible explanations include limitation of larval dispersal by water temperature due to the influence of the Arabian Sea upwelling outside the Red Sea and by turbidity in the south Red Sea (Sheppard et al. 1992; DiBattista et al. 2016b). Notably, 10.5% of the coral species occurring in the Arabian region are absent from the Red Sea (Fig. 7.2m–o). These include Arabian endemics such as *Sclerophyllia maxima* (Fig. 7.2n), the sister species of the Red Sea endemic *S. margariticola*, and *Acropora arabensis*. Also absent from the Red Sea are some common Indo-Pacific corals found everywhere else in the Arabian region, such as *Turbinaria peltata*, (Fig. 7.2m) and some species exclusively recorded in Socotra (Fig. 7.2o).

Within the Arabian region, the Red Sea zooxanthellate coral fauna has greatest affinity with that of Socotra Island in terms of diversity and composition with an overall similarity of almost 70% (Fig. 7.3c, d). Although Socotra has no coral endemics, it features species absent elsewhere in the Arabian region (like *Astrea devantieri*, Fig. 7.2o). Socotra's high coral diversity may be the result of a superimposition of the Red Sea, Gulf of Aden, Arabian Sea, and western Indian Ocean coral faunas, a pattern described for reef fish (DiBattista et al. 2015a).

Table 7.3 Species of scleractinian corals recorded in the different regions of the Red Sea by different authors in the last 30 years

Reference	Gulf of Suez	Gulf of Aqaba	North	Central	South
Head (1987)[a]	47	130	128	143	74
Sheppard and Sheppard (1991)	137			149	115
Veron (1995)[a]	139			150	115
Veron et al. (2009)	289				297
DiBattista et al. (2016a)	283			176	241
This study	290			201	263

[a]Only reports zooxanthellate species

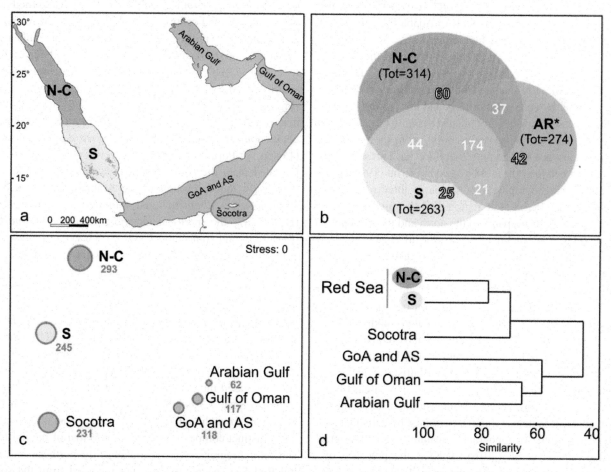

Fig. 7.3 Composition and affinities of the scleractinian coral fauna in different marine regions around the Arabian Peninsula. (**a**) Marine regions bordering the Arabian Peninsula (modified from Roberts et al. 2016, and DiBattista et al. 2016a), northern and central Red Sea (N-C), southern Red Sea (S), the Gulf of Aden and Arabian Sea (GoA and AS), Socotra Island, the Gulf of Oman, and the Arabian Gulf; (**b**) schematic representation of the azooxanthellate and zooxanthellate coral fauna affinities between and among N-C, S, and the remainder of the Arabian region excluding the Red Sea (AR*). Within each partition the total number of scleractinian coral species recorded (Tot) and of those recorded exclusively within that partition (bordered in black) is given. The number of species shared by partitions is placed in their respective intersections; (**c**) Multi-Dimensional Scaling analysis plot obtained from the data set of the zooxanthellate coral species found in the regions shown in panel a (Euclidean distance). Circle diameter is proportional to the total number of species recorded (given below the region name); (**d**) Hierarchical Cluster analysis plot (Bray-Curtis similarity) of the same data set analyzed in panel c

Biogeographic subdivisions of the Red Sea, largely congruent with a latitudinal gradient, have been proposed and discussed by different authors (see Table 7.3). Ormond et al. (1984) initially subdivided the Red Sea into four biogeographic sub-zones based on various taxa, including corals. Head (1987) differentiated the Gulf of Aqaba and Gulf of Suez from the north-central and southern Red Sea regions. Sheppard et al. (1992) recognized a group of at least seven Indo-Pacific corals occurring in the Gulf of Aqaba but nowhere else in the Red Sea. Possible explanations besides misidentification (which the authors explicitly ruled out) included the role of the Gulf of Aqaba as a refuge in the last glaciation and the fact that these species had gone undetected despite the sampling effort by different authors. Subsequently, each of these seven species has been recorded in the south Red Sea, thus confirming the latter hypothesis and the importance of the exploration effort of different coral reef environments in the southern Red Sea. Therefore, we did not attempt to create a distinction of the Gulf of Aqaba from the rest of the Red Sea for biogeographic purposes. Sheppard and Sheppard (1991, Figure 6 therein) subdivided the Red Sea into northern, central, and southern regions. However, Sheppard et al. (1992) argued that for corals this subdivision was justified by ecological criteria and the proportion composition of each species in the subdivisions, but also noted that all subdivisions contain most reef species. Strict presence-absence criteria may, therefore, have presented less distinction between these subdivisions. Veron (1995) maintained the biogeographic subdivision in three regions and remarked that the central region had the highest species diversity of coral compared to the north and the even poorer south (Table 7.3), but did not discuss the differences in detail. Spalding et al. (2007) recognized two distinct ecoregions (Marine Ecoregions of the World, or MEOWs) in the Red Sea, namely MEOW 87 (Northern and Central Red Sea) and 88 (Southern Red Sea), with a division occurring at 20°N latitude. More recently, following the subdivision by Spalding et al. (2007), most authors have considered the north and central Red Sea together and distinct from the south region (Veron et al. 2009; DiBattista et al. 2016a; Roberts et al. 2016). MEOWs were defined as "areas of relatively homogeneous species composition, clearly distinct from adjacent systems…[and] determined by a distinct suite of oceanographic or topographic features" (Spalding et al. 2007). Roberts et al. (2016) explored the presence of a within- Red Sea ecological boundary at 20°N through multivariate analyses of fish abundance (215 spp.) and benthic cover data (90 categories, 59 of which were scleractinians identified at genus or genus/growth form level) collected between 26.8°N and 18.6°N latitude in the Saudi Arabian Red Sea. The authors did not find evidence supporting the division of the Red Sea into two separate regions and provided evidence of a relative homogeneity of coral reef communities within their study area. However, they did not sample coral reef communities occurring at lower latitudes in the Red Sea, thus excluding the large island systems below 17° of latitude, namely the Farasan and the Kamaran Islands (Saudi Arabia and Yemen, respectively) in the east and the Dahlak Islands (Eritrea) in the west. These archipelagos provide extensive reef habitats occurring in environmental conditions different from those occurring at higher latitudes in the Red Sea (see Sect. 7.4 below and Davis et al., Chap. 2 in this volume). Roberts et al. (2016) also note that the majority of reefs surveyed were offshore reefs, whereas the Sheppard et al. (1992) surveys primarily considered fringing reefs; it is possible that there is less homogeneity in the near-coastal and fringing reefs.

Within the Red Sea, a total of 314 scleractinians are recorded from the north and central region and 263 from the south with 60.7% (n = 218) of the species known from both regions (Table 7.2). Of the remainder, 27% are recorded only from the north and central region (84 zooxanthellate and 12 azooxanthellate spp.) (e.g., Fig. 7.2i) and 12.5% only from the south (36 zooxanthellate and 9 azooxanthellate spp.) (e.g., Fig. 7.2j–l). The 25 corals exclusively found in the southern Red Sea and nowhere else in the Arabian region (Fig. 7.3b) are either poorly known and likely undersampled deep-water species (6 spp.) or they are from families in need of taxonomic revision, as mentioned above (e.g., Acroporidae (15 spp.) or Poritidae (2 spp.)). However, other coral species recorded from the southern Red Sea only include species commonly found throughout the Arabian region, such as *Psammocora albopicta* and *Anomastraea irregularis* (Fig. 7.2j), the regional endemic *Goniopora tantillus* (Fig. 7.2l), and some Indo-Pacific species only represented in the region in Socotra (e.g. *Euphyllia paradivisa*, Fig. 7.2k). To date, no southern Red Sea endemics are known (Table 7.2). Various authors have remarked on the relatively low coral diversity in the south of the Red Sea (Table 7.3). Head (1987) argued that the low scleractinian diversity recorded in the south Red Sea was likely a sampling artefact and that further collecting effort would increase the species count. Although he was certainly right, and thirty years of sampling effort have increased the count, the overall trend has not changed (see also Sect. 7.3 below). Nevertheless, the southern Red Sea coral fauna, though less diverse and devoid of endemics, has a distinct composition compared to that of the northern and central regions. More than half of the northern and central Red Sea species that are not recorded from the south belong to the Acroporidae (30 spp.), to currently unrevised Merulinidae genera (19 spp.), or to the genus *Porites* (3 spp.). It is thus likely that once the study of these species-rich and currently-unrevised taxa is finalized, differences in species richness between the sub-regions could be revised. Among the 60 coral species exclusively found in the northern and central Red Sea and nowhere else in the Arabian region, 14 are endemic (Table 7.2, Fig. 7.2h). Another 11 are azooxanthellate, such as *Javania insignis* and *Eguchipsammia fistula*, and their absence from the south could be explained by uneven sampling effort or habitat availability. The remainder are Acroporidae (14 spp.) and other taxa awaiting revision. Other records included some Pacific taxa (e.g., *Montipora niugini*), some of which could not be verified.

7.2 Coral Communities in the Red Sea

Coral communities thrive along both sides of the Red Sea, with some latitudinal variability from north to south as well as some cross-shelf variability where different habitats are available (Sheppard and Sheppard 1991; Roberts et al. 2016; Khalil et al. 2017). At the north of the Red Sea, the shallow Gulf of Suez harbors corals only at its southern end, while the Gulf of Aqaba, with depths reaching 1800 m, is surrounded by narrow fringing reefs that support corals down to the limits of the photic zone (Sheppard and Sheppard 1991). Further south, the northern and central Red Sea coasts are bordered by fringing reefs of variable width, protected in the central Red Sea by a number of elongated patch and barrier reefs along its shores. In the southern Red Sea, fringing reefs are progressively replaced by mangrove stands, with corals mostly found around islands and patch reefs (Sheppard and Sheppard 1991).

On a latitudinal gradient, the northern and central Red Sea share similar coral communities, although the central Red Sea harbors additional communities that correspond to the additional habitats provided by the patch and barrier reefs (Sheppard and Sheppard 1991). The southern Red Sea shows the most coral community differentiation with a potential community break between the Farasan Banks and the Farasan Islands around 17.5°N latitude (Sheppard and Sheppard 1991; Roberts et al. 2016). This break, south of which turbidity and productivity levels are much higher than in the rest of the Red Sea (Raitsos et al. 2013), has not yet been investigated in detail for coral communities; however, this habitat transition marks an important change corresponding to an apparent gene flow barrier for the clownfish *Amphiprion bicinctus* (Nanninga et al. 2014) and for the sponge *Stylissa carteri* (Giles et al. 2015). The environmental shift may therefore effectively act as a kind of barrier for a number of taxa with resultant impacts on the community composition on either side of the barrier.

Along a cross-shelf gradient, offshore and midshore communities are relatively homogenous while inshore reefs show a different coral community, with lower coral cover and sometimes lower coral diversity, potentially due to the increased presence of macroalgae and turf algae on those reefs (Ellis et al. 2017; Khalil et al. 2017). A cross-shelf study of benthic and fish communities in the central Red Sea showed that inshore reefs generally contained few branching species (e.g., *Acropora* spp., *Pocillopora* spp., *Stylophora* spp.) and were dominated by massive *Porites* species (Fig. 7.4; Khalil et al. 2017). However, some cross-shelf differentiation could also be due to inshore environments being more susceptible to stress such as thermal stress, leading to short- to long-term shifts in coral communities (van Woesik et al. 2011; Furby et al. 2013).

7.3 Red Sea Corals in an Indian Ocean Perspective

The Coral Triangle (for definition see Hoeksema 2007) represents the geographic area hosting the maximum species richness of scleractinian corals. Although diversity of reef-building coral species decreases in all directions from the Coral Triangle (Veron 1993; Wallace 1999; Hoeksema 2007; Veron et al. 2009), the Indian Ocean has at least two regions of high biodiversity and endemism, the Red Sea (Scheer and Pillai 1983; Sheppard and Sheppard 1991; DiBattista et al. 2016a) and the Northern Mozambique Channel (Obura 2016). The Red Sea hosts 359 coral species while coral diversity is moderately low in the seas around the Arabian Peninsula, with the minimum value occurring in the Arabian Gulf (66 species) and the maximum peak in Socotra (233 species) (Table 7.2). A combination of extreme physical and environmental conditions, including, for example, monsoon-induced upwelling as well as extreme water temperature and salinity, appears to limit the diversity of corals in the Gulf of Aden, Arabian Sea, Gulf of Oman, and Arabian Gulf (Glynn 1993; Sheppard et al. 2000). Conversely, Socotra harbors a diverse stony coral fauna likely due to the aforementioned overlap of fauna from several biogeographic regions (DeVantier et al. 2004; DiBattista et al. 2015a). Concerning the Northern Mozambique Channel, the maximum species richness occurs in Nacala (Mozambique), with a total of 297 recorded scleractinian species (Obura 2012), while the diversity decreases radially from this area. Genetic material seems to be transported on the South Equatorial Current and retained by eddies within the Mozambique Channel, with north and south export of diversity along linear transport corridors (New et al. 2004; Obura 2012).

In the last 5 years, evidence from molecular studies on corals seems to corroborate the importance of the Western Indian Ocean and the Red Sea as source of evolutionary novelty (Bowen et al. 2013). Genetic surveys in the Red Sea have allowed the formal description of several species of reef corals (Terraneo et al. 2014; Arrigoni et al. 2016a, 2017) and the resurrection of *Sclerophyllia* from the seas around the Arabian Peninsula (Arrigoni et al. 2015), suggesting that the previous taxonomic approach (i.e., reliance solely on traditional morphological characters) may have underestimated

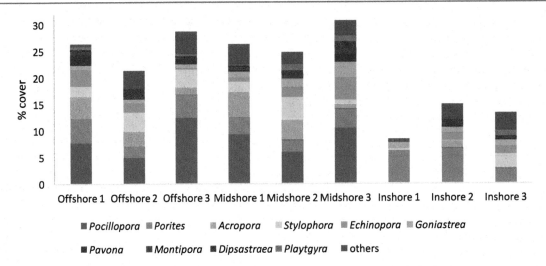

Fig. 7.4 Coral communities in offshore, midshore, and inshore reef environments in the central Red Sea, off the coast of Thuwal, Saudi Arabia. Vertical bars indicate the mean percentage cover of benthic substrate (determined by 10 m line-intercept transects) for each of the ten most abundant coral genera are listed. Other hard coral cover is included in "others". Offshore reefs are typically >10-15 km from the coast and surrounded by deep water (> 200 m). Midshore reefs are typically <15 km from the coast and located on the continental shelf (surrounding water <100 m depth). Inshore reefs are typically <5 km from the coast (surrounding water <40 m depth). (Data is extracted from Khalil et al. 2017)

coral biodiversity in this region. A combined morphomolecular approach led to the formal restoration of *Craterastrea* from the Western Indian Ocean and the Red Sea as well as the introduction of the predominantly Indian Ocean family Coscinaraeidae (Benzoni et al. 2012a). Moreover, genetic investigations revealed the existence of several distinct evolutionary lineages of *Pocillopora* and *Stylophora* that are restricted to the seas around the Arabian Peninsula and/or the Western Indian Ocean (Flot et al. 2011; Stefani et al. 2011; Pinzon et al. 2013; Keshavmurthy et al. 2013; Arrigoni et al. 2016c; Gélin et al. 2017). Finally, a growing body of molecular data indicated cases of deep intraspecific divergence between the Pacific and the Indian Ocean populations in several species, suggesting that morphological convergence of skeletal features might have occurred (Ladner and Palumbi 2012; Arrigoni et al. 2012, 2016b; Huang et al. 2014a; b; Kitano et al. 2014; Richards et al. 2016).

Recent findings suggest that the distribution of extant scleractinian corals is influenced more by geological features, such as tectonic plates and mantle plume tracks (Keith et al. 2013), than by variations in contemporary environmental conditions and habitats. The high levels of diversity of the Red Sea coral fauna may indeed be explained by the peculiar geological history of this basin and its tectonic activity during the Neogene (Bosworth et al. 2005). The establishment of the Red Sea reef fauna dates to the Pliocene and Pleistocene (4–3 Ma) (DiBattista et al. 2016b). After this period, the Red Sea experienced strong variations in environmental conditions (notably, temperature and salinity). Additionally, fluctuating isolation and connectivity of the Red Sea and Arabian Gulf occurred in the Pliocene and Pleistocene, all of which likely promoted speciation processes, contributing to high level of coral diversity of the Red Sea (Sheppard et al. 1992).

The Red Sea harbors the highest level of endemism in the Indian Ocean region. Specifically, 23 of the 359 Red Sea species (i.e., 6.4% endemism) are so far known only from the Red Sea (Table 7.3), whereas levels of endemism are less than 3% for all the other areas of the Indian Ocean (Veron et al. 2015; Obura 2016). For example, the Western Indian Ocean hosts only 7 endemic species (about 2% of the species present). In total, there are 60 coral species with distributions restricted to the Indian Ocean (the 23 Red Sea endemics and an additional 37 other Indian Ocean endemics (Obura 2012)). Although the Red Sea and the Northern Mozambique Channel represent two separate coral communities with distinct endemism and coral assemblages, they are closely related and several similarities can be found between them. For example, seven coral genera are endemic to the western and northern Indian Ocean, including *Anomastrea*, *Coscinaraea*, *Craterastrea*, *Ctenella*, *Gyrosmilia*, *Horastrea*, and *Sclerophyllia* (Veron 2000; Obura 2012; Benzoni et al. 2012a; Arrigoni et al. 2015). While *Sclerophyllia* seems to be restricted to the seas around the Arabian Peninsula (Arrigoni et al. 2015) and *Horastrea* and *Ctenella* have a restricted distribution within the Western Indian Ocean, *Anomastrea*, *Craterastrea*, and *Gyrosmilia* occur in both the Red Sea and the Western Indian Ocean (Veron 2000). Moreover, a new cryptic genus of Lobophylliidae from the Red Sea, the Gulf of Aden, Mayotte, and Madagascar was recently described (Arrigoni et al. 2018), further contributing to the peculiarity of the Indian Ocean coral fauna. Notably, three of these seven genera, i.e. *Ctenella*, *Anomastrea*, and *Horastrea*, are included in the top-20 list of evolutionary distinct and globally endangered corals (Huang 2012) and the extinction of these endangered lineages could therefore result in disproportionate loss within the coral tree of life.

The relatively high number of endemic species in the Red Sea may reflect its unusual environmental conditions, such as high temperature and salinity (Ngugi et al. 2012; see also Chap. 1 in this volume), and may be due to its isolated and peripheral geographical position. Unfortunately, the origin of Red Sea coral endemism is far from being definitively understood. This is due in part to a lack of genetic data coupled with estimated divergence times for Red Sea corals. The sole Red Sea work including this type of analyses, to our knowledge, focuses on *Stylophora* (Arrigoni et al. 2016a) and showed that the two endemic species (*S. wellsi* and *S. mamillata*) originated recently within the Red Sea (during the last 2 Ma). Moreover, these genetic analyses questioned the validity of these two species, suggesting that they could be recent species characterized by incomplete lineage sorting or they could simply be ecomorphs of *S. pistillata*, one of the most common scleractinian coral species of the Red Sea (Arrigoni et al. 2016a). Although there is limited information for corals, a growing amount of data is now available from reef fishes. If the case studies from fishes are indicative of the evolutionary histories of corals, then the overall picture may be rather complicated. Genetic evidence has revealed that there have been multiple evolutionary origins for Red Sea endemic fishes (Fernandez-Silva et al. 2015; DiBattista et al. 2015b, 2016b, 2017; Ahti et al. 2016; Coleman et al. 2016; Priest et al. 2016; Waldrop et al. 2016). Some endemic fishes diverged from their Indian Ocean relatives long before the most recent glaciations and apparently survived glacial cycles in low-salinity refugia, while other groups are younger, with likely origins within the Red Sea. There may be an equally diverse number of evolutionary histories of Red Sea coral endemics.

7.4 Climate Change and Red Sea Corals

The world's changing climate is threating the oceans at different scales. Global warming and associated increased sea surface temperatures pose major threats to reef ecosystems around the world (Hughes et al. 2003; Heron et al. 2016). Seawater temperature is a key factor controlling the distribution and diversity of zooxanthellate corals. Temperature tolerance in corals is species-specific as well as location-specific. Some corals adapt to local environmental conditions, so temperature tolerance thresholds can vary (e.g., Howells et al. 2012). Together with Arabian Gulf inhabitants, for example, Red Sea corals evolved thermal stress tolerance mechanisms that allow them to live in one of the hottest seas on earth (Davis et al. 2011). Nonetheless, climate change poses challenges to the most resistant corals. Overall, when temperature limits are exceeded several physiological processes are affected. One of the most striking example is the symbiosis between corals and the microscopic algae (zooxanthellae) that live within their tissues. Providing most of the corals' nutrition through photosynthesis, the zooxanthellae are essential for the well-being of the corals. This association is extremely vulnerable to temperature stress, with most coral species living in areas generally not exceeding 30 °C (Gates et al. 1992; Tanaka et al. 2014). As previously mentioned, this threshold can vary at different localities and under different conditions, but prolonged temperature fluctuations of even 1 °C above the local tolerance limit can lead to the breakdown of this association, causing the expulsion of the algae – a phenomenon known as bleaching. Corals can recover from mild bleaching and regain the zooxanthellae; severe or long term bleaching, on the other hand, can ultimately cause death of the coral and subsequently affect the community composition and reef diversity (Glynn and D'Croz 1990; Goreau and Hayes 1994).

Red Sea waters are characterized by extreme temperature gradients, seasonal temperature fluctuations that can exceed 10 °C, and sea surface temperatures often above 32 °C during the summer (e.g., Davis et al. 2011; Roik et al. 2016) (also see Chap. 1 in this volume). Despite these harsh environmental factors, the Red Sea harbors vibrant coral reefs that remain among the most diverse ecosystems on the planet. Indeed, Red Sea corals appear to be regionally adapted to live in these extreme conditions and are currently regarded as model organisms to help understanding the future of reefs in a changing climate scenario (ReFuGe 2020 Consortium 2015).

Nonetheless, as is the case for the majority of the world's reefs, Red Sea corals are already living close to their thermal tolerance limit (Roik et al. 2016) and are thus fragile and vulnerable to ocean warming. Since the mid 1990's, there has been evidence of pantropical sea surface temperature warming (0.4–1 °C) (Morice et al. 2012), and severe thermal stress on Red Sea corals has already been documented. Historically hot years have been shown to correspond with reduced coral calcification rates in Red Sea *Diploastrea* corals (Cantin et al. 2010) and, when prolonged over several weeks, to result in coral bleaching and coral mortality (DeVantier et al. 2000).

Temperature-driven bleaching was first reported from the central Red Sea in 1998. Following an El Niño Southern Oscillation (ENSO) event in 1997–1998, a prolonged heatwave of 1–2 °C above mean monthly SST averages hit coral communities in the central and southern Red Sea, causing patchy bleaching and high levels of mortality in both shallow water reefs and deeper slopes (De Vantier et al. 2000; Goreau et al. 2000). In 2010, following a heatwave that lasted more than 10 weeks, another bleaching event affected parts of the central Saudi Arabian Red Sea, with major consequences on shallow water and in shore reefs. Some of these communities, previously dominated by fast growing corals (e.g., the genus *Acropora*), experienced up to 95% mortality and shifted to a *Porites*-dominated system (Furby et al. 2013). In October 2015, the United States National Oceanic and Atmospheric Administration (NOAA) declared the third global mass bleaching event to be underway. This global bleaching event has been attributed to the coupled effect of ongoing climate change and a strong ENSO event, and is now considered the longest (i.e., greatest temporal duration) bleaching event on record (Heron et al. 2016). The Great Barrier Reef (GBR) was carefully monitored before, during, and after this bleaching event. The proportion of reefs categorized as "extremely bleached" following the heatwave was up to 4 times higher compared to prior bleaching events (i.e., 1998 and 2002) (Hughes et al. 2017). In the Red Sea, in-water surveys revealed that widespread bleaching occurred, albeit at different levels of severity (e.g., Monroe et al. 2018). In the central-northern Saudi Arabian Red Sea (around 22° N), inshore and shallow reefs were the most impacted by bleaching, with an average of 56% of total hard coral cover experiencing bleaching and with some reefs showing up to 95% of hard corals bleached at 5 m (Monroe et al. 2018). The pattern in the central Red Sea appeared to be very similar to that documented by Furby et al. (2013) in the 2010 bleaching event, with the majority of mid-shelf and offshore reefs only slightly affected by bleaching and remaining in good condition (Monroe et al. 2018). In the central-southern region (20° N), on the other hand, the effects of the heat stress were much broader, with reefs as far as 50 km from shore and at depths down to 30 m experiencing high mortality as a result of the bleaching (Lozano-Cortes et al. 2016). Initial observations in the region indicated widespread mortality on most of the reefs in the Farasan Banks region and probably extended into the Farasan Islands in the southern Red Sea. Lasting impacts on the health and resilience of these reefs, and flow-on effects for reef fish and invertebrate communities, are expected.

Overall, the causes shaping the geographic pattern of the bleaching events in the Red Sea can be attributed to the intensity and span of the heat stress in the different regions. The northern Red Sea has so far escaped mass bleaching events, leading to the suggestion that these reefs may serve as an important thermal refuge (Krueger et al. 2017; Osman et al. 2018). Nonetheless, small-scale spatial variability in patterns of bleaching remains rarely studied. As recently shown on the GBR, local weather conditions can strongly influence the likelihood of bleaching; for example, the remnants of a tropical storm created cloudy conditions in one section of the GBR which likely kept temperatures from reaching bleaching levels (Hughes et al. 2017). However, local conditions can also have the opposite effect. During a summer heatwave in 2015, an unusual wind pattern in the central-northern Red Sea seems to have limited water mixing, leading to a deadlock of warm water in inshore-shallow sites (Furby et al. 2013). This likely contributed to the high mortality at inshore sites whereas deeper sites and sites further offshore exhibited less bleaching and greater recovery rates. With regards to the bleaching in the southern Red Sea in 2015, there are ongoing efforts to understand the physical and environmental conditions that led to such extensive bleaching. One area of investigation is related to the input of Indian Ocean waters through the Bab al Mandeb, particularly as there was a strong ENSO event (Dreano et al. 2016). Alterations in the exchange of Red Sea and Gulf of Aden waters may have altered temperature-at-depth patterns or nutrient levels in the southern half of the Red Sea (e.g., Raitsos et al. 2015) and may have further exacerbated the heat stress condition. The presence of excess nutrients has been linked to reduced thermal tolerance in zooxanthellate organisms (Wooldridge et al. 2016), but on the GBR, water quality did not significantly affect the overall bleaching impact in 2016 (Hughes et al. 2017). Whatever the cause, the resultant mortality in the Farasan Banks and other areas of the southern Red Sea will likely necessitate a lengthy recovery period. In other systems, recovery rates are typically on the order of at least 10-15 years (Sweatman et al. 2011; Hughes et al. 2018). Unfortunately, the Red Sea may be susceptible to the same increasing global pressures that are leading to the recurrence of bleaching events on a shorter timescale (Osborne et al. 2017; Hughes et al. 2018), potentially precluding full recovery between bleaching events. While some Red Sea regions have so far escaped thermal stress events (Osman et al. 2018), the fate of the Red Sea's reefs remains in question in the face of global climate change.

Appendices

Appendix 1

Checklist of coral species from the Arabian region updated from DiBattista et al. (2016a). Crosses indicate presence in the major subdivisions of Arabian waters. New records and distributions in **bold**, references and/or specimens are provided in the corresponding footnote. Cl. = major phylogenetic clade (following Kitahara et al. (2016)); E = Red Sea endemic; zoox = zooxanthellate species; RS = Red Sea; GoA = Gulf of Aqaba; AS = Arabian Sea. "RS" in a species' row indicates that the species was recorded in the Red Sea but without indication of a specific locality.

Cl.	Family	Genus	Species	Taxonomic authority	zoox	North RS	Central RS	South RS	GoA and AS	Gulf of Oman	Socotra	Arabian Gulf	Endemism	Reference
II	Dendrophylliidae	Balanophyllia	cf cumingii	Milne Edwards and Haime 1848		x								
II	Dendrophylliidae	Balanophyllia	diffusa	Harrison and Poole 1909				x						1
II	Dendrophylliidae	Balanophyllia	gemmifera	Klunzinger 1879		x	x	x						
II	Dendrophylliidae	Balanophyllia	gigas	Moseley 1881				x		x				
II	Dendrophylliidae	Balanophyllia	rediviva	Moseley 1881		x								
II	Dendrophylliidae	Dendrophyllia	arbuscula	van der Horst 1922		x	x	x		x				
II	Dendrophylliidae	Dendrophyllia	cf minuscula	Bourne 1905		x								
II	Dendrophylliidae	Dendrophyllia	cf cornigera	Lamarck 1816		x								
II	Dendrophylliidae	Dendrophyllia	robusta	Bourne 1905		x	x	x	x	x				
II	Eguchipsammia		fistula	Alcock 1902		x	x							
II	Dendrophylliidae	Heteropsammia	cochlea	Spengler 1781					x	x				
II	Dendrophylliidae	Rhizopsammia	compacta	Sheppard and Sheppard 1991					x					
II	Dendrophylliidae	Rhizopsammia	wettsteini	Scheer and Pillai 1983		x	x	x	x	x				2
II	Dendrophylliidae	Tubastraea	coccinea	Lesson 1829		x	x	x	x	x		x		
II	Dendrophylliidae	Tubastraea	diaphana	Dana 1846		x	x	x						
II	Dendrophylliidae	Tubastraea	micranthus	Ehrenberg 1834		x	x	x	x	x	x			3
II	Dendrophylliidae	Turbinaria	frondens	Dana 1846	+	x					x			
II	Dendrophylliidae	Turbinaria	irregularis	Bernard 1896	+	x					x			
II	Dendrophylliidae	Turbinaria	mesenterina	Lamarck 1816	+	x	x	x		x	x	x		
II	Dendrophylliidae	Turbinaria	peltata	Esper 1794	+									
II	Dendrophylliidae	Turbinaria	reniformis	Bernard 1896	+	x		x	x	x	x	x		3
II	Dendrophylliidae	Turbinaria	stellulata	Lamarck 1816	+	x		x	x	x	x	x		
III	Poritidae	Goniopora	albiconus	Veron 2000	+		x	x	x	x	x			4
III	Poritidae	Goniopora	burgosi	Nemenzo 1971	+	x			x	x	x			
III	Poritidae	Goniopora	ciliatus	Veron 2000	+	x	x		x	x	x			
III	Poritidae	Goniopora	columna	Dana 1846	+	x	x	x		x	x			
III	Poritidae	Goniopora	djiboutiensis	Vaughan 1907	+	x	x	x	x	x	x			4
III	Poritidae	Goniopora	gracilis	(Milne Edwards and Haime, 1849)	+	x	x	x	x	x	x			4
III	Poritidae	Goniopora	lobata	Milne Edwards 1860	+	x	x	x	x	x	x	x		3, 4
III	Poritidae	Goniopora	minor	Crossland, 1952	+	x	x	x	x	x				3, 4
III	Poritidae	Goniopora	pedunculata	Quoy and Gaimard 1833	+	x	x	x		x	x			
III	Poritidae	Goniopora	planulata	Ehrenberg 1834	+	x	x	x						
III	Poritidae	Goniopora	savignyi	Dana 1846	+	x	x	x	x	x				
III	Poritidae	Goniopora	somaliensis	Vaughan 1907	+	x	x	x	x	x	x			
III	Poritidae	Goniopora	stokesi	Milne Edwards and Haime 1851	+	x	x	x	x	x				
III	Poritidae	Goniopora	sultani	Veron, DeVantier and Turak 2000	+	x							E	

(continued)

Cl.	Family	Genus	Species	Taxonomic authority	zoox	North RS	Central RS	South RS	GoA and AS	Gulf of Oman	Socotra	Arabian Gulf	Endemism	Reference
III	Poritidae	Goniopora	tantillus	Claereboudt and Al Amri 2004	+			x	x	x				5
III	Poritidae	Goniopora	tenella	Quelch 1886	+		x	x						
III	Poritidae	Goniopora	tenuidens	Quelch 1886	+	x		x	x	x	x			3
III	Poritidae	Poritidae	annae	Crossland 1952	+			x	x	x	x			
III	Poritidae	Porites	columnaris	Klunzinger 1879	+	x		x						
III	Poritidae	Porites	compressa	Dana 1846	+	x	x	x		x	x	x		
III	Poritidae	Porites	cf cumulatus	Nemenzo 1975	+			x		x	x			
III	Poritidae	Porites	cylindrica	Dana 1846	+		x							
III	Poritidae	Porites	echinulata	Klunzinger 1879	+	x	x	x	x		x			
III	Poritidae	Porites	fontanesii	Benzoni and Stefani 2012	+	x	x	x	x		x			6
III	Poritidae	Porites	harrisoni	Veron 2000	+			x	x	x	x	x		3
III	Poritidae	Porites	cf latistella	Quelch 1886	+			x			x			
III	Poritidae	Porites	lichen	Dana 1846	+	x	x	x	x	x	x			3
III	Poritidae	Porites	lobata	Dana 1846	+		x	x	x	x	x	x		
III	Poritidae	Porites	lutea	Quoy and Gaimard 1833	+	x	x	x	x	x	x	x		
III	Poritidae	Porites	monticulosa	Dana 1846	+		x	x	x		x			
III	Poritidae	Porites	nigrescens	Dana 1846	+			x	x					
III	Poritidae	Porites	nodifera	Klunzinger 1879	+		x	x	x	x	x	x		3
III	Poritidae	Porites	ornata	Nemenzo 1971	+	x								
III	Poritidae	Porites	profundus	Rehberg 1892	+			x		x	x			
III	Poritidae	Porites	rus	Forskal 1775	+	x	x	x	x	x				
III	Poritidae	Porites	solida	Forskal 1775	+	x	x	x	x	x	x			3
III	Poritidae	Porites	somaliensis	Gravier 1911	+		x							
III	Poritidae	Stylaraea	punctata	Linnaeus 1758	+		RS	x		x	x			7
IV	Agariciidae	Pachyseris	inattesa	Benzoni and Terraneo 2014	+	x	x	x					E	
IV	Agariciidae	Pachyseris	rugosa	Lamarck 1801	+	x								
IV	Agariciidae	Pachyseris	speciosa	Dana 1846	+	x	x	x			x			
V	Euphylliidae	Euphyllia	divisa	Veron and Pichon 1980	+					x				
V	Euphylliidae	Euphyllia	glabrescens	Chamisso and Eysenhardt 1821	+	x		x	x		x			
V	Euphylliidae	Euphyllia	paradivisa	Veron 1990	+						x			8
V	Euphylliidae	Galaxea	astreata	Lamarck 1816	+	x	x	x	x		x			9
V	Euphylliidae	Galaxea	fascicularis	Linnaeus 1767	+	x	x	x	x	x	x			
V	Euphylliidae	Gyrosmilia	interrupta	Ehrenberg 1834	+	x	x	x			x			
VI	Acroporidae	Acropora	abrotanoides	Lamarck 1816	+	x	x	x						
VI	Acroporidae	Acropora	aculeus	Dana 1846	+		x	x						
VI	Acroporidae	Acropora	acuminata	Verrill 1864	+	x	x	x						

	Family	Genus	species	Author									
VI	Acroporidae	*Acropora*	*anthocercis*	Brook 1893	+	x							10
VI	Acroporidae	*Acropora*	*arabensis*	Hodgson and Carpenter 1995	+				x	x			11 E
VI	Acroporidae	*Acropora*	*aspera*	Dana 1846	+				x				
VI	Acroporidae	*Acropora*	*austera*	Dana 1846	+	x	x						
VI	Acroporidae	*Acropora*	*capillaris*	Klunzinger 1879	+	x	x	x					
VI	Acroporidae	*Acropora*	*cerealis*	Dana 1846	+		x	x					
VI	Acroporidae	*Acropora*	*clathrata*	Brook 1891	+	x	x	x	x	x			
VI	Acroporidae	*Acropora*	*cytherea*	Dana 1846	+	x	x	x					
VI	Acroporidae	*Acropora*	*digitifera*	Dana 1846	+	x		x					
VI	Acroporidae	*Acropora*	*divaricata*	Dana 1846	+			x					
VI	Acroporidae	*Acropora*	*donei*	Veron and Wallace 1984	+			x					
VI	Acroporidae	*Acropora*	*downingi*	Wallace 1999	+	x	x	x	x	x			3
VI	Acroporidae	*Acropora*	*elseyi*	Brook 1982	+			x					
VI	Acroporidae	*Acropora*	*eurystoma*	Klunzinger 1879	+	x		x					
VI	Acroporidae	*Acropora*	*gemmifera*	Brook 1892	+	x		x					
VI	Acroporidae	*Acropora*	*grandis*	Brook 1892	+			x					
VI	Acroporidae	*Acropora*	*granulosa*	Milne Edwards 1860	+	x		x	x				
VI	Acroporidae	*Acropora*	*hemprichii*	Ehrenberg 1834	+	x	x	x	x				
VI	Acroporidae	*Acropora*	*horrida*	Dana 1846	+	x	x	x	x				
VI	Acroporidae	*Acropora*	*humilis*	Dana 1846	+	x		x					
VI	Acroporidae	*Acropora*	*hyacinthus*	Dana 1846	+	x	x	x					
VI	Acroporidae	*Acropora*	*intermedia*	Brook 1891	+			x					
VI	Acroporidae	*Acropora*	*lamarcki*	Veron 2000	+			x		x			
VI	Acroporidae	*Acropora*	*latistella*	Brook 1892	+	x		x					
VI	Acroporidae	*Acropora*	*listeri*	Brook 1893	+	x		x					
VI	Acroporidae	*Acropora*	*loripes*	Brook 1892	+			x					
VI	Acroporidae	*Acropora*	*lukeni*	Crossland 1952	+	x		x	x				12
VI	Acroporidae	*Acropora*	*microphthalma*	Verrill 1869	+	x		x	x				
VI	Acroporidae	*Acropora*	*millepora*	Ehrenberg 1834	+			x					
VI	Acroporidae	*Acropora*	*muricata*	Linnaeus 1758	+	x	x	x	x				
VI	Acroporidae	*Acropora*	*nasuta*	Dana 1846	+	x	x	x	x	x			
VI	Acroporidae	*Acropora*	*cf natalensis*	Riegl 1995	+			x					
VI	Acroporidae	*Acropora*	*parapharaonis*	Veron 2000	+			x					
VI	Acroporidae	*Acropora*	*pharaonis*	Milne Edwards 1860	+	x	x	x	x	x			E
VI	Acroporidae	*Acropora*	*polystoma*	Brook 1891	+	x		x					
VI	Acroporidae	*Acropora*	*pulchra*	Brook 1891	+			x					
VI	Acroporidae	*Acropora*	*robusta*	Dana 1846	+	x		x					
VI	Acroporidae	*Acropora*	*rufa*	Veron 2000	+			x	x				
VI	Acroporidae	*Acropora*	*samoensis*	Brook 1891	+	x		x	x				
VI	Acroporidae	*Acropora*	*secale*	Studer 1878	+	x	x	x	x	x			13
VI	Acroporidae	*Acropora*	*selago*	Studer 1878	+			x					
VI	Acroporidae	*Acropora*	*spicifera*	Dana 1846	+			x	x				

(continued)

Cl.	Family	Genus	Species	Taxonomic authority	zoox	North RS	Central RS	South RS	GoA and AS	Gulf of Oman	Socotra	Arabian Gulf	Endemism	Reference
VI	Acroporidae	Acropora	squarrosa	Ehrenberg 1834	+	x	x						E	
VI	Acroporidae	Acropora	subulata	Dana 1846	+			x						
VI	Acroporidae	Acropora	valenciennesi	Milne Edwards 1860	+	x	x	x		x	x	x		
VI	Acroporidae	Acropora	valida	Dana 1846	+	x	x	x	x	x				
VI	Acroporidae	Acropora	variolosa	Klunzinger 1879	+	x								
VI	Acroporidae	Acropora	vaughani	Wells 1954	+			x						
VI	Acroporidae	Acropora	yongei	Veron and Wallace 1984	+			x						
VI	Acroporidae	Alveopora	allingi	Hoffmeister 1925	+	x	x	x			x			
VI	Acroporidae	Alveopora	daedalea	Forskal 1775	+	x	x	x						
VI	Acroporidae	Alveopora	fenestrata	Lamarck 1816	+	x								
VI	Acroporidae	Alveopora	marionensis	Veron and Pichon 1982	+						x			
VI	Acroporidae	Alveopora	ocellata	Wells 1954	+	x	x	x			x			
VI	Acroporidae	Alveopora	spongiosa	Dana 1846	+	x	x	x			x			
VI	Acroporidae	Alveopora	superficialis	Pillai and Scheer 1976	+		x							
VI	Acroporidae	Alveopora	tizardi	Bassett-Smith 1890	+		x	x			x			
VI	Acroporidae	Alveopora	verrilliana	Dana 1846	+	x	x							
VI	Acroporidae	Alveopora	viridis	Quoy and Gaimard 1833	+	x	x	x						
VI	Acroporidae	Anacropora	spumosa	Veron, Turak and DeVantier 2000	+	x					x		E	
VI	Acroporidae	Astreopora	cucullata	Lamberts 1980	+	x					x			
VI	Acroporidae	Astreopora	expansa	Brueggemann 1877	+		x	x		x				
VI	Acroporidae	Astreopora	explanata	Veron 1985	+					x				
VI	Acroporidae	Astreopora	gracilis	Bernard 1896	+	x					x			
VI	Acroporidae	Astreopora	listeri	Bernard 1896	+	x					x			
VI	Acroporidae	Astreopora	myriophthalma	Lamarck 1816	+	x	x	x	x	x	x			3
VI	Acroporidae	Isopora	cf brueggemanni	Brook 1893	+						x			
VI	Acroporidae	Isopora	palifera	Lamarck 1816	+						x			
VI	Acroporidae	Montipora	aequituberculata	Bernard 1897	+	x	x	x		x	x	x	E	
VI	Acroporidae	Montipora	aspergillus	Veron, DeVantier and Turak 2000	+	x	x	x			x			
VI	Acroporidae	Montipora	circumvallata	Ehrenberg 1834	+	x	x	x		x	x	x	E	14
VI	Acroporidae	Montipora	cryptus	Veron 2000	+	x					x			
VI	Acroporidae	Montipora	danae	Milne Edwards and Haime 1851	+	x	x	x		x				
VI	Acroporidae	Montipora	digitata	Dana 1846	+	x			x		x			
VI	Acroporidae	Montipora	echinata	Veron, DeVantier and Turak 2000	+	x		x					E	
VI	Acroporidae	Montipora	edwardsi	Bernard 1897	+	x	x							
VI	Acroporidae	Montipora	efflorescens	Bernard 1897	+	x								
VI	Acroporidae	Montipora	effusa	Dana 1846	+		x							
VI	Acroporidae	Montipora	ehrenbergi	Verrill 1872	+	x	x	x						
VI	Acroporidae	Montipora	floweri	Wells 1954	+		x	x						

	Family	Genus	species	Author	+	1	2	3	4	5	6	Notes
VI	Acroporidae	*Montipora*	*foliosa*	Pallas 1766	+	x						
VI	Acroporidae	*Montipora*	*gracilis*	Klunzinger 1879	+	x						
VI	Acroporidae	*Montipora*	*grisea*	Bernard 1897	+	x						
VI	Acroporidae	*Montipora*	*hemispherica*	Veron 2000	+	x						E
VI	Acroporidae	*Montipora*	*hispida*	Dana 1846	+	x				x		
VI	Acroporidae	*Montipora*	*informis*	Bernard 1897	+	x	x			x		
VI	Acroporidae	*Montipora*	*meandrina*	Ehrenberg 1834	+	x	x	x				
VI	Acroporidae	*Montipora*	*millepora*	Crossland 1952	+	x	x					
VI	Acroporidae	*Montipora*	*mollis*	Bernard 1897	+	x						
VI	Acroporidae	*Montipora*	*monasteriata*	Forskal 1775	+	x	x	x	x			
VI	Acroporidae	*Montipora*	*niugini*	Veron 2000	+	x		x				
VI	Acroporidae	*Montipora*	*nodosa*	Dana 1846	+	x			x			
VI	Acroporidae	*Montipora*	*orientalis*	Nemenzo 1967	+	x			x			
VI	Acroporidae	*Montipora*	*pachytuberculata*	Veron, DeVantier and Turak 2000	+	x	x					E
VI	Acroporidae	*Montipora*	*peltiformis*	Bernard 1897	+	x	x					
VI	Acroporidae	*Montipora*	*saudii*	Turak, DeVantier and Veron 2000	+	x	x					E
VI	Acroporidae	*Montipora*	*spongiosa*	Ehrenberg 1834	+	x	x	x	x	x		3
VI	Acroporidae	*Montipora*	*spongodes*	Bernard 1897	+	x	x		x	x		
VI	Acroporidae	*Montipora*	*spumosa*	Lamarck 1816	+	x	x		x			
VI	Acroporidae	*Montipora*	*stellata*	Bernard 1897	+	x	x					
VI	Acroporidae	*Montipora*	*stilosa*	Ehrenberg 1834	+	x	x	x	x	x		
VI	Acroporidae	*Montipora*	*tuberculosa*	Lamarck 1816	+	x	x	x	x	x		
VI	Acroporidae	*Montipora*	*turgescens*	Bernard 1897	+	x	x	x	x	x		
VI	Acroporidae	*Montipora*	*venosa*	Ehrenberg 1834	+	x	x	x	x	x		
VI	Acroporidae	*Montipora*	*verrilli*	Vaughan 1907	+	x	x					
VI	Acroporidae	*Montipora*	*verrucosa*	Lamarck 1816	+	x	x	x	x	x		
VII	Agariciidae	*Dactylotrochus*	*cervicornis*	Moseley 1881	+	x		x				
VII	Agariciidae	*Gardineroseris*	*planulata*	Dana 1846	+	x	x	x	x	x		
VII	Agariciidae	*Leptoseris*	*explanata*	Yabe and Sugiyama 1941	+	x	x	x	x	x		
VII	Agariciidae	*Leptoseris*	*foliosa*	Dinesen 1980	+	x	x	x	x	x		
VII	Agariciidae	*Leptoseris*	*fragilis*	Milne Edwards and Haime 1849	+	x	x	x	x	x		
VII	Agariciidae	*Leptoseris*	*gardineri*	van der Horst 1921	+	x	x	x				
VII	Agariciidae	*Leptoseris*	*hawaiiensis*	Vaughan 1907	+	x	x	x	x	x		
VII	Agariciidae	*Leptoseris*	*incrustans*	Quelch 1886	+	x	x	x				
VII	Agariciidae	*Leptoseris*	*mycetoseroides*	Wells 1954	+	x	x	x	x	x		3
VII	Agariciidae	*Leptoseris*	*scabra*	Vaughan 1907	+	x	x	x	x	x		
VII	Agariciidae	*Leptoseris*	*solida*	Quelch 1886	+	x	x	x				
VII	Agariciidae	*Leptoseris*	*yabei*	Pillai and Scheer 1976	+	x	x			x		
VII	Agariciidae	*Pavona*	*bipartita*	Nemenzo 1980	+	x						
VII	Agariciidae	*Pavona*	*cactus*	Forskal 1775	+	x	x	x	x	x		

(continued)

Cl.	Family	Genus	Species	Taxonomic authority	zoox	North RS	Central RS	South RS	GoA and AS	Gulf of Oman	Socotra	Arabian Gulf	Endemism	Reference
VII	Agariciidae	Pavona	clavus	Dana 1846	+	x	x		x		x			3
VII	Agariciidae	Pavona	danai	Milne Edwards 1860	+	x	x					x		
VII	Agariciidae	Pavona	decussata	Dana 1846	+	x	x	x	x	x	x	x		
VII	Agariciidae	Pavona	diffluens	Lamarck 1816	+		x		x	x	x	x		
VII	Agariciidae	Pavona	divaricata	Lamarck 1816	+	x	x	x			x			
VII	Agariciidae	Pavona	duerdeni	Scheer and Pillai 1974	+	x	x	x			x			3
VII	Agariciidae	Pavona	explanulata	Lamarck 1816	+	x	x	x	x	x	x	x		
VII	Agariciidae	Pavona	frondifera	Lamarck 1816	+	x		x	x		x			3
VII	Agariciidae	Pavona	maldivensis	Gardiner 1905	+	x	x	x	x		x			
VII	Agariciidae	Pavona	minuta	Wells 1954	+					x				
VII	Agariciidae	Pavona	varians	Verrill 1864	+	x	x	x	x	x	x	x		
VII	Agariciidae	Pavona	venosa	Ehrenberg 1834	+	x		x		x	x			
IX	Siderastreidae	Pseudosiderastrea	tayamai	Yabe and Sugiyama 1935	+	x					x			
IX	Siderastreidae	Siderastrea	savignyana	Milne Edwards and Haime 1850	+	x	x	x	x	x	x	x		
X	Pocilloporidae	Madracis	interjecta	Marenzeller 1907	+	x								
X	Pocilloporidae	Madracis	kirbyi	Veron and Pichon 1976	+	x	x	x		x	x	x		15
X	Pocilloporidae	Stylocoeniella	armata	Ehrenberg 1834	+	x	x	x		x	x			
X	Pocilloporidae	Stylocoeniella	guentheri	Bassett-Smith 1890	+	x	x	x		x	x			
X	Pocilloporidae	Pocillopora	damicornis	Linnaeus 1758	+	x	x	x	x	x	x	x		
X	Pocilloporidae	Pocillopora	grandis	Dana 1846	+						x			
X	Pocilloporidae	Pocillopora	verrucosa	Ellis and Solander 1786	+	x	x	x	x	x	x			
X	Pocilloporidae	Seriatopora	caliendrum	Ehrenberg 1834	+	x	x	x			x			
X	Pocilloporidae	Seriatopora	hystrix	Dana 1846	+	x	x	x			x			
X	Pocilloporidae	Seriatopora	octoptera	Ehrenberg 1834	+	x								
X	Pocilloporidae	Stylophora	danae	Milne Edwards and Haime 1850	+	x	x	x	x	x	x			
X	Pocilloporidae	Stylophora	kuehlmanni	Scheer and Pillai 1983	+	x	x	x		x	x			16
X	Pocilloporidae	Stylophora	mamillata	Scheer and Pillai 1983	+	x	x	x			x		E	
X	Pocilloporidae	Stylophora	pistillata	Esper 1797	+	x	x	x	x	x	x	x		
X	Pocilloporidae	Stylophora	subseriata	Ehrenberg 1834	+	x	x	x	x	x	x	x		
X	Pocilloporidae	Stylophora	wellsi	Scheer 1964	+	x	x	x		x	x			
X	Pocilloporidae	Stylophora	madagascarensis	Veron 2000	+				x					
XI	Coscinareidae	Anomastraea	irregularis	Marenzeller 1901	+	x		x	x	x	x	x		17, 18
XI	Coscinareidae	Coscinaraea	columna	Dana 1846	+		x	x	x	x	x			
XI	Coscinareidae	Coscinaraea	monile	Forskal 1775	+	x	x	x	x	x	x	x		
XI	Coscinareidae	Craterastrea	levis	Head 1983	+	x	x	x	x	x	x			19
XI	Fungiidae	Cantharellus	doederleini	von Marenzeller 1907	+	x	x		x	x	x	x	E	
XI	Fungiidae	Cantharellus	noumeae	Hoeksema and Best 1984	+	x					x			
XI	Fungiidae	Ctenactis	crassa	Dana 1846	+	x	x				x			
XI	Fungiidae	Ctenactis	echinata	Pallas 1766	+	x	x							

7 Corals of the Red Sea

	Family	Genus	species	Author									Notes
XI	Fungiidae	*Cycloseris*	*costulata*	Ortmann 1889	+		x	x					
XI	Fungiidae	*Cycloseris*	*curvata*	Hoeksema 1989	+			x			x		
XI	Fungiidae	*Cycloseris*	*cyclolites*	Lamarck 1815	+	x	x	x	x		x		
XI	Fungiidae	*Cycloseris*	*distorta*	Michelin 1842	+	x	x	x	x			x	
XI	Fungiidae	*Cycloseris*	*explanulata*	Van der Horst 1922	+	x	x	x	x	x			
XI	Fungiidae	*Cycloseris*	*fragilis*	Alcock 1893	+	x	x	x	x				
XI	Fungiidae	*Cycloseris*	*somervillei*	Gardiner 1909	+			x					
XI	Fungiidae	*Cycloseris*	*tenuis*	Dana 1846	+	x		x	x				
XI	Fungiidae	*Cycloseris*	*vaughani*	Boschma 1923	+	x	x	x	x				3
XI	Fungiidae	***Cycloseris***	***wellsi***	(Veron and Pichon, 1980)	+	x	x	x	x				20
XI	Fungiidae	*Danafungia*	*horrida*	Dana 1846	+	x	x	x	x				
XI	Fungiidae	*Danafungia*	*scruposa*	Klunzinger 1879	+	x	x	x	x				
XI	Fungiidae	*Fungia*	*fungites*	Linnaeus 1758	+	x	x	x	x				
XI	Fungiidae	*Herpolitha*	*limax*	Esper 1797	+	x	x	x	x				3
XI	Fungiidae	*Lithophyllon*	*concinna*	Verrill 1864	+	x	x	x					
XI	Fungiidae	*Lithophyllon*	*repanda*	Dana 1846	+	x	x	x					
XI	Fungiidae	*Lithophyllon*	*scabra*	Doederlein 1901	+				x				
XI	Fungiidae	*Lobactis*	*scutaria*	Lamarck 1801	+	x	x		x				
XI	Fungiidae	*Pleuractis*	*granulosa*	Klunzinger 1879	+	x	x		x				
XI	Fungiidae	*Pleuractis*	*moluccensis*	Van der Horst 1919	+	x			x				
XI	Fungiidae	*Pleuractis*	*paumotensis*	Stutchbury 1833	+	x	x		x				
XI	Fungiidae	*Pleuractis*	*seychellensis*	Hoeksema 1993	+		x						
XI	Fungiidae	*Podabacia*	*crustacea*	Pallas 1766	+	x	x						
XI	Fungiidae	*Podabacia*	*sinai*	Veron 2000	+	x							
XI	incertae sedis	*Leptastrea*	*aequalis*	Veron 2000	+						x		
XI	incertae sedis	*Leptastrea*	*bottae*	Milne Edwards and Haime 1849	+	x		x	x				
XI	incertae sedis	*Leptastrea*	*inaequalis*	Klunzinger 1879	+	x	x	x	x				3
XI	incertae sedis	*Leptastrea*	*pruinosa*	Crossland 1952	+	x	x	x	x				
XI	incertae sedis	*Leptastrea*	*purpurea*	Dana 1846	+	x	x	x	x				
XI	incertae sedis	*Leptastrea*	*transversa*	Klunzinger 1879	+	x	x	x	x		x		
XI	Psammocoridae	*Psammocora*	*albopicta*	Benzoni 2006	+	x	x	x	x		x		3, 21
XI	Psammocoridae	*Psammocora*	*contigua*	Esper 1794	+	x		x	x		x		
XI	Psammocoridae	*Psammocora*	*profundacella*	Gardiner 1898	+	x		x	x		x		22
XI	Psammocoridae	*Psammocora*	*stellata*	Verrill 1866	+	x		x	x		x		3, 21, 23
XI	Psammocoridae	*Psammocora*	*nierstraszi*	Vaughan 1907	+	x	x	x	x				24
XIV	incertae sedis	*Blastomussa*	*loyae*	Head 1978	+	x	x	x	x				
XIV	incertae sedis	*Blastomussa*	*merleti*	Wells 1961	+	x		x	x		x		
XIV	incertae sedis	*Blastomussa*	*omanensis*	Sheppard and Sheppard 1991	+				x		x		
XIV	incertae sedis	*Blastomussa*	*wellsi*	Wijsman-Best 1973	+	x							
XIV	incertae sedis	*Physogyra*	*lichtensteini*	Milne Edwards and Haime 1851	+		x						

(continued)

Cl.	Family	Genus	Species	Taxonomic authority	zoox	North RS	Central RS	South RS	GoA and AS	Gulf of Oman	Socotra	Arabian Gulf	Endemism	Reference
XIV	incertae sedis	Plerogyra	sinuosa	Dana 1846	+	x	x	x			x			25
XIV	incertae sedis	Plesiastrea	versipora	Lamarck 1816	+	x	x	x	x	x	x	x		
XIV	Incertae sedis	Cyathelia	axillaris	(Ellis and Solander, 1786)	+						x			
XV	Diploastreidae	Diploastrea	heliopora	Lamarck 1816	+	x	x	x			x			
XVII	Merulinidae	Astrea	curta	Dana 1846	+	x	x	x			x			18
XVII	Merulinidae	Astrea	devantieri	Veron 2000	+						x			
XVII	Merulinidae	Caulastraea	connata	Ortmann 1892	+	x		x						
XVII	Merulinidae	Caulastraea	tumida	Matthai 1928	+	x		x			x			
XVII	Merulinidae	Coelastrea	aspera	Verrill 1866	+			x			x			
XVII	Merulinidae	Cyphastrea	chalcidicum	Forskal 1775	+	x	x	x			x			
XVII	Merulinidae	Cyphastrea	hexasepta	Veron, DeVantier and Turak 2000	+	x		x			x		E	
XVII	Merulinidae	Cyphastrea	microphthalma	Lamarck 1816	+	x	x	x	x	x	x	x		
XVII	Merulinidae	Cyphastrea	serailia	Forskal 1775	+	x	x	x	x	x	x	x		26
XVII	Merulinidae	Cyphastrea	kausti	Bouwmeester and Benzoni 2015	+	x	x	x					E	26
XVII	Merulinidae	Cyphastrea	magna	Benzoni and Arrigoni 2017	+	x	x	x					E	27
XVII	Merulinidae	Dipsastraea	albida	Veron 2000	+	x					x			
XVII	Merulinidae	Dipsastraea	amicorum	Milne Edwards and Haime 1849	+	x					x			
XVII	Merulinidae	Dipsastraea	danai	Milne Edwards 1857	+	x	x	x			x			
XVII	Merulinidae	Dipsastraea	favus	Forskal 1775	+	x	x	x	x	x	x	x		3
XVII	Merulinidae	Dipsastraea	helianthoides	Wells 1954	+	x	x							
XVII	Merulinidae	Dipsastraea	lacuna	Veron, Turak and DeVantier 2000	+	x					x			
XVII	Merulinidae	Dipsastraea	laxa	Klunzinger 1879	+	x	x	x						
XVII	Merulinidae	Dipsastraea	lizardensis	Veron, Pichon and Wijsman-Best 1977	+	x	x	x						
XVII	Merulinidae	Dipsastraea	maritima	Nemenzo 1971	+	x								
XVII	Merulinidae	Dipsastraea	matthai	Vaughan 1918	+	x			x	x	x			3
XVII	Merulinidae	Dipsastraea	maxima	Veron, Pichon and Wijsman-Best 1977	+	x			x	x	x			
XVII	Merulinidae	Dipsastraea	pallida	Dana 1846	+	x	x	x		x	x	x		
XVII	Merulinidae	Dipsastraea	rotumana	Gardiner 1899	+	x	x		x	x	x	x		3
XVII	Merulinidae	Dipsastraea	speciosa	Dana 1846	+	x	x		x	x	x	x		
XVII	Merulinidae	Dipsastraea	veroni	Moll and Best 1984	+					x	x			
XVII	Merulinidae	Dipsastraea	wisseli	Scheer and Pillai 1983	+		x				x		E	
XVII	Merulinidae	Echinopora	forskaliana	Milne Edwards and Haime 1849	+	x	x	x			x			28
XVII	Merulinidae	Echinopora	fruticulosa	Klunzinger 1879	+	x	x	x		x				
XVII	Merulinidae	Echinopora	gemmacea	Lamarck 1816	+	x	x	x		x	x			

7 Corals of the Red Sea

	Family	Genus	species	Author										
XVII	Merulinidae	Echinopora	hirsutissima	Milne Edwards and Haime 1849	+			x		x				
XVII	Merulinidae	Echinopora	irregularis	Veron, Turak and DeVantier 2000	+	x							E	
XVII	Merulinidae	Echinopora	lamellosa	Esper 1795	+	x	x	x		x				
XVII	Merulinidae	Echinopora	mammiformis	Nemenzo 1959	+	x							E	
XVII	Merulinidae	Echinopora	tiranensis	Veron, Turak and DeVantier 2000	+	x								
XVII	Merulinidae	Eryhrastrea	flabellata	Pichon, Scheer and Pillai 1983	+	x		x						
XVII	Merulinidae	Favites	abdita	Ellis and Solander 1786	+	x	x	x		x				
XVII	Merulinidae	Favites	acuticollis	Ortmann 1889	+	x	x				x			
XVII	Merulinidae	Favites	chinensis	Verrill 1866	+	x	x	x		x	x			
XVII	Merulinidae	Favites	complanata	Ehrenberg 1834	+	x	x	x		x	x			
XVII	Merulinidae	Favites	flexuosa	Dana 1846	+	x	x	x		x	x			
XVII	Merulinidae	Favites	halicora	Ehrenberg 1834	+	x	x	x		x				3
XVII	Merulinidae	Favites	magnistellata	Chevalier 1971	+		x			x				
XVII	Merulinidae	Favites	micropentagonus	Veron 2000	+					x				
XVII	Merulinidae	Favites	paraflexuosus	Veron 2000	+	x			x	x				
XVII	Merulinidae	Favites	pentagona	Esper 1795	+	x	x	x		x	x			
XVII	Merulinidae	Favites	rotundata	Veron, Pichon and Wijsman-Best 1977	+	x	x			x				
XVII	Merulinidae	Favites	spinosa	Klunzinger 1879	+		RS							29
XVII	Merulinidae	Favites	vasta	Klunzinger 1879	+	x				x				
XVII	Merulinidae	Goniastrea	columella	Crossland 1948	+	x		x		x				
XVII	Merulinidae	Goniastrea	edwardsi	Chevalier 1971	+	x	x		x					
XVII	Merulinidae	Goniastrea	pectinata	Ehrenberg 1834	+	x	x	x		x				
XVII	Merulinidae	Goniastrea	retiformis	Lamarck 1816	+	x	x	x	x	x				30
XVII	Merulinidae	Goniastrea	stelligera	Dana 1846	+	x	x	x		x				
XVII	Merulinidae	Goniastrea	thecata	Veron, DeVantier and Turak 2000	+	x	x	x		x				
XVII	Merulinidae	Hydnophora	exesa	Pallas 1766	+	x	x	x		x	x			
XVII	Merulinidae	Hydnophora	microconos	Lamarck 1816	+	x	x	x		x				
XVII	Merulinidae	Hydnophora	pilosa	Veron 1985	+	x		x		x	x			
XVII	Merulinidae	Leptoria	irregularis	Veron 1990	+									
XVII	Merulinidae	Leptoria	phrygia	Ellis and Solander 1786	+	x	x	x	x	x				
XVII	Merulinidae	Merulina	ampliata	Ellis and Solander 1786	+	x	x	x		x				
XVII	Merulinidae	Merulina	scheeri	Head 1983	+	x	x		x	x				
XVII	Merulinidae	Mycedium	elephantotus	Pallas 1766	+	x	x	x		x				
XVII	Merulinidae	Mycedium	umbra	Veron 2000	+	x	x						E	
XVII	Merulinidae	Oulophyllia	bennettae	Veron, Pichon and Best 1977	+		x	x		x				
XVII	Merulinidae	Oulophyllia	crispa	Lamarck 1816	+	x	x	x		x				
XVII	Merulinidae	Paragoniastrea	australensis	Milne Edwards 1857	+	x	x	x		x				

(continued)

Cl.	Family	Genus	Species	Taxonomic authority	zoox	North RS	Central RS	South RS	GoA and AS	Gulf of Oman	Socotra	Arabian Gulf	Endemism	Reference
XVII	Merulinidae	Paragoniastrea	deformis	Veron 1990	+						x			
XVII	Merulinidae	Paragoniastrea	russelli	Wells 1954	+	x		x		x	x			
XVII	Merulinidae	Paramontastraea	peresi	Faure and Pichon 1978	+	x	x	x	x		x			
XVII	Merulinidae	Pectinia	africana	Veron 2000	+			x			x			
XVII	Merulinidae	Platygyra	acuta	Veron 2000	+	x					x			
XVII	Merulinidae	Platygyra	contorta	Veron 1990	+						x			
XVII	Merulinidae	Platygyra	crosslandi	Matthai 1928	+	x	x	x		x	x			
XVII	Merulinidae	Platygyra	daedalea	Ellis and Solander 1786	+	x	x	x		x	x	x		
XVII	Merulinidae	Platygyra	lamellina	Ehrenberg 1834	+	x	x	x		x	x	x		
XVII	Merulinidae	Platygyra	pini	Chevalier	+	x								
XVII	Merulinidae	Platygyra	sinensis	Milne Edwards and Haime 1849	+		x	x		x	x	x		
XVII	Merulinidae	Trachyphyllia	geoffroyi	Audouin 1826	+	x	x	x	x					31
XVIII	Lobophylliidae	Acanthastrea	hillae	Wells 1955	+				x	x		x		32
XVIII	Lobophylliidae	Micromussa	indiana	Benzoni and Arrigoni 2016	+	x		x	x	x	x			33
XIX	Lobophylliidae	Cynarina	lacrymalis	Milne Edwards and Haime 1849	+	x	x	x		x	x			
XIX	Lobophylliidae	Echinophyllia	aspera	Ellis and Solander 1786	+	x	x	x	x	x	x	x		
XIX	Lobophylliidae	Echinophyllia	echinata	Saville-Kent 1871	+			x			x			
XIX	Lobophylliidae	Echinophyllia	orpheensis	Veron and Pichon 1980	+	x								34
XIX	Lobophylliidae	Echinophyllia	bulbosa	Arrigoni, Benzoni and Berumen 2016	+	x	x						E	
XIX	Lobophylliidae	Lobophyllia	corymbosa	Forskal 1775	+	x	x	x	x					3
XIX	Lobophylliidae	Lobophyllia	hattai	Yabe and Sugiyama 1936	+	x	x	x	x		x			
XIX	Lobophylliidae	Lobophyllia	hemprichii	Ehrenberg 1834	+	x	x	x	x		x			
XIX	Lobophylliidae	Lobophyllia	pachysepta	Chevalier 1975	+						x			
XIX	Lobophylliidae	Lobophyllia	robusta	Yabe and Sugiyama 1936	+	x		x			x			
XIX	Lobophylliidae	Oxypora	crassispinosa	Nemenzo 1980	+	x								
XIX	Lobophylliidae	Oxypora	glabra	Nemenzo 1959	+	x								
XIX	Lobophylliidae	Oxypora	lacera	Verrill 1864	+	x	x	x	x	x	x			3
XIX	Lobophylliidae	Parascolymia	vitiensis	Brueggemann 1877	+		x							
XIX	Lobophylliidae	Sclerophyllia	margariticola	Klunzinger 1879	+	x	x	x						36
XIX	Lobophylliidae	Sclerophyllia	maxima	Sheppard and Salm 1988	+				x	x	x	x	E	
XIX	Lobophylliidae	Symphyllia	agaricia	Milne Edwards and Haime 1849	+	x		x		x				
XIX	Lobophylliidae	Symphyllia	erythraea	Klunzinger 1879	+	x	x	x		x	x			37
XIX	Lobophylliidae	Symphyllia	cf hassi	Pillai and Scheer 1976	+						x			
XIX	Lobophylliidae	Symphyllia	radians	Edwards and Haime 1849	+			x	x	x	x	x		

7 Corals of the Red Sea

Order	Family	Genus	species	Author							Fig.
XIX	Lobophylliidae	Symphyllia	recta	Dana 1846	+			x			
XIX	Lobophylliidae	Symphyllia	valenciennesii	Milne Edwards and Haime 1849	+	x		x	x	x	38
XX	Lobophylliidae	Acanthastrea	brevis	Milne Edwards and Haime 1849	+	x			x		
XX	Lobophylliidae	Acanthastrea	echinata	Dana 1846	+	x	x	x	x	x	
XX	Lobophylliidae	Acanthastrea	faviaformis	Veron 2000	+	x			x		
XX	Lobophylliidae	Acanthastrea	hemprichii	Ehrenberg 1834	+	x		x	x		
XX	Lobophylliidae	Acanthastrea	ishigakiensis	Veron 1990	+				x		
XX	Lobophylliidae	Acanthastrea	lordhowensis	Veron and Pichon 1982		x			x		
XX	Lobophylliidae	Acanthastrea	rotundoflora	Chevalier 1975	+				x		
	Caryophylliidae	Anomocora	fecunda	Pourtales 1871		x					
	Caryophylliidae	Caryophyllia	paradoxus	Alcock 1898		x					
	Caryophylliidae	Caryophyllia	sewelli	Gardiner and Waugh 1938		x					
	Caryophylliidae	Dasmosmilia	valida	Marenzeller 1907		x				E	
	Caryophylliidae	Heterocyathus	aequicostatus	Milne Edwards and Haime 1848	+	x	x	x	x		39
	Caryophylliidae	Paracyathus	stokesii	Milne Edwards and Haime 1848			x				
	Caryophylliidae	Paracyathus	conceptus	Gardiner and Waugh 1938		x					
	Caryophylliidae	Polycyathus	fuscomarginatus	Klunzinger 1879		x		x			40
	Caryophylliidae	Solenosmilia	variabilis	Duncan 1873		x				x	
	Caryophylliidae	Tethocyathus	virgatus	Alcock 1902		x					
	Caryophylliidae	Trochocyathus	oahensis	Vaughan 1907		x					
	Flabellidae	Javania	insignis	Duncan 1876		x					
	Flabellidae	Rhizotrochus	typus	Milne Edwards and Haime 1848		x		x			
	Rhizangiidae	Culicia	cuticulata	Klunzinger 1879		x					
	Rhizangiidae	Culicia	rubeola	Quoy and Gaimard 1833		x		x	x	x	41
	Turbinoliidae	Deltocyathoides	orientalis	Duncan 1876		x					

[1]Specimen KAUST SA1732, Shib Farasan (16°45.286′ N; 41°29.552′ E);
[2]Specimen KAUST SA1547, Ghurob (17°06.620′ N; 42°04.052′ E);
[3]Record for the NW Gulf of Aden from Pichon et al. (2010);
[4]Record from Terraneo et al. (2016);
[5]Specimen KAUST SA333, Abulad Island (16°47.456′ N; 42°11.920′ E);
[6]Specimens KAUST SA725, Ras Al-Ubayd (26°44.167′ N; 36°02.659′ E), SA766, Jaz'air Silah (27°38.302′ N; 35°18.369′ E);
[7]Specimen KAUST SA0218 (18°13.028′ N; 41°29.716′ E);
[8]Specimens KAUST SA1807, SA1808, SA1809, Hindiyah (16°34.602′ N; 42°14.379′ E);
[9]Specimen SA1212, Shark reef (22°25.810′ N; 38°59.569′ E);
[10]Specimen KAUST SA1255, Fsar (22°13.779′ N; 39°01.730′ E);
[11]Specimens KAUST SA1262, Fsar (22°13.779′ N, 39°01.730′ E), SA399, Shib Nizar (22°19.500′ N, 38°51.000′ E), SA411, Al Fahal (22°18.427′ N, 38°57.842′ E);
[12]Specimens KAUST SA226 (18°11.350′ N, 41°06.658′ E), SA69 (18°39.571′N, 40°49.618′ E);
[13]Specimens KAUST SA1265, Fsar (22°13.779′ N, 39°01.730′ E), SA510, Shib Nizar (22°20.391′ N, 38°50.872′ E);

[14] Type locality is Sharm al-Sheikh, Sinai Peninsula, Egypt (Veron 2002);
[15] Specimen KAUST SA1179, Yanbu (24°26.561' N; 37°14.860' E);
[16] Record from Arrigoni et al. (2016a);
[17] Record for the NW GoA from Stefani et al. (2011);
[18] Record for Socotra from Benzoni et al. (2012a);
[19] Specimens KAUST SA1725, SA1726, SA1728, SA1730, Shib Farasan (16°45.286' N; 41°29.552' E);
[20] Records from Benzoni et al. (2012c);
[21] Record for the south Red Sea (Yemen) and Socotra in Benzoni et al. (2012b);
[22] Specimen KAUST SA767, Jaz'air Silah (27°38.302' N; 35°18.369' E);
[23] Specimen KAUST SA1456, Aminah (16°46.240' N; 42°27.875' E);
[24] Following Benzoni et al. (2010) *Psammocora verrilli* Vaughan, 1907 is considered a junior synonym of *P. nierstraszi*. Records for this species in DiBattista et al. (2016a) are revised here. Additional evidence for the presence of this species in the north and central Red Sea is provided by specimens KAUST SA0768, Jaz'air Silah (27°38.302' N; 35°18.369' E), and SA0392, Palace Reef (22°18.427' N; 38° 57.842' E), respectively. Specimen NHM 1991.6.4.65 collected from Yanbu and figured in Sheppard and Sheppard (1991, Fig. 67c) also belong to this species (Benzoni et al. 2010);
[25] Specimen from Socotra figured in Benzoni et al. (2011);
[26] Records from Arrigoni et al. (2017);
[27] Specimens KAUST SA1452, Aminah (16°46.240' N; 42°27.875 E), SA1519, Ghurob (17°06.620' N; 42°04.052' E);
[28] Specimen KAUST SA0391, Shi'b Nazar (22°28.199'N; 38°53.755' E);
[29] DiBattista et al. (2016a) provide no record for this species in the Red Sea which is, however, the region where the material described by Klunzinger (1879) comes from. The record is, therefore, added here;
[30] Record for the on Iranian islands in the Straits of Hormuz from Riegl et al. (2012);
[31] Specimen KAUST SA0328, Abulad Island (16°47.456' N; 42°11.920' E);
[32] Previously recorded in the region as *Micromussa amakusensis* Veron 1990 (Arrigoni et al. 2016b);
[33] Specimens KAUST SA0445, Qita al Kirsh (22°25.597' N; 38°59.769'E), SA0879, Jazirat Burcan (27°54.356' N; E35°03.555' E);
[34] Possibly a record of *E. aspera* (see discussion in Arrigoni et al. 2016c);
[35] Records from Arrigoni et al. (2016c);
[36] Specimens KAUST SA1805, SA1806, SA1817, Hindiyah (16°34.602' N; 42°14.379' E);
[37] Record from Mayotte Island in Arrigoni et al. (2014);
[38] Specimen KAUST SA1823, Hindiyah (16°34.602' N; 42°14.379' E);
[39] Specimen KAUST SA1851, Zara Durah (16°50.123' N; 42°18.386' E);
[40] Specimen KAUST SA1733, Shib Farasan (16°45.286' N; 41°29.552' E);
[41] Specimen KAUST SA0740, Ras Al-Ubayd (26°44.167' N; 36°02.659' E).

Appendix 2

In situ images of representatives of the scleractinian coral genera from the Red Sea.

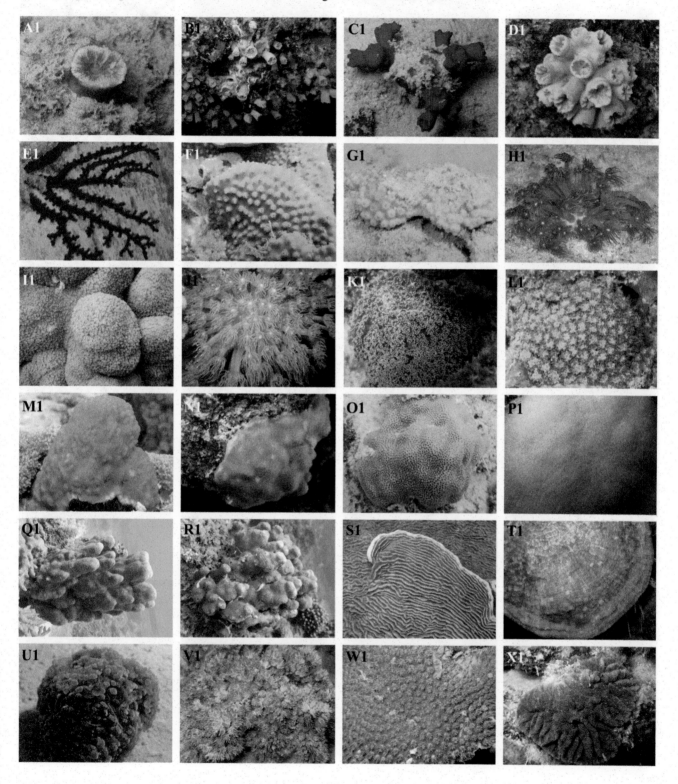

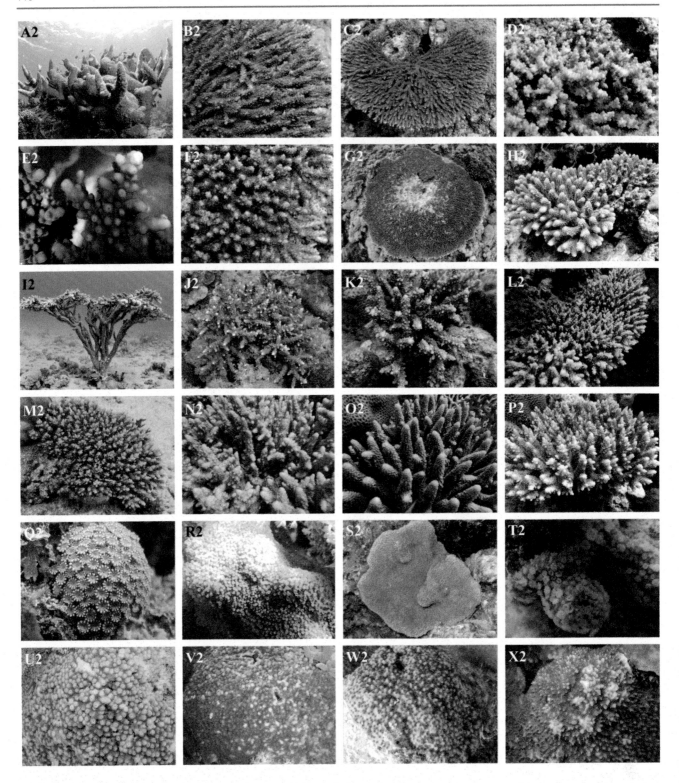

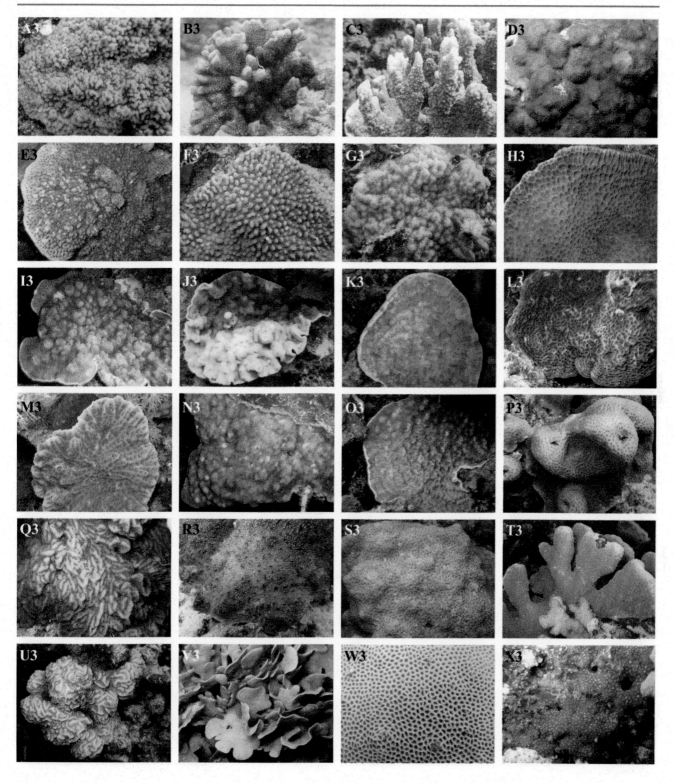

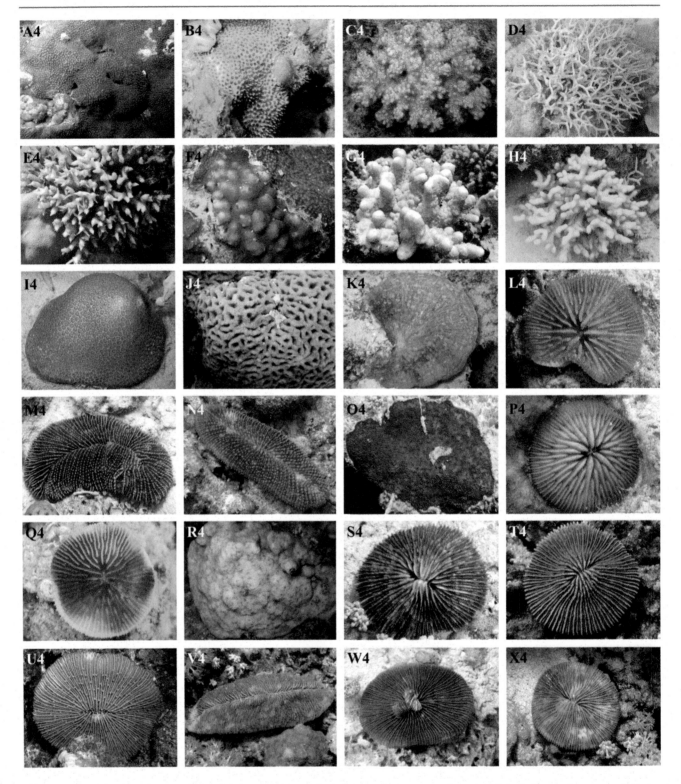

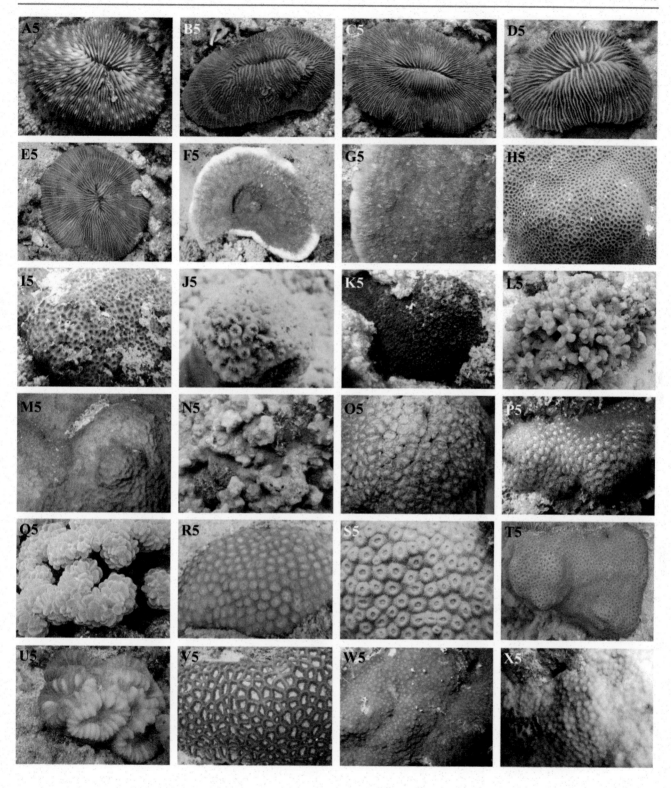

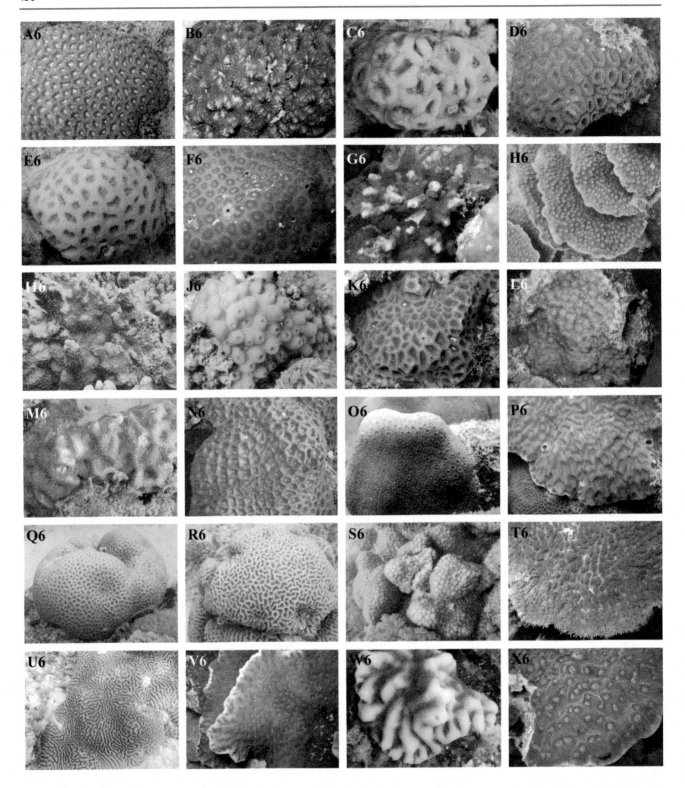

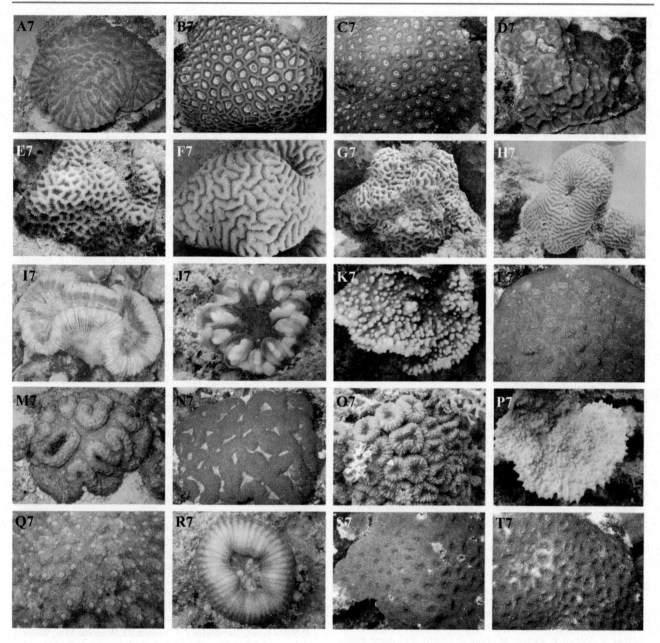

In situ images of representatives of the scleractinian coral genera from the Red Sea. (A1) *Balanophyllia*; (B1) *Rhizopsammia*; (C1-E1) *Tubastraea*; (F1-G1) *Turbinaria*; (H1-L1) *Goniopora*; (M1-R1) *Porites*; (S1-T1) *Pachyseris*; (U1) *Euphyllia*; (V1-W1) *Galaxea*; (X1) *Gyrosmilia*; (A2-P2) *Acropora*; (Q2-T2) *Alveopora*; (U2-V2) *Astreopora*; (W2-G3) *Montipora*; (H3) *Gardineroseris*; (I3-O3) *Leptoseris*; (P3-V3) *Pavona*; (W3) *Siderastrea*; (X3) *Madracis*; (A4-B4) *Stylocoeniella*; (C4) *Pocillopora*; (D4-E4) *Seriatopora*; (F4-H4) *Stylophora*; (I4) *Anomastraea*; (J4) *Coscinaraea*; (K4) *Craterastrea*; (L4) *Cantharellus*; (M4-N4) *Ctenactis*; (O4-R4) *Cycloseris*; (S4-T4) *Danafungia*; (U4) *Fungia*; (V4) *Herpolitha*; (W4-X4) *Lithophyllon*; (A5) *Lobactis*; (B5-E5) *Pleuractis*; (F5-G5) *Podabacia*; (H5-K5) *Leptastrea*; (L5-N5) *Psammocora*; (O5-P5) *Blastomussa*; (Q5) *Plerogyra*; (R5) *Plesiastrea*; (S5) *Diploastrea*; (T5) *Astrea*; (U5) *Caulastraea*; (V5) *Coelastrea*; (W5-X5) *Cyphastrea*; (A6-F6) *Dipsastraea*; (G6-J6) *Echinopora*; (K6-N6) *Favites*; (O6-R6) *Goniastrea*; (S6-T6) *Hydnophora*; (U6) *Leptoria*; (V6-W6) *Merulina*; (X6) *Mycedium*; (A7-B7) *Oulophyllia*; (C7) *Paragoniastrea*; (D7) *Paramontastraea*; (E7-H7) *Platygyra*; (I7) *Trachyphyllia*; (J7) *Cynarina*; (K7-L7) *Echinophyllia*; (M7-O7) *Lobophyllia*; (P7-Q7) *Oxypora*; (R7) *Sclerophyllia*; (S7-T7) *Acanthastrea*. Photos A1-K4, H5-T7 by F Benzoni, photos L4-G5 by BW Hoeksema.

References

Ahti PA, Coleman RR, DiBattista JD, Berumen ML, Rocha LA, Bowen BW (2016) Phylogeography of Indo-Pacific reef fishes: sister wrasses *Coris gaimard* and *C. cuvieri* in the Red Sea, Indian Ocean and Pacific Ocean. J Biogeogr 43:1103–1115

Arrigoni R, Stefani F, Pichon M, Galli P, Benzoni F (2012) Molecular phylogeny of the Robust clade (Faviidae, Mussidae, Merulinidae, and Pectiniidae): an Indian Ocean perspective. Mol Phylogenet Evol 65:183–193

Arrigoni R, Terraneo TI, Galli P, Benzoni F (2014) Lobophylliidae (Cnidaria, Scleractinia) reshuffled: pervasive non-monophyly at genus level. Mol Phylog Evol 73:60–74

Arrigoni R, Berumen ML, Terraneo TI, Caragnano A, Bouwmeester J, Benzoni F (2015) Forgotten in the taxonomic literature: resurrection of the scleractinian coral genus *Sclerophyllia* (Scleractinia, Lobophylliidae) from the Arabian Peninsula and its phylogenetic relationships. Syst Biodivers 13:140–163

Arrigoni R, Benzoni F, Terraneo TI, Caragnano A, Berumen ML (2016a) Recent origin and semi-permeable species boundaries in the scleractinian coral genus *Stylophora* from the Red Sea. Sci Rep 6:34612

Arrigoni R, Benzoni F, Huang D, Fukami H, Chen CA, Berumen ML, Hoogenboom M, Thomson DP, Hoeksema BW, Budd AF, Zayasu Y, Terraneo TI, Kitano YF, Baird AH (2016b) When forms meet genes: revision of the scleractinian genera *Micromussa* and *Homophyllia* (Lobophylliidae) with a description of two new species and one new genus. Contrib Zool 85:387–422

Arrigoni R, Berumen ML, Chen CA, Terraneo TI, Baird AH, Payri C, Benzoni F (2016c) Species delimitation in the reef coral genera *Echinophyllia* and *Oxypora* (Scleractinia, Lobophylliidae) with a description of two new species. Mol Phylogenet Evol 105:146–159

Arrigoni R, Berumen ML, Huang D, Terraneo TI, Benzoni F (2017) *Cyphastrea* (Cnidaria: Scleractinia: Merulinidae) in the Red Sea: phylogeny and a new reef coral species. Invertebr Syst 31:141–156

Arrigoni R, Berumen ML, Stolarski J, Terraneo TI, Benzoni F (2018) Uncovering hidden coral diversity: a new cryptic lobophylliid scleractinian from the Indian Ocean. Cladistics https://doi.org/10.1111/cla.12346

Benzoni F (2006) *Psammocora albopicta* sp. nov., a new species of Scleractinian Coral from the Indo-West Pacific (Scleractinia; Siderastreidae). Zootaxa 1358:49–57

Benzoni F, Stefani F (2012) *Porites fontanesii*, a new species of hard coral (Scleractinia, Poritidae) from the southern Red Sea, the Gulf of Tadjoura, and the Gulf of Aden, and its phylogenetic relationships within the genus. Zootaxa 3447:56–68

Benzoni F, Stefani F, Pichon M, Galli P (2010) The name game: morpho-molecular species boundaries in the genus *Psammocora* (Cnidaria, Scleractinia). Zool J Linnean Soc 160:421–456

Benzoni F, Arrigoni R, Stefani F, Pichon M (2011) Phylogeny of the coral genus *Plesiastrea* (Cnidaria, Scleractinia). Contr Zool 80:231–249

Benzoni F, Arrigoni R, Stefani F, Stolarski J (2012a) Systematics of the coral genus *Craterastrea* (Cnidaria, Anthozoa, Scleractinia) and description of a new family through combined morphological and molecular analyses. Syst Biodivers 10:417–433

Benzoni F, Pichon M, Dutrieux E, Chaîneau CH (2012b) The scleractinian fauna of Yemen: diversity and species distribution patterns. Proceedings of the 12th international coral reef symposium, Cairns, pp 11–16. http://eprints.jcu.edu.au/22394/

Benzoni F, Arrigoni R, Stefani F, Reijnen BT, Montano S, Hoeksema BW (2012c) Phylogenetic position and taxonomy of *Cycloseris explanulata* and *C. wellsi* (Scleractinia: Fungiidae): lost mushroom corals find their way home. Contrib Zool 81:125–146

Bosworth W, Huchon P, McClay K (2005) The Red Sea and Gulf of Aden Basins. J Afr Earth Sci 43:334–378

Bouwmeester J, Benzoni F, Baird AH, Berumen ML (2015) *Cyphastrea kausti* sp. n. (Cnidaria, Anthozoa, Scleractinia), a new species of reef coral from the Red Sea. ZooKeys 496:1–13

Bowen BW, Rocha LA, Toonen RJ, Karl SA (2013) The origins of tropical marine biodiversity. Trends Ecol Evol 8:359–366

Budd AF, Fukami H, Smith ND, Knowlton N (2012) Taxonomic classification of the reef coral family Mussidae (Cnidaria: Anthozoa: Scleractinia). Zool J Linnean Soc 166:465–529

Cantin NE, Cohen AL, Karnauskas KB, Tarrant AM, McCorkle DC (2010) Ocean warming slows coral growth in the central Red Sea. Science 329:322–325

Claereboudt MR (2006) Reef corals and coral reefs of the Gulf of Oman. Historical Society of Oman, Oman

Claereboudt MR, Al-Amri IS (2004) *Calathiscus tantillus*, a new genus and new species of scleractinian coral (Scleractinia, Poritidae) from the Gulf of Oman. Zootaxa 532:1–8

Coleman RR, Eble JA, DiBattista JD, Rocha LA, Randall JE, Berumen ML, Bowen BW (2016) Regal phylogeography: Range-wide survey of the marine angelfish *Pygoplites diacanthus* reveals evolutionary partitions between the Red Sea, Indian Ocean, and Pacific Ocean. Mol Phylogenet Evol 100:243–253

Davis KA, Lentz SJ, Pineda J, Farrar JT, Starczak VR, Churchill JH (2011) Observations of the thermal environment on Red Sea platform reefs: a heat budget analysis. Coral Reefs 30:25–36

DeVantier L, De'Ath G, Klaus R, Al-Moghrabi S, Abdulaziz M, Reinicke GB, Cheung C (2004) Reef-building corals and coral communities of the Socotra Archipelago, a zoogeographic crossroads' in the Arabian Sea. Fauna of Arabia 20:117–168

DeVantier L, Turak E, Al-Shaikh K, De'ath G (2010) Coral communities of the central-northern Saudi Arabian Red Sea. Fauna of Arabia 18:23–66

DiBattista JD, Rocha LA, Hobbs JP, He S, Priest MA, Sinclair-Taylor TH, Bowen BW, Berumen ML (2015a) When biogeographical provinces collide: hybridization of reef fishes at the crossroads of marine biogeographical provinces in the Arabian Sea. J Biogeogr 42:1601–1614

DiBattista JD, Waldrop E, Rocha LA, Craig MT, Berumen ML, Bowen BW (2015b) Blinded by the bright: a lack of congruence between colour morphs, phylogeography and taxonomy for a cosmopolitan Indo-Pacific butterflyfish, *Chaetodon auriga*. J Biogeogr 42:1919–1929

DiBattista JD, Roberts MB, Bouwmeester J, Bowen BW, Coker DJ, Lozano-Cortés DF, Choat JH, Gaither MR, Hobbs J-PA, Khalil MT, Kochzius M, Myers RF, Paulay G, Robitzch VSN, Saenz-Agudelo P, Salas E, Sinclair-Taylor TH, Toonen RJ, Westneat MW, Williams ST, Berumen ML (2016a) A review of contemporary patterns of endemism for shallow water reef fauna in the Red Sea. J Biogeogr 43:423–439

DiBattista JD, Howard Choat J, Gaither MR (2016b) On the origin of endemic species in the Red Sea. J Biogeogr 43:13–30

DiBattista JD, Gaither MR, Hobbs JP, Saenz-Agudelo P, Piatek MJ, Bowen BW, Rocha LA, Choat JH, McIlwain JH, Priest MA, Sinclair-Taylor TH, Berumen ML (2017) Comparative phylogeography of reef fishes from the Gulf of Aden to the Arabian Sea reveals two cryptic lineages. Coral Reefs 36:623–638

Ditlev H (2003) New scleractinian corals (Cnidaria: Anthozoa) from Sabah, North Borneo. Description of one new genus and eight new species, with notes on their taxonomy and ecology. Zoologische Mededelingen Leiden 77:193–219

Dreano D, Raitsos DE, Gittings J, Krokos G, Hoteit I (2016) The Gulf of Aden intermediate water intrusion regulates the southern Red Sea summer phytoplankton blooms. PLoS One 11:e0168840

Ellis J, Anlauf H, Kürten S, Lozano-Cortés D, Alsaffar Z, Cúrdia J, Jones B, Carvalho S (2017) Cross shelf benthic biodiversity patterns in the Southern Red Sea. Sci Rep 7:437

Fernandez-Silva I, Randall JE, Coleman RR, DiBattista JD, Rocha LA, Reimer JD, Meyer CG, Bowen BW (2015) Yellow tails in the Red Sea: phylogeography of the Indo-Pacific goatfish *Mulloidichthys flavolineatus* reveals isolation in peripheral provinces and cryptic evolutionary lineages. J Biogeogr 42:2402–2413

Flot JF, Blanchot J, Charpy L, Cruaud C, Licuana WY, Nakano Y, Payri C, Tillier S (2011) Incongruence between morphotypes and genetically delimited species in the coral genus *Stylophora*: phenotypic plasticity, morphological convergence, morphological stasis or interspecific hybridization? BMC Ecol 11:22

Furby KA, Bouwmeester J, Berumen ML (2013) Susceptibility of central Red Sea corals during a major bleaching event. Coral Reefs 32:505–513

Gates RD, Baghdasarian G, Muscatine L (1992) Temperature stress causes host cell detachment in symbiotic cnidarians: implications for coral bleaching. Biol Bull 182:324–332

Gélin P, Postaire B, Fauvelot C, Magalon H (2017) Reevaluating species number, distribution and endemism of the coral genus *Pocillopora* Lamarck, 1816 using species delimitation methods and microsatellites. Mol Phylogenet Evol 109:430–446

Giles EC, Saenz-Agudelo P, Hussey NE, Ravasi T, Berumen ML (2015) Exploring seascape genetics and kinship in the reef sponge *Stylissa carteri* in the Red Sea. Ecol Evol 5:2487–2502

Glynn PW (1993) Monsoonal upwelling and episodic *Acanthaster* predation as probable controls of coral reef distribution and community structure in Oman, Indian Ocean. Atoll Res Bull 379:1–66

Glynn PW, D'Croz L (1990) Experimental evidence for high temperature stress as the cause of El Nino-coincident coral mortality. Coral Reefs 8:181–191

Goreau TJ, Hayes RL (1994) Coral bleaching and ocean hot spots. Ambio 23:176–180

Goreau T, McClanahan T, Hayes R, Strong AL (2000) Conservation of coral reefs after the 1998 global bleaching event. Conser Biol 14:5–15

Head SM (1980) The ecology of corals in the Sudanese Red Sea. PhD thesis. University of Cambridge

Heron F, Maynard JA, van Hooidonk R, Eakin CM (2016) Warming trends and bleaching stress of the world's coral reefs 1985–2012. Sci Rep 6:38402

Hoeksema BW (2007) Delineation of the Indo-Malayan centre of maximum marine biodiversity: The coral triangle. In: Renema W (ed) Biogeography, time and place: Distributions, barriers and islands. Springer, Dordrecht, pp 117–178

Hoeksema B (2014) *Acropora* Oken, 1815. In: Fautin, Daphne G. 2013. Hexacorallians of the world. http://geoportal.kgs.ku.edu/hexacoral/anemone2/index.cfm

Howells EJ, Beltran VH, Larsen NW et al (2012) Coral thermal tolerance shaped by local adaptation of photosymbionts. Nat Clim Chang 2:116–120

Huang D (2012) Threatened reef corals of the world. PLoS One 7:e34459

Huang D, Benzoni F, Arrigoni R, Baird AH, Berumen ML, Bouwmeester J, Chou LM, Fukami H, Licuanan WY, Lovell ER, Meier R, Todd PA, Budd AF (2014a) Towards a phylogenetic classification of reef corals: the Indo-Pacific genera *Merulina*, *Goniastrea* and *Scapophyllia* (Scleractinia, Merulinidae). Zool Scripta 43:531–548

Huang D, Benzoni F, Fukami H, Knowlton N, Smith ND, Budd AF (2014b) Taxonomic classification of the reef coral families Merulinidae, Montastraeidae, and Diploastraeidae (Cnidaria: Anthozoa: Scleractinia). Zool J Linnean Soc 171:277–355

Huang D, Arrigoni R, Benzoni F, Fukami H, Knowlton N, Smith ND, Stolarski J, Chou LM, Budd AF (2016) Taxonomic classification of the reef coral family Lobophylliidae (Cnidaria: Anthozoa: Scleractinia). Zool J Linnean Soc 178:436–481

Hughes TP, Bellwood DR, Connolly SR (2002) Biodiversity hotspots, centres of endemicity, and the conservation of coral reefs. Ecol Lett 5:775–784

Hughes TP, Baird AH, Bellwood DR, Card M, Connolly SR, Folke C, Grosberg R, Hoegh-Guldberg O, Jackson JBC, Kleypas J, Lough JM, Marshall P, Nyström M, Palumbi SR, Pandolfi JM, Rosen B, Roughgarden J (2003) Climate change, human impacts, and the resilience of coral reefs. Science 301(5635):929–933

Hughes TP, Kerry JT, Álvarez-Noriega M, Álvarez-Romero JG, Anderson KD, Baird AH, Babcock RC, Beger M, Bellwood DR, Berkelmans R, Bridge TC, Butler IR, Byrne M, Cantin NE, Comeau S, Connolly SR, Cumming GS, Dalton SJ, Diaz-Pulido G, Eakin CM, Figueira WF, Gilmour JP, Harrison HB, Heron SF, Hoey AS, Hobbs J-PA, Hoogenboom MO, Kennedy EV, Kuo C-Y, Lough JM, Lowe RJ, Liu G, McCulloch MT, Malcom HA, McWilliam MJ, Pandolfi JM, Pears RJ, Pratchett MS, Schoepf V, Simpson T, Skirving WJ, Sommer B, Torda G, Wachenfeld DR, Willis BL, Wilson SK (2017) Global warming and recurrent mass bleaching of corals. Nature 543:373–377

Hughes TP, Anderson KD, Connolly SR, Heron SF, Kerry JT, Lough JM, Baird AH, Baum JK, Berumen ML, Bridge TC, Claar DC, Eakin CM, Gilmour JP, Graham NAJ, Harrison H, Hobbs J-PA, Hoey AS, Hoogenboom M, Lowe RJ, McCulloch MT, Pandolfi JM, Pratchett M, Schoepf V, Torda G, Wilson SK (2018) Spatial and temporal patterns of mass bleaching of corals in the Anthropocene. Science 359:80–83

Keith SA, Baird AH, Hughes TP, Madin JS, Connolly SR (2013) Faunal breaks and species composition of Indo-Pacific corals: the role of plate tectonics, environment and habitat distribution. Proc R Soc Biol Sci Ser B280:20130818

Keshavmurthy S, Yang SY, Alamaru A, Chuang Y-Y, Pichon M, Obura D, Fontana S, De Palmas S, Stefani F, Benzoni F, MacDonald A, Noreen AME, Chen C, Wallace CC, Pillay RM, Denis V, Amri AY, Reimer JD, Mezaki T, Sheppard C, Loya Y, Abelson A, Mohammed MS, Baker AC, Mostafavi PG, Suharsono BA, Chen CA (2013) DNA barcoding reveals the coral "laboratory-rat", *Stylophora pistillata* encompasses multiple identities. Sci Rep 3:1520

Khalil MT, Bouwmeester J, Berumen ML (2017) Spatial variation in coral reef fish and benthic communities in the central Saudi Arabian Red Sea. PeerJ 5:e3410

Kitahara MV, Cairns SD, Miller DJ (2010) Monophyletic origin of *Caryophyllia* (Scleractinia, Caryophylliidae), with descriptions of six new species. Syst Biodivers 8:91–118

Kitahara MV, Fukami H, Benzoni F, Huang D (2016) The new systematics of scleractinia: integrating molecular and morphological evidence. In: Goffredo S, Dubinsky Z (eds) The cnidaria, past, present and future: the world of Medusa and her sisters. Springer, Heidelberg

Kitano YF, Benzoni F, Arrigoni R, Shirayama Y, Wallace CC, Fukami H (2014) A phylogeny of the family Poritidae (Cnidaria, Scleractinia) based on molecular and morphological analyses. PLoS One 9:e98406

Klunzinger CB (1879) Die Korallenthiere des Rothen Meeres, 3. Theil: Die Steinkorallen. Zweiter Abschnitt: Die Asteraeaceen und Fungiaceen. 100pp.

Krueger T, Horwitz N, Bodin J, Giovani M-E, Escrig S, Meibom A, Fine M (2017) Common reef-building coral in the Northern Red Sea resistant to elevated temperature and acidification. R Soc Open Sci 4:170038

Ladner JT, Palumbi SR (2012) Extensive sympatry, cryptic diversity and introgression throughout the geographic distribution of two coral species complexes. Mol Ecol 21:2224–2238

Lozano-Cortes D, Robitzch V, Abdulkader K, Kattan Y, Elyas A, Berumen M (2016) Coral bleaching report-Saudi Arabia (The Red Sea and The Arabian Gulf). Reef Encounter 31:50–52

Monroe AA, Ziegler M, Roik A, Röthig T, Hardenstine RS, Emms MA, Jensen T, Voolstra CR, Berumen ML (2018) *In situ* observations of coral bleaching in the central Saudi Arabian Red Sea during the 2015/2016 global coral bleaching event. PLoS One 13:e0195814

Morice CP, Kennedy JJ, Rayner NA, Jones PD (2012) Quantifying uncertainties in global and regional temperature change using an ensemble of observational estimates: the HadCRUT4 data set. J Geophys Res 117:D08101

Nanninga GB, Saenz-Agudelo P, Manica A, Berumen ML (2014) Environmental gradients predict the genetic population structure of a coral reef fish in the Red Sea. Mol Ecol 23:591–602

New AL, Stansfield K, Smythe-Wright D, Smeed DA, Evans AJ, Alderson SG (2004) Physical and biochemical aspects of the flow across the Mascarene plateau in the Indian Ocean. Phil Trans Royal Soc (A) 363:151–168

Ngugi DK, Antunes A, Brune A, Stingl U (2012) Biogeography of pelagic bacterioplankton across an antagonistic temperature–salinity gradient in the Red Sea. Mol Ecol 21:388–405

Obura D (2012) The diversity and biogeography of Western Indian Ocean reef-building corals. PLoS One 7:e45013

Obura D (2016) An Indian Ocean centre of origin revisited: Palaeogene and Neogene influences defining a biogeographic realm. J Biogeogr 43:229–242

Ormond RFG, Dawson Shepherd AR, Price ARG, Pitts JR (1984) Management of Red Sea coastal resources: recommendations for marine protected areas. IUCN/MEPA, Kingdom of Saudi Arabia

Osborne K, Thompson AA, Cheal AJ, Emslie MJ, Johns KA, Jonker MJ, Logan M, Miller IR, Sweatman HPA (2017) Delayed coral recovery in a warming ocean. Glob Chang Biol 23:3869–3881

Osman EO, Smith DJ, Zielger M, Kürten B, Conrad C, El-Haddad KM, Voolstra CR (2018) Thermal regufia against coral bleaching throughout the northern Red Sea. Glob Change Biol 24:e474–e484

Pichon M, Benzoni F, Chaineu CH, Dutrieux E (2010) Field guide to the hard corals of the southern coast of Yemen. BIOTOPE Parthenope, Paris

Pinzón JH, Sampayo E, Cox E, Chauka LJ, Chen CA, Voolstra CR, LaJeunesse TC (2013) Blind to morphology: genetics identifies several widespread ecologically common species and few endemics among Indo-Pacific cauliflower corals (*Pocillopora*, Scleractinia). J Biogeogr 40:1595–1608

Priest MA, DiBattista JD, McIlwain JL, Taylor BM, Hussey NE, Berumen ML (2016) A bridge too far: dispersal barriers and cryptic speciation in an Arabian Peninsula grouper (*Cephalopholis hemistiktos*). J Biogeogr 43:820–832

Raitsos DE, Pradhan Y, Brewin RJW, Stenchikov G, Hoteit I (2013) Remote sensing the phytoplankton seasonal succession of the Red Sea. PLoS One 8:e64909

Raitsos DE, Yi X, Platt T, Racault M-F, Brewin RJW, Pradhan Y, Papadopoulos VP, Sathyendranath S, Hoteit I (2015) Monsoon oscillations regulate fertility of the Red Sea. Geophys Res Lett 42:855–862

ReFuGe 2020 Consortium (2015) The ReFuGe 2020 consortium–using "omics" approaches to explore the adaptability and resilience of coral holobionts to environmental change. Front Mar Sci 2: 68

Richards ZT, Berry O, van Oppen MJ (2016) Cryptic genetic divergence within threatened species of *Acropora* coral from the Indian and Pacific Oceans. Conserv Genet 17:577–591

Riegl B, Benzoni F, Sheppard CRC, Samimi-Namin K (2012) The hermatypic scleractinian (hard) coral fauna of the Gulf. In: Riegl B, Purkins S (eds) Coral Reefs of the Gulf: Adaptation to Climatic Extremes, Coral Reefs of the World 3. Dordrecht, Springer, pp 187–224

Roberts MB, Jones GP, McCormick MI, Munday PL, Neale S, Thorrold S, Robitzch VSN, Berumen ML (2016) Homogeneity of coral reef communities across 8 degrees of latitude in the Saudi Arabian Red Sea. Mar Poll Bull 105:558–565

Rocha LA, Aleixo A, Allen G, Almeda F, Baldwin CC, Barclay MVL, Bates JM, Bauer AM, Benzoni F, Berns CM, Berumen ML, Blackburn DC, Blum S, Bolaños F, Bowie RCK, Britz R, Brown RM, Cadena CD, Carpenter K, Ceríaco LM, Chakrabarty P, Chaves G, Choat JH, Clement KD, Collette BB, Collins A, Coyne J, Cracraft J, Daniel T, de Carvalho MR, Queiroz K, Di Dario F, Drewes R, Dumbacher JP, Engilis Jr A, Erdmann MV, Eschmeyer W, Feldman CR, Fisher BL, Fjeldså J, Fritsch PW, Fuchs J, Getahun A, Gill A, Gomon M, Gosliner T, Graves GR, Griswold CE, Guralnick R, Hartel K, Helgen KM, Ho H, Iskandar DT, Iwamoto T, Jaafar Z, James HF, Johnson D, Kavanaugh D, Knowlton N, Lacey E, Larson HK, Last P, Leis JM, Lessios H, Liebherr J, Lowman M, Mahler DL, Mamonekene V, Matsuura K, Mayer GC, Mays Jr H, McCosker J, McDiarmid RW, McGuire J, Miller MJ, Mooi R, Mooi RD, Moritz C, Myers P, Nachman MW, Nussbaum RA, Foighil DÓ, Parenti LR, Parham JF, Paul E, Paulay G, Pérez-Emán J, Pérez-Matus A, Poe S, Pogonoski J, Rabosky DL, Randall JE, Reimer JD, Robertson DR, Rödel M-O, Rodrigues MT, Roopnarine P, Rüber L, Ryan MJ, Sheldon F, Shinohara G, Short A, Simison WB, Smith-Vaniz WF, Springer VG, Stiassny M, Tello JG, Thompson CW, Trnski T, Tucker P, Valqui T, Vecchione M, Verheyen E, Wainwright PC, Wheeler TA, White WT, Will K, Williams JT, Williams G, Wilson EO, Winker K, Winterbottom R, Witt CC (2014) Specimen collection: an essential tool. Science 344:814–815

Roik A, Röthig T, Roder C, Ziegler M, Kremb SG, Voolstra CR (2016) Year-long monitoring of physico-chemical and biological variables provide a comparative baseline of coral reef functioning in the central Red Sea. PLoS One 11:e0163939

Scheer G, Pillai CG (1983) Report on the stony corals from the Red Sea. Zoologica 45:1–198

Schmidt-Roach S, Miller KJ, Lundgren P, Andreakis N (2014) With eyes wide open: a revision of species within and closely related to the *Pocillopora damicornis* species complex (Scleractinia; Pocilloporidae) using morphology and genetics. Zool J Linnean Soc 170:1–33

Sheppard CRC, Sheppard ALS (1991) Corals and coral communities of Saudi Arabia. Fauna Saudi Arabia 12:1–170

Sheppard CRC, Price A, Roberts C (1992) Marine ecology of the Arabian region: patterns and processes in extreme tropical environments. Academic, London

Sheppard CRC, Wilson SC, Salm RV, Dixon D (2000) Reefs and coral communities of the Arabian Gulf and Arabian Sea. In: McClanahan TR, Sheppard CRC, Obura DO (eds) Coral reefs of the Indian Ocean: their ecology and conservation. Oxford University Press, Oxford, pp 259–293

Spalding MD, Fox HE, Allen GR, Davidson N, Ferdaña ZA, Finlayson M, Halpern BS, Jorge MA, Lombana A, Lourie SA, Martin KD, McManus E, Molnar J, Recchia C, Robertson J (2007) Marine ecoregions of the world: a bioregionalization of coastal and shelf areas. Bioscience 57:573–583

Stefani F, Benzoni F, Sung Yin Y, Pichon M, Galli P, Chen CA (2011) Comparison of morphological and genetic analyses reveal cryptic divergence and morphological plasticity in *Stylophora* (Cnidaria, Scleractinia). Coral Reefs 30:1033–1049

Sweatman H, Delean S, Syms C (2011) Assessing loss of coral cover in Australia's Great Barrier Reef over two decades, with implications for longer-term trends. Coral Reefs 30:521–531

Tanaka Y, Inoue M, Nakamura T, Suzuki A, Sakai K (2014) Loss of zooxanthellae in a coral under high seawater temperature and nutrient enrichment. J Exo Mar Biol Ecol 457:220–225

Terraneo TI, Berumen ML, Arrigoni R, Waheed Z, Bouwmeester J, Caragnano A, Stefani F, Benzoni F (2014) *Pachyseris inattesa* sp. n. (Cnidaria, Anthozoa, Scleractinia): a new reef coral species from the Red Sea and its phylogenetic relationships. ZooKeys 433:1–30

Terraneo TI, Benzoni F, Arrigoni R, Berumen ML (2016) Species delimitation in the coral genus *Goniopora* (Scleractinia, Poritidae) from the Saudi Arabian Red Sea. Mol Phylogenet Evol 102:278–294

Terraneo TI, Arrigoni R, Benzoni F, Tietbohl MD, Berumen ML (2017) Exploring genetic diversity of shallow-water Agariciidae (Cnidaria: Anthozoa) from the Saudi Arabian Red Sea. Mar Biodivers 47:1065–1078

Todd PA (2008) Morphological plasticity in scleractinian corals. Biol Rev 83:315–337

Turak E, Brodie J, DeVantier L (2007) Reef-building corals and coral communities of the Yemen Red Sea. Fauna of Arabia 23:1–40

Van Oppen MJH, Willis BL, Van Vugt HWJA, Miller DJ (2000) Examination of species boundaries in the *Acropora cervicornis* group (Scleractinia, Cnidaria) using nuclear DNA sequence analyses. Mol Ecol 9:1363–1373

van Woesik R, Sakai K, Ganase A, Loya Y (2011) Revisiting the winners and the losers a decade after coral bleaching. Mar Ecol Prog Ser 434:67–76

Veron JEN (1993) A biogeographic database of hermatypic corals: species of the Central Indo-Pacific genera of the world. Australian Institute of Marine Science, Townsville

Veron JEN (1995) Corals in space and time, the biogeography and evolution of the Scleractinia. Australian Institute of Marine Science, Townsville

Veron JEN (2000) Corals of the world. Australian Institute of Marine Science, Townsville

Veron JEN (2002) New species described in *Corals of the World*. Australian Institute of Marine Science, Townsville

Veron JEN, Devantier LM, Turak E, Green A, Kininmonth S, Stafford-Smith M, Peterson N (2009) Delineating the coral triangle. Galaxea 11:91–100

Veron J, Stafford-Smith M, DeVantier L, Turak E (2015) Overview of distribution patterns of zooxanthellate Scleractinia. Front Mar Sci 1:81

Waldrop E, Hobbs JP, Randall JE, DiBattista JD, Rocha LA, Kosaki RK, Berumen ML, Bowen BW (2016) Phylogeography, population structure and evolution of coral-eating butterflyfishes (Family Chaetodontidae, genus *Chaetodon*, subgenus *Corallochaetodon*). J Biogeogr 43:1116–1129

Wallace CC (1999) Staghorn corals of the world: a revision of the genus *Acropora*. CSIRO publishing, Victoria

Wallace CC, Turak E, DeVantier L (2011) Novel characters in a conservative coral genus: three new species of *Astreopora* (Scleractinia: Acroporidae) from West Papua. J Nat Hist 45:1905–1924

Wallace CC, Done BJ, Muir PR (2012) Revision and catalogue of worldwide staghorn corals *Acropora* and *Isopora* (Scleractina: Acroporidae) in the Museum of Tropical Queensland. Queensland Museum, Australia

Wheeler QD (2004) Taxonomic triage and the poverty of phylogeny. Phil Trans Roy Soc Lond B 359:571–583

Wolstenholme JK, Wallace CC, Chen CA (2003) Species boundaries within the *Acropora humilis* species group (Cnidaria; Scleractinia): a morphological and molecular interpretation of evolution. Coral Reefs 22:155–166

Wooldridge SA, Heron SF, Brodie JE, Done TJ, Masiri I, Hinrichs S (2016) Excess seawater nutrients, enlarged algal symbiont densities and bleaching sensitive reef locations: 1. Identifying thresholds of concern for the Great Barrier Reef, Australia. Mar Pollut Bull 114:343–354

Fishes and Connectivity of Red Sea Coral Reefs

Michael L. Berumen, May B. Roberts, Tane H. Sinclair-Taylor, Joseph D. DiBattista, Pablo Saenz-Agudelo, Stamatina Isari, Song He, Maha T. Khalil, Royale S. Hardenstine, Matthew D. Tietbohl, Mark A. Priest, Alexander Kattan, and Darren J. Coker

Abstract

The coral reefs of the Red Sea are host to a diverse fish fauna. Ichthyofauna studies began in the Red Sea during expeditions undertaken by some of the earliest European naturalists. In the more than 200 years that have passed, much has been learned about Red Sea fishes. Nonetheless, many knowledge gaps remain. Although it is a relatively young sea, the geologic history of the Red Sea provides an interesting context for many evolutionary biology studies. The strong environmental gradients within the Red Sea and the broader Arabian region may play a role in structuring some observed biodiversity patterns, perhaps most notably in the context of high numbers of Arabian and Red Sea endemics. As such, Red Sea fishes provide ideal opportunities for connectivity studies, both based on adult movement and larval dispersal patterns. These studies are increasingly important as multiple modern "mega-developments" are planned on Red Sea shores in locations where a lack of scientific information may still hinder conservation efforts and planning for sustainable development. Coupled with increasing pressures from global climate change, each of the Red Sea countries faces unique challenges for the preservation of the rich biological resources for which their reefs are historically known.

Keywords

Biodiversity · Connectivity · Conservation · Endemism · Fisheries · Ichthyofauna · Movement ecology

8.1 Red Sea Ichthyofauna and Movement Ecology

8.1.1 Early Natural Historians and Red Sea Taxonomy

The ichthyofauna of the Red Sea attracted the attention of some of the earliest naturalist historians; several of them spent a great deal of time in the Red Sea, or at least working with material collected from the Red Sea. Peter Forsskål, a Swedish explorer and naturalist, may have the most unfortunate story, dying near the end of a 7-year journey to what is now called Yemen, but not before sending many preserved specimens back to his mentor, Carl Linnaeus (Hansen 1962). Several fishes bear scientific epithets honoring these naturalists, including *Parapeneus forsskali*, *Thalassoma rueppellii* (named for the German Eduard Rüppell, one of the first European naturalists to reach the Gulf of Aqaba), and *Lutjanus ehrenbergi* (named after Christian Gottfried Ehrenberg, another German naturalist / explorer among the earliest Europeans to study northern Red Sea fauna) (Fig. 8.1). Notably, many of the species were given scientific names derived from local Arabic names for the fishes, such as *Acanthurus gahhm*, *Acanthurus sohal*, *Hipposcarus harid*, *Carangoides bajad*, *Pomacanthus asfur*, *Lethrinus harak*, *Lutjanus bohar*, and the genus *Abudefduf* (Fig. 8.2).

As the Red Sea is home to more than 1000 species of fishes (DiBattista et al. 2016b), it is a daunting task to create a field or pictorial guide for the taxonomic diversity of the Red Sea. This chapter is not intended to serve as a field guide, particularly as good examples already exist. Lieske and Myers (2004) offer a very good treatment for most conspicuous reef fishes, including many species from the Gulf of Aden. Users should take care to note that not all species included in this book are found in the Red Sea (i.e., it includes species found in the Gulf of Aden or other parts of Arabia but not found in the Red Sea). Perhaps the most comprehensive and most recent checklist is that provided by Golani and Bogorodsky (2010). Instead of attempting to provide a field guide or a checklist, this chapter instead seeks to review the

Fig. 8.1 A selection of some Red Sea reef fishes given scientific names honoring early European natural historians who explored and cataloged Red Sea ichthyofauna: (**a**) *Thalassoma rueppellii*, (**b**) *Parapeneus forsskali*, and (**c**) *Lutjanus ehrenbergii*

Fig. 8.2 A selection of Red Sea fishes bearing scientific names derived from Arabic: (**a**) the genus *Abudefduf*, represented here by *Abudefduf vaigiensis*, (**b**) *Pomacanthus asfur*, with the specific epithet named after the Arabic word for "yellow", (**c**) *Carangoides bajad*, taking a specific epithet named after the Arabic word used for most trevallies, (**d**) *Lutjanus bohar*, (**e**) *Acanthurus sohal*, bearing a specific epithet derived from the Arabic word used for most *Acanthurus* surgeonfishes, (**f**) *Acanthurus gahhm*, (**g**) *Lethrinus harak*, and (**h**) *Hipposcarus harid*, with a specific epithet named after the Arabic word used for most parrotfishes

state of knowledge of Red Sea coral reef fish work, particularly with respect to recent work conducted outside the Gulf of Aqaba (where scientific knowledge has traditionally been more developed than in the main body of the Red Sea (Berumen et al. 2013)).

8.1.2 Fishes and Movement Ecology

Fishes provide many ideal model systems for investigations in the broad domain of movement ecology. For the study of basic biogeographic patterns, the state of knowledge in many other organisms is not yet sufficient even to describe basic distribution patterns. Nonetheless, a review by DiBattista et al. (2016b) assembled information from the few Red Sea groups for which sufficient checklists were available. Recent efforts to understand Red Sea fishes in a broader context were captured in a special issue of the Journal of Biogeography highlighting numerous studies (not exclusively of fishes) from the Red Sea and western Indian Ocean (Berumen et al. 2017). However, the taxonomy of Red Sea fishes is far from perfect. In fact, detailed studies of less-conspicuous groups (e.g., blennies and gobies) are very few; unsurprisingly, the few works to delve into these groups indicate that the Red Sea ichthyofauna may yet hold much more diversity (more on this in Sect. 8.3.3).

On the subject of Red Sea evolutionary biology, fishes again provide one of the most useful study systems. DiBattista et al. (2016a) provides perspective on the potential origins of Red Sea fauna, and particularly the potential reasons that high levels of endemism emerged in the region. The Red Sea's unique conditions (see Chap. 1) create an important opportunity to investigate adaptation mechanisms to climate change; modern-day conditions in the Red Sea may reflect future scenarios in other oceans, and Red Sea fauna may therefore provide insights (particularly genetic) to the adaptive capacity of reef fauna elsewhere (ReFuGe 2020 Consortium 2015).

In general reef ecology, fishes are also frequent model organisms. When considering the movement ecology of fishes, temporal and spatial scales are important. For many species of reef fishes, the largest distances that individuals will move are realized during the larval phase (Green et al. 2015). Unfortunately, acquiring empirical measurements of the movement patterns of larval fishes poses major practical challenges, primarily due to their small size, the quantities typically produced, and the naturally high mortality rates larvae experience during their pelagic dispersal phase (Thorrold et al. 2007). The movements of many adult fishes can be studied using a variety of techniques and off-the-shelf equipment, although these are typically time-intensive and expensive endeavors. In the Red Sea, there are examples of ecological studies at most scales, although the work may only have taken place with a limited number of species or in a limited number of places. This chapter will touch on various ecological aspects of Red Sea fishes in three broad areas (biodiversity patterns, genetic connectivity, and ecological work) and will conclude with comments on conservation and associated challenges in the region.

8.2 Biodiversity Patterns

8.2.1 Latitudinal/Longitudinal Gradients

Despite the Red Sea's strong environmental gradients (see Chap. 1) and a long history of research on fishes in the Red Sea, there are few publications examining fish assemblages from a latitudinal perspective. While fish community composition does seem to gradually change along most gradients of the Red Sea, there is likely more difference between reefs across the continental shelf (Khalil et al. 2017) than observed along latitudinal gradients, which is a well-established pattern seen in other reef systems (e.g., Aguilar-Perera and Appeldoorn 2008; Malcolm et al. 2010).

Surveys covering conspicuous fish species on offshore reefs from Al Wajh (26.8°N latitude) to the southern Farasan Banks (18.6°N latitude) (see Fig. 8.3) suggest that overall fish community assemblages do not differ greatly among reefs at the edge of the continental shelf across this span (Roberts et al. 2016). A slight shift in community composition in the central-northern portion of the Red Sea was attributed, in part, to the influence of few taxa with narrow range limits and with relatively low abundances. The butterflyfishes (Chaetodontidae) and angelfishes (Pomacanthidae) are good examples of groups with species following this pattern. Surveys of inshore reef crests from the Gulf of Aqaba (29°N latitude) to the Gulf of Aden (12°N latitude) revealed a shift in these taxa in the central Red Sea (around 20°N latitude) (Roberts et al. 1992). Two species of butterflyfish, *Chaetodon paucifasciatus* and *Chaetodon austriacus,* were present only on central and northern reefs while *Chaetodon trifasciatus, Chaetodon melannotus, Chaetodon fasciatus, Chaetodon auriga,* and *Pygoplytes diacanthus* all showed marked decreases in abundance towards the south. Other species, including *Chaetodon mesoleucos, Chaetodon larvatus, Pomacanthus asfur,* and *Pomacanthus maculosus* showed the opposite trend.

These patterns may have influenced the demarcation of the Red Sea into two Marine Ecoregions of the World by Spalding et al. (2007), splitting the Red Sea roughly in half at ~20°N, although subsequent community analyses suggest the appropriate division may be closer to 17°N. A comparison of coastal coral reef communities (including corals, benthic invertebrates, and fishes) found that sites between the Gulf of Aqaba and the cen-

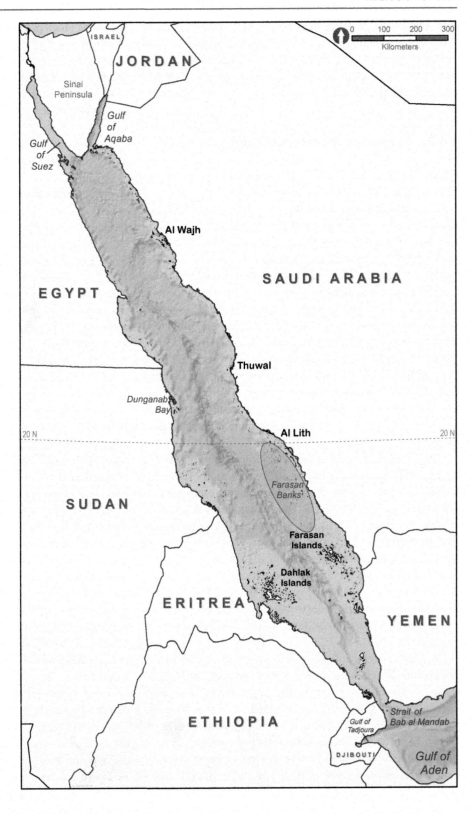

Fig. 8.3 Map of the Red Sea highlighting key features referenced in this chapter. Aquatic features are indicated in blue text; terrestrial features are indicated in black text. The circle drawn around the Farasan Banks indicates the approximate location of the extensive network of more than 100 reefs spread through this area. (Map data sources are ESRI and M. Campbell)

tral Red Sea were relatively uniform, while Farasan Island communities were distinctly separate (Sawall et al. 2014). These community differences were attributed to a greater abundance of predators and herbivores and lower abundance of small planktivorous fishes in the lower latitudes. The shallow, turbid, and patchy reef structure of the reefs in the Farasan Islands area likely supports a distinctly different assemblage of fishes than the more uniform reefs found in the central and northern Red Sea (Roberts et al. 1992).

A recent study comparing cryptobenthic fishes between the central Red Sea and the Farasan Islands found marked differences in fish abundance and species richness driven by habitat characteristics and productivity (assessed using chlorophyll a values) (Coker et al. 2018). The widening of the continental shelf in the southern part of the Red Sea results in expansive shallow patchy reef systems across the shelf, similar to habitats found on inshore and midshelf reefs of the central Red Sea. Coupled with the influence of Indian Ocean water influx through the Strait of Bab al Mandab, these conditions make the Farasan Islands a distinctly different habitat among Red Sea regions.

A broader analysis by Khalaf & Kochzius (2002), including detailed surveys in the Gulf of Aqaba, supported the suggestion that there are gradual latitudinal shifts in reef fish assemblage from north to south, identifying the clear difference between the Red Sea and the Gulf of Aden / southern Arabian regions. However, Roberts et al. (2016) suggest that the latitudinal shifts become less well-defined with increased distance from shore, possibly indicating that the factors structuring fish assemblage (e.g., habitat variables) have greater change from inshore to offshore sites than they do from north to south (at a given distance from shore). Patterns of prevalent cross-shelf effects have been found in other reef systems (Aguilar-Perera and Appeldoorn 2008; Malcolm et al. 2010). This is also seen in Red Sea reefs, characterized by an increase in herbivore and planktivorous fish diversity in the offshore reefs compared to inshore reefs (Khalil et al. 2017). Coker et al. (2018) compared cryptobenthic fish assemblages across inshore, midshelf, and offshore reefs. The authors found that differences in fish assemblages were driven by proximity to shore, likely due to the change in habitat quality along this gradient. Mechanisms driving the fish assemblage changes are likely associated with distance from shore (Khalil et al. 2017; Coker et al. 2018).

From these few studies, we can say that fish assemblages are not distinctly different from the Gulf of Aqaba to the central Red Sea (Khalaf and Kochzius 1992; Sawall et al. 2014; Roberts et al. 2016). There is indication that the Farasan Islands support an assemblage most different to the rest of the Red Sea, though more investigation is necessary. The well-established pattern of more pronounced differences in assemblages across reefs longitudinally than across latitudes and as seen in reefs such as the Great Barrier Reef, also hold true in the Red Sea thus far.

8.2.2 Understudied Regions of the Red Sea

Many parts of the Red Sea remain poorly studied. For example, the southernmost reaches of the Red Sea contain perhaps the most unique reef habitats (see Chap. 1), but these are among the least-represented among Red Sea reef fish publications. This includes the Farasan Islands in the southern Saudi Arabian Red Sea (and extending into Yemeni waters) and the Dahlak Archipelago in the Eritrean Red Sea. Combined, these two coastal and offshore systems contain more than 200 islands and host a variety of marine biota. Many of the islands are fringed with shallow reefs. Some of the islands, particularly to the far west of the Farasan Islands, have well-developed coral reefs. Multiple groups have conducted surveys of the reef habitats in this region, including the Living Oceans Foundation (Bruckner et al. 2011), and have arrived at the conclusion that the reef communities are unique among Saudi Arabian reef systems (e.g., Sheppard and Sheppard 1991; Sheppard et al. 1992). The reefs are subject to far more sedimentation than most other Saudi Red Sea reefs, the water is consistently more turbid, and remote sensing data indicates very high productivity in this region (Raitsos et al. 2013; Racault et al. 2015). Some of the reefs in this area are largely dominated by macroalgae. In these respects, the Farasan Islands region has greater affinities with the Gulf of Aden region reefs than with the remainder of Red Sea reefs (Sheppard and Sheppard 1991). In some ways, the southern Red Sea islands may functionally be more like inshore, coastal reefs, even though they are relatively distant (>100 km) from the mainland coast.

The southern Red Sea also hosts the largest area of shallow soft-bottom habitats in Saudi Arabia, and is home to some of the only major trawling operations in the country. Although these trawling operations primarily target shrimp, there is some catch of bottom fishes. In recent years, the armed conflict in Yemen has severely hindered any scientific progress in the Yemeni Red Sea. Border tensions exist between Eritrea and most of its neighboring countries, resulting in similarly restricted access (or no access) to its territorial waters. The active geological fault in the southern Red Sea has even given rise to new islands, which could be the subject of fascinating study (to observe primary colonization, etc.), but due to their location in Yemeni waters, work to date has been limited to satellite observations (Xu et al. 2015).

8.3 Genetic Connectivity

8.3.1 Genetic Barriers in the Red Sea

As discussed in Sect. 8.2.1 above, there is mixed evidence for a strong faunal change at the proposed 20°N boundary of Spalding et al. (2007). Few genetic surveys have directly tested for the presence of this barrier, but they provide equally mixed results. Clear signals of a genetic break at 20°N have been shown for an anemonefish (Nanninga et al. 2014; Saenz-Agudelo et al. 2015) and a sponge (Giles et al. 2015). These same patterns seem to exist also for an anemone (*Entacmea quadricolor* (Emms 2015)) and two damselfishes (*Dascyllus marginatus* (Robitzch 2017) and *Dascyllus trimaculatus* (Salas De la Fuente 2017)). However, work in other species has failed to detect this signal, including *Chaetodon* species and *Ctenochaetus striatus* (JD DiBattista et al. unpublished), *Dascylllus aruanus* (Robitzch 2017), two anemone species (*Heteractis magnifica* and *Stichodactyla haddoni* (Emms 2015)), and a coral (*Pocillopora verrucosa* (Robitzch et al. 2015)). Taken together, there does not seem to be a clear connection between the presence of a genetic break at 20°N and biological traits such as pelagic larval duration or spawning mode. At least two of the studied species suggest that environmental characteristics play an important role in shaping gene flow near 20°N (Nanninga et al. 2014; Giles et al. 2015; Saenz-Agudelo et al. 2015), but recent modeling work suggests that oceanographic patterns are tightly linked with genetic similarity among populations (Raitsos et al. 2017).

While there are some species that exhibit this genetic break, it is not typically reflected in a presence / absence change (see Sect. 8.2.1). A more interesting barrier is perhaps the Strait of Bab al Mandab, the narrow opening dividing the Red Sea from the Gulf of Aden (and the wider Indian Ocean), which is the most common range limit for the majority of Arabian endemics (DiBattista et al. 2016a). Approximately half of the species investigated so far have shown signatures of restricted gene flow between the Red Sea and the Gulf of Aden. Recent and unpublished data suggest that this structure is explained to some extent by historical interruption of gene flow, followed by secondary contact. As with the putative barrier at 20°N, evidence for disruption of gene flow between populations at either side of this strait is divided. Several species of fish show genetic structure between populations in Djibouti and populations in the southern or central Red Sea (Saenz-Agudelo et al. 2015; Salas De la Fuente 2016; DiBattista et al. 2017; Robitzch 2017), and some anemone species show similar genetic structure (Emms 2015). However, there are also several fishes for which this pattern is not the case (DiBattista et al. 2017). Although there is limited data available, no consistent pattern has emerged, and it appears that there is not a single biological characteristic that can explain the observed patterns. Indeed, the evolutionary history of Red Sea fauna may be rather complicated and each species may have a unique story (DiBattista et al. 2013, 2016a). Further work with additional species (and application of next-generation sequencing technologies) may reveal common histories for some groups of fishes.

8.3.2 East-West Connectivity

While questions about genetic connectivity along the latitudinal gradient of the Red Sea have received limited attention (Sect. 8.3.1 above), even fewer studies have explicitly tested whether connectivity across the Red Sea (east-west connectivity) is occurring. The geography and oceanography of the Red Sea make this a reasonable possibility; the typical width of the Red Sea is ~200–300 km, and the Red Sea is characterized by periodic basin-width eddies hypothetically capable of facilitating the transport of larvae across these distances (Zhan et al. 2014; Yao et al. 2014). A recent modeling study confirmed the potential for cross-sea connections of larval particles and found correspondence with available genetic data for clownfish (Raitsos et al. 2017). This work demonstrates that the eddies and cross-basin currents should be sufficient to link reefs on opposite sides of the Red Sea on a regular basis. The eddies are somewhat ephemeral (Zhan et al. 2014), and the timing of spawning in most Red Sea reef fishes is not clear (see Sect. 8.4.4 below), but the 'average' oceanography appears to be conducive to genetic mixing, even for species with a short pelagic larval duration, such as clownfish (Nanninga et al. 2014; Saenz-Agudelo et al. 2015; Raitsos et al. 2017). Some groupers, which have longer pelagic larval durations than clownfish, also exhibit genetic patterns suggesting east-west connectivity (Priest et al. 2016). The timing of spawning and interactions with the hydrodynamic conditions present during the larval dispersal phase (as opposed to time-averaged conditions) can have substantial influence over specific dispersal potential. Empirical measurements of specific dispersal events are uncommon, but application of genetic parentage analysis has proven to be powerful in this regard (e.g., Harrison et al. 2012; Almany et al. 2017). To our knowledge, parentage analysis has only been conducted in one study in the Red Sea (Nanninga et al. 2015). Based on modeled hydrodynamics of the inferred spawning dates, most of the clownfish larvae would have been advected out of the study area, corresponding with the lack of parent-offspring matches in the study (Nanninga et al. 2015). However, additional modeling work suggested that if the study had focused on a reef located further inshore, a greater portion of the larvae may have been locally retained and self-recruitment may have been more prominent (Nanninga 2013). The potential for connections across the width of the Red Sea, especially if they occur on a regular basis, has important implications for conservation as healthier populations could reseed heavily exploited populations (see Sect. 8.5.2) on opposite sides of the Red Sea.

8.3.3 Genetic Identification of Cryptobenthic Species

Cryptobenthic fishes are generally characterized as fishes that have a proximate association with the benthos and attain body lengths ≤50 mm (Ackerman and Bellwood 2000; Depczynski and Bellwood 2003). These fishes are often cryptic in nature and coloration, hence they are often overlooked or excluded during standard visual reef fish censuses.

Despite their small size, this group can be strikingly abundant and diverse across coral reefs. By some estimates this group contributes approximately 50% of the individual fish abundance and 10% of the overall reef fish biomass on coral reefs (Ackerman and Bellwood 2000). Additionally, a large proportion of these fishes exhibit high fecundity, growth, and metabolic rates (Hernaman and Munday 2005, 2007; Depczynski and Bellwood 2006; Depczynski et al. 2007). Due to the rates at which these fishes are preyed upon, they play a disproportionate role in the transfer of energy in reef food webs. In addition to being prey items for larger fishes, cryptobenthic fishes may also play other important functional roles (Goatley and Brandl 2017). However, logistical constraints limit the number of studies that include or focus on cryptobenthic fishes and subsequently impact our understanding of their ecology.

The Red Sea is no exception; few studies have examined cryptobenthic fishes in Arabian waters. The family Gobiidae has been the subject of some study (Herler and Hilgers 2007; Herler 2007), but only recently studies have begun to investigate community-level composition of this assemblage among different habitats (Troyer et al. 2018; Coker et al. 2018). Importantly, the application of molecular tools to identify species (DNA barcoding) has enabled community-level and ecological investigations even though morphology-based taxonomy remains problematic for these fishes (Troyer 2017; Coker et al. 2018; see also Tornabene et al. 2013) There are many undescribed species and very few morphological identification keys are available for Red Sea specimens (Troyer 2017). Fortunately, each new study that combines morphology and genetic analyses steadily contributes to global genetic databases (such as GenBank) and helps to slowly fill some of the many gaps in coverage of Red Sea species (Troyer 2017; Robitzch 2017; Isari et al. 2017a, b; Coker et al. 2018). Barcoding is not a panacea (Rubinoff 2006), but the technique can be a valuable component of an integrated approach (DeSalle 2006).

Standardized field sampling suggests that cryptobenthic fish communities differ along a latitudinal gradient and with distance from shore (Coker et al. 2018). The Red Sea's environmental gradients (Raitsos et al. 2013; see also Chap. 1) and are predicted to influence species composition and abundance through direct (e.g., temperature, salinity, productivity) and indirect variables (habitat availability, predation pressure). Given the size of individuals within this group, microhabitat is likely to explain finer-scale spatial patterns (see Troyer 2017) while environmental variables are likely driving larger-scale patterns. Given the importance of this group, future work is needed in the Red Sea to better understand biodiversity, spatial patterns, and ecosystem processes. The work so far on these fishes, and particularly the molecular barcoding work, suggests that there are many new fishes (some not yet recorded from the Red Sea and many others probably new to science) to be discovered.

8.3.4 Inter-Species Genetic Variation and Cryptic Speciation

The uniqueness of the Red Sea fauna is only apparent in comparison to the fauna of the seas outside of the Red Sea. The Red Sea is undoubtedly an important biodiversity hotspot among the entire western Indian Ocean (DiBattista et al. 2016b), but there are important unanswered questions as to why this is the case (DiBattista et al. 2016a). Many Red Sea populations may have colonized the Red Sea and then had to adapt to its unique environmental conditions, effectively diverging from the "parent" populations in the Indian Ocean, but there is also evidence that some Indian Ocean species have their origins within the Red Sea, challenging the historical assumption that peripheral seas rarely "export" biodiversity (Bowen et al. 2013). What exactly drives the generation of diversity within the Red Sea is still not well understood, but it could be that novel genes and adaptations emerge to cope with typical Red Sea conditions (see ReFuGe 2020 Consortium 2015), which might otherwise be considered "harsh" in other parts of the Indo-Pacific.

When widespread species have been examined with samples both from within the Red Sea and outside the Red Sea, the patterns of intra-specific genetic variation are unpredictable. In some species, the Red Sea populations appear to show evidence of contemporary genetic exchange with other western Indian Ocean populations (e.g., *Abudefduf vaigiensis*, DiBattista et al. 2017), while other species show unexpected divergence dating far beyond recent sea level minima (when the Red Sea would have been quite, but not completely, isolated from the Gulf of Aden and the rest of the Indian Ocean (DiBattista et al. 2016a)). Examples of the latter case include *Chaetodon melannotus* (DiBattista et al. 2017) and *Mulloidichthys flavolineatus* (Fernandez-Silva et al. 2015, 2016). In some cases, the isolation appears to be so complete that the species should likely be considered separate species yet to be described, such as *Pygoplites diacanthus* (DiBattista et al. 2013; Coleman et al. 2016) and *Cephalopholis hemistiktos* (Priest et al. 2016). In the context of Red Sea fishes, there can therefore be some semantic confusion with regards to "cryptic species". One definition applies to the preceding examples, and is taxonomic in nature, wherein populations have species-level divergence but have evaded detection by taxonomists because the morphology has not diverged (at least obviously enough to have been recognized). Another definition of "cryptic species" is functional or ecological in nature; the Red Sea has many fishes that, due to their size, coloration, or behavior, are difficult to detect in visual surveys, and are often overlooked or understudied (see Sect. 8.3.3).

While the aforementioned studies have examined a small number of species in some detail, the results indicate that there is no single explanation for the evolutionary history of Red Sea ichthyofauna (DiBattista et al. 2016a). We therefore thought it would be useful to broadly assess the genetic "connectedness" of Red Sea fishes using samples from within the Red Sea compared to samples from outside the Red Sea (using Indian Ocean sites when available). For species endemic to the Red Sea, we included samples from sister species (or at least congeners). For many species, prior genetic data was publicly available in the NCBI GenBank repository (http://www.ncbi.nlm.nih.gov/genbank/) (specifically, the mitochondrial gene cytochrome oxidase I (COI) "barcoding" marker), but for many species that were not available, we sequenced new samples. For these species, we used a small (~2mm^2) piece of fin tissue and extracted DNA following the "HotSHOT" protocol (Truett et al. 2000; Meeker et al. 2007). The COI barcoding fragment was amplified using the primers FishF2/FishR2 (Ward et al. 2005). PCR products were sequenced in the forward direction with fluorescently labeled dye terminators following the manufacturer's protocols (BigDye, Applied Biosystems Inc., Foster City, CA, USA) and were analyzed using an ABI 3130XL Genetic Analyzer (Applied Biosystems) in the King Abdullah University of Science and Technology (KAUST) Biosciences Core Laboratory. (Details of the samples used, including accession numbers for existing and newly-generated sequences, are available in Table 8.1 and Appendix 1) Sequences were aligned using Geneious R8 (Biomatters Ltd., Auckland, New Zealand) and divergence was calculated using the Kimura 2-parameter model (K2P) in MEGA 6.0 (Tamura et al. 2013). The results of this comparison show that there do not seem to be any obvious family-specific or genus-specific patterns of genetic relatedness. For species that occur inside and outside of the Red Sea (Appendix 1), there were varying levels of differentiation, and there were a few species exhibiting quite high values (e.g., *Bothus pantherinus*). Several explanations are possible: among other possibilities, the values may be the result of as-yet undetected cryptic speciation, the samples could have been misidentified, or intraspecific variation may be quite high in general. Values for sister species comparisons (Table 8.1) were, as expected, generally higher than the intraspecific comparisons. There were several interesting species pairs for which the K2P values were very low (e.g., *Pseudochromis fridmani* + *Pseudochromis sankeyi* and *Chaetodon austriacus* + *Chaetodon melapterus*). These pairs may be in the early stages of speciation (e.g., Waldrop et al. 2016).

8.4 Ecology

8.4.1 Application of Stable Isotope Techniques to Red Sea Fishes

Stable isotope analyses have been traditionally used to track the movements of fishes through natural isotope gradients, or isoscapes, via analysis of the calcified earbones ("otoliths") (Campana and Thorrold 2001; Thorrold et al. 2001; Kennedy et al. 2002; Elsdon et al. 2008). While these studies have provided useful insights on the movements of marine organisms, there are some notable challenges to using stable isotope from fish otoliths. Bulk isotope values can be affected by fish metabolism (Kalish 1991; Stephenson et al. 2001), environmental conditions (Mulcahy et al. 1979), and changes in dissolved inorganic carbon $\delta^{13}C$ values (Schwarcz et al. 1998). There is also difficulty with associating any changes in otolith $\delta^{13}C$ values with either changes in basal resource use or trophic shifts (Post 2002), and this is particularly apparent when working with species that undergo ontogenetic shifts in habitat use, as many coral reef fishes do (Cocheret de la Morinière et al. 2002; Kimirei et al. 2013). However, the use of compound-specific stable isotope analysis (CSIA) of essential amino acids (EAAs) may help to circumvent these complexities. Essential amino acids are those that most animals, including fishes, have lost the ability to synthesize at sufficient rates for survival (Borman et al. 1946; Reeds 2000), therefore EAAs must be assimilated through the fishes' diets. Once taken up, EAAs remain virtually unaltered biochemically, so that fractionation factors between food and consumers are essentially zero (Hare et al. 1991; McMahon et al. 2011a). This means $\delta^{13}C$ values of a consumer's EEAs represent the isotopic signatures of the primary producers (e.g. plants, algae, and microbes) at the bottom of the food web. When this information is combined with known isoscapes across marine environments, it allows for the possibility to track movements through habitats, provided the fish is present long enough to incorporate the isotopic signature of its habitat. The use of CSIA-EAA to investigate residency, ontogenetic movement, and food web ecology has been pioneered in studies of fishes from the Red Sea (McMahon et al. 2011a, b, 2012, 2016).

The analysis of $\delta^{13}C$ values of essential amino acids in Red Sea fishes has expanded our understanding of fish residency and ontogenetic movements. CSIA-EAA has been utilized to study residency patterns of coral reef fish in the Red Sea, providing information applicable to coastal ecosystems across the globe. McMahon et al. (2011b) documented the advantage of CSIA-EAA compared to traditional bulk analysis for determining habitat use in economically important fishes. Although isotopic differences between mangrove and seagrass habitats have been previously documented (Marguillier et al. 1997; Layman 2007), McMahon et al. (2011b) failed to find any clear relationship between habitat residency and bulk isotope $\delta^{13}C$ values. Bulk isotope values were only able to distinguish between ocean basins rather than specific habitats, while EAA $\delta^{13}C$ values provided sufficient resolution to reliably distinguish between mangrove and seagrass habitats (McMahon et al. 2011b), including across ocean basins.

Table 8.1 Genetic relatedness (K2P, based on COI, see Sect. 8.3.4) of Red Sea / Gulf of Aden (RS/GoA) endemic fish species with their nominal sister species (or congener). Sampling sites for the samples used are indicated (KSA = Saudi Arabia). For samples sequenced at KAUST, the internal tissue library reference is included. GenBank accession numbers are provided for each sample

Family	Red Sea / Gulf of Aden species	Sampling site (Red Sea / Gulf of Aden)	KAUST library ref.	Accession number (RS/GoA sample)	Sister species (Outside of Red Sea / Gulf of Aden)	Sampling site	KAUST library ref.	Accession number	K2P
Acanthuridae	*Acanthurus gahhm*	Thuwal, KSA	RS7771	MH331650	*Acanthurus nigracauda*	Reunion Island	–	JQ349655.1	*0.082*
Balistidae	*Sufflamen albicaudatus*	Thuwal, KSA	RS4725	MH331875	*Sufflamen chrysopterus*	Maldives	RS7150	MH331877	*0.025*
Balistidae	*Sufflamen albicaudatus*	Obock, Djibouti	RS1993	MH331876	*Sufflamen chrysopterus*	Maldives	RS7150	MH331877	*0.025*
Caesionidae	*Caesio suevica*	South Farasan Banks, KSA	RS3471	MH331681	*Caesio xanthonota*	Madagascar	–	JQ349810	*0.028*
Chaetodontidae	*Chaetodon austriacus*	South Farasan Banks, KSA	RS3552	MH331699	*Chaetodon melapterus*	Obock, Djibouti	RS1857	MH331708	*0.002*
Chaetodontidae	*Chaetodon larvatus*	Thuwal, KSA	RS6624	MH331701	*Chaetodon triangulum*	Maldives	RS6893	MH331714	*0.077*
Chaetodontidae	*Chaetodon fasciatus*	Thuwal, KSA	Cfa16	MH331700	*Chaetodon lunula*	Socotra, Yemen	RS5959	MH331704	*0.013*
Chaetodontidae	*Chaetodon paucifasciatus*	Thuwal, KSA	Cpa15	MH331711	*Chaetodon madagaskariensis*	Madagascar	Cma11	MH331705	*0.007*
Clupeidae	*Etrumeus golanii*	Gulf of Suez, Egypt	Eter63	MH331756	*Etrumeus wongratanai*	Durban, South Africa	Eter60	MH331757	*0.023*
Holocentridae	*Myripristis xanthacra*	Farasan Islands, KSA	RS7560	MH331804	*Myripristis hexagona*	Reunion Island	–	JQ350120.1	*0.072*
Labridae	*Chlorurus gibbus*	Farasan Islands, KSA	RS7588	MH331722	*Chlorurus strongylocephalus*	Socotra, Yemen	RS6194	MH331725	*0.004*
Labridae	*Cheilinus abudjubbe*	Socotra, Yemen	RS6268	MH331715	*Cheilinus cf. chlororus*	Maldives	RS7391	MH331716	*0.091*
Labridae	*Hemigymnus sexfasciatus*	South Farasan Banks, KSA	RS3743	MH331776	*Hemigymnus fasciatus*	Socotra, Yemen	RS5945	MH331773	*0.074*
Mullidae	*Parupeneus forsskali*	Shi'b Al Karrah, KSA	RS6752	MH331820	*Parupeneus barberinus*	Maldives	RS6915	MH331818	*0.118*
Pomacanthidae	*Amblyglyphidodon flavilatus*	Farasan Islands, KSA	ROT205	MH331657	*Amblyglyphidodon indicus*	Yanbu, KSA	RS6721	MH331658	*0.052*
Pomacentridae	*Chromis dimidiata*	Thuwal, KSA	RS2461	MH331728	*Chromis fieldi*	Socotra, Yemen	RS6158	MH331729	*0.020*
Pseudochromidae	*Pseudochromis sankeyi*	Socotra, Yemen	RS8876	MH331846	*Pseudochromis fridmani*	Al Lith, KSA	ROT484	MH331844	*0.008*
Serranidae	*Diploprion drachi*	Yanbu, KSA	RS6696	MH331748	*Diploprion bifasciatum*	Lizard Island, Australia	–	KP194601.1	*0.208*
Serranidae	*Epinephelus geoffroyi*	Farasan Islands, KSA	RS3965	MH331752	*Epinephelus chlorostigma*	India	–	KT835686.1	*0.113*
Serranidae	*Epinephelus summana*	Thuwal, KSA	A516	MH331754	*Epinephelus caeruleopunctatus*	Mozambique	–	JF493438	*0.044*
Serranidae	*Epinephelus summana*	Obock, Djibouti	RS1973	MH331753	*Epinephelus caeruleopunctatus*	Mozambique	–	JF493438	*0.046*
Siganidae	*Siganus rivulatus*	Al Halliniyat Islands, Oman	RS4383	MH331868	*Siganus sutor*	Masirah Island, Oman	RS4461	MH331871	*0.081*
Siganidae	*Siganus stellatus laqueus*	Shi'b Al Baydah, KSA	RS6776	MH331870	*Siganus stellatus*	Maldives	RS6927	MH331869	*0.002*
Tetraodontidae	*Arothron diadematus*	Thuwal, KSA	RS6617	MH331668	*Arothron nigropunctatus*	Philippines	–	FJ582890.1	*0.014*

The application of CSIA to otolith EAAs has also revealed plasticity in the ontogenetic movements between coastal ecosystems of reef fish. Past studies have documented the importance of coastal habitats (e.g., mangrove and seagrass beds) as nurseries for coral reef fishes (Adams et al. 2006; Nagelkerken et al. 2008), though most of these studies have inferred this relationship by analyzing size-frequency distributions and relative densities of juvenile fishes (Nagelkerken et al. 2000; Cocheret de la Moriniére et al. 2002). A Red Sea study was the first to quantify the contribution of different juvenile habitats to adult fish populations via CSIA-EAA of otoliths (McMahon et al. 2012). By assessing EEAs in material from the core of the otoliths (i.e., the material deposited as a juvenile), McMahon et al. (2012) assigned adult fishes into different juvenile habitats. In addition to documenting movements of economically important snappers among coastal habitats in the Red Sea, the study more generally emphasized the importance of seascape configuration as a factor driving ontogenetic movement patterns.

Densities of Ehrenberg's snapper (*Lutjanus ehrenbergii*) were found to be highest on shelf reefs near shore, which also happened to have the greatest levels of connectivity between coastal wetland habitats and other shelf reefs. This finding lends empirical support to others that have found higher fish biomass on reefs closer to coastal habitats (Nagelkerken et al. 2000; Mumby et al. 2004). While these snapper are able to migrate from coastal habitats to shelf reefs, there does appear to be a break in connectivity at the shelf edge, where snappers cannot or will not migrate beyond. Red Sea oceanic reefs were dominated by snappers that had settled directly onto these types of reefs, despite the complete absence of juveniles from extensive visual surveys. A small portion (<30%) of snappers on offshore reefs were also found to have migrated from a large island near the shelf edge, crossing deep water and making horizontal movements of at least 30 km. McMahon et al. (2012) demonstrated not only a plasticity in ontogenetic movements of snappers, but also the ability to migrate large distances between coastal wetlands and reef habitats. The role of seascape configuration plays an important role in structuring how snapper, or any fish, may be able to move between coastal habitats. In light of planned coastal developments in the Red Sea (see Sect. 8.5.3), understanding linkages between coastal habitats and nearby reef fish populations will be important to consider.

While isotopic studies from the Red Sea have demonstrated patterns of residency and connectivity in coral reef fishes, the more traditional use of isotopic analyses has been to tease apart information about resource usage. Isotopic approaches have been especially useful in reconstructing the diets of important fishery species (e.g., cod, Hanson and Chouinard 2002). Several isotopic studies have documented reliance on microbially-processed carbon in mangrove ecosystems (Bouillon et al. 2002; Kieckbusch et al. 2004; Kristensen et al. 2017), raising interesting questions about the structure of some marine food webs. In the Red Sea, mangrove-derived carbon contributes little to the diets of coastal snappers compared to other locations (e.g., the Pacific coast of Panama and the Caribbean) (McMahon et al. 2011b). The reduced reliance on mangrove-derived carbon in the Red Sea is potentially due to the relatively diminutive mangrove stands that typically exist on a narrow strip of coastal land, as opposed to the more extensive forests found at some non-Red Sea sites that spend more time submerged and accessible for fishes (McMahon et al. 2011b).

In addition to documenting differences in food webs between broad ocean basins, CSIA is revealing how resource use can change among reefs in the Red Sea. Using a CSIA-EAA analysis of fish muscle samples, McMahon et al. (2016) documented changes in the basal nutrient source that supports Red Sea coral reef fishes. Some functional groups of Red Sea fishes exhibited consistency in their nutritional ecology while other groups appeared to be flexible. Highly specialized functional groups, including obligate corallivorous butterflyfish (*Chaetodon trifascialis*), algal-farming damselfish (*Stegastes nigricans*), and detritivorous surgeonfish (*Ctenochaetus striatus*) show little change in the main nutrient source they rely on across the seascape from shelf to oceanic reefs. Several species were more variable in their resource usage across reefs, though they were generally reliant on mostly a single basal food source. Planktivorous damselfish (*Amblyglyphidodon indicus*) were found to rely almost equally on carbon sources from macroalgae and phytoplankton on shelf reefs, while these fish on oceanic reefs sourced nearly all their carbon from phytoplankton production. *Lutjanus ehrenbergii* also showed a similar pattern, being reliant mostly on macroalgae production on shelf reefs and switching to phytoplankton carbon on oceanic reefs (see Figure 4 in McMahon et al. 2016). Giant moray eels (*Gymnothorax javanicus*) relied mostly on phytoplankton-derived carbon on both shelf and oceanic reefs, though they had a greater phytoplankton reliance on oceanic reefs. The pattern for many species to increase reliance on phytoplankton-derived carbon on oceanic reefs is likely not unique to the Red Sea (e.g., Wyatt et al. 2012; Letourneur et al. 2013). Given the lack of terrestrial/freshwater input into the Red Sea, the patterns documented by McMahon et al. (2016) are likely to be slightly different in other reef systems as runoff and riverine outflow can alter food web nutrient dynamics (e.g., Dromard et al. 2013; Letourneur et al. 2013; Docmac et al. 2017). CSIA-EAA represents a powerful technique for determining broad differences in the nutrient sources supporting coral reefs in oligotrophic systems such as the Red Sea.

While McMahon et al. (2016) have demonstrated the utility of CSIA-EEA for determining broad differences in highly dissimilar functional groups, the approach also has the potential to identify subtler nutritional differences within functional groups than previous techniques. Robust differences have been shown in the $\delta^{13}C$ isotope values of EEAs from basal food sources, including various tropical marine algae (Larsen et al. 2009, 2012, 2013). The technique is sensitive enough to discriminate isotope values between similar algae and bacterial species, indicating that CSIA-EEA could be used to determine fine-scale differences in the nutritional ecology of functional groups that may normally be missed in traditional feeding observation or stomach content analysis (Bearhop et al. 2004; Larsen et al. 2012). Indeed, in the Red Sea, preliminary analysis of fishes within the functional group of herbivores have found discreet differences in the nutritional ecology of herbivorous fishes (Tietbohl 2016). Fishes that appear to have nearly identical feeding habits show robust and distinct clustering from other species. The approach even clearly separates scraping and excavating parrotfish species, which implies these fish are actually using different nutritional sources within the turf algae they feed in together. Distinctions among functional (sub)groups of parrotfishes have been previously suggested (Clements et al. 2016); CSIA-EAA of Red Sea parrotfishes may be able to definitively show these differences and further attribute the differences to the use of isotopically distinct food sources. Broader application in other geographic regions will provide important comparisons and determine the generality of Red Sea trends for reef systems in other parts of the world.

8.4.2 Megafauna Movements

Reports of whale sharks (*Rhincodon typus*) impaled on the bows of steams ships, including four incidents from the Red Sea and Gulf of Aden, make up some of the earliest published records of these sharks in the Arabian region (Gudger 1940). These instances resulted in Gudger concluding, "whale sharks must surely abound in this region" (Gudger 1938). Following these reports, sporadic sightings of whale sharks were recorded throughout the region, but research was limited. Whale shark research within the region began to increase with the discovery of a juvenile male dominated whale shark aggregation in the Gulf of Tadjura, Djibouti (Rowat et al. 2007). Several years later, a juvenile whale shark aggregation with sexual parity was described within the Red Sea along the central Saudi Arabian coast approximately 200 km south of Jeddah (Berumen et al. 2014; Cochran et al. 2016). Historically, work on sharks in the Red Sea has been sparse and generally concentrated in the Gulf of Aqaba (Spaet et al. 2012), but efforts over the last decade have begun to fill in vital knowledge gaps for select elasmobranch species.

The only known Saudi Arabian whale shark aggregation takes place at a nearshore reef, locally known as Shib Habil, which lies approximately 4 km from the coast of the small town of Al Lith. Whale sharks are commonly encountered here from March through May (Berumen et al. 2014). In addition, reef mantas (*Mobula alfredi*, following the taxonomic synonymization of the genus *Manta* (White et al. 2017)) are occasionally encountered alongside whale sharks and commonly at the surrounding nearshore reefs (Braun et al. 2014; Berumen et al. 2014). Despite their similar habitat use near Al Lith during the spring, the two species show distinct differences in movement patterns the rest of the year. *Mobula alfredi* movements were restricted to coastal areas and reefs primarily within the Al Lith region, which was confirmed by acoustic monitoring (Braun et al. 2014; Braun et al. 2015). Similar restricted coastal movements of *M. alfredi* have been documented using satellite tags at a large manta aggregation in Dunganab Bay along the Sudanese coast (Kessel et al. 2017). One manta at this location was the first (and currently the only) documented *M. alfredi* x *Manta* (now *Mobula*) *birostris* hybrid (Walter et al. 2014; Kessel et al. 2017).

In contrast to the mantas, whale sharks leave the Al Lith region outside of the aggregation season. Most satellite-tagged sharks (39 of 47) made basin-scale movements throughout the southern Red Sea. Seasonal variation was present, with sharks preferring the central Red Sea in the spring and shifting to the south-central and far southern Red Sea during the summer, fall, and into the winter months (Berumen et al. 2014). These high-use areas include waters of multiple countries including Saudi Arabia, Sudan, Yemen, and Eritrea, highlighting the need for international cooperation to protect such highly mobile species. Only three of the whale sharks moved into the northern Red Sea, but tagged sharks ventured as far north as Sharm el-Sheikh on the Sinai Peninsula (see Fig. 8.3). Five sharks left the Red Sea and passed through the Gulf of Aden into the northwestern Indian Ocean (Berumen et al. 2014). On-going photo identification efforts and monitoring of the aggregation site have not identified these sharks as returning to the Al Lith region after exiting the Red Sea. Limited satellite tagging data is available from the Djibouti aggregation, with only one track showing short term movements of a single individual around the Gulf of Tadjoura (Rowat et al. 2007). On the other side of the Arabian Peninsula, a presumed pregnant female shark was tagged in Qatari waters and was tracked moving toward the Gulf of Aden. The shark traveled at least 2640 km over 37 days, with the tag detaching between the Somali coast and the main island of Socotra (Robinson et al. 2017).

Photo-identification of whale sharks from 2010 through 2017 at the Shib Habil aggregation has resulted in the identification of 147 unique individuals in the Al Lith region. Cochran et al. (2016) described the population structure at Shib Habil using the 136 individuals identified from 305 encounters between 2010 and 2015. The population exhibited sexual parity and all individuals were immature based on size estimate and male clasper morphology. Daily abundances at the aggregation site were estimated as 15 to 34 individuals with individual residence times of 4–44 days (Cochran et al. 2016).

An international database, *Wildbook for Whale Sharks* (whaleshark.org), invites researchers and citizen scientists to submit photos of whale sharks from anywhere in the world. Suitable images are used for photo-identification and are then cross-referenced against the entire database. At the end of 2017, *Wildbook* had a total of 585 Red Sea whale shark encounters submitted from dive companies, tourists, and researchers. There are sightings from all Red Sea nations except for Eritrea. Shib Habil has the most encounter records with 318, an expected result considering the area has been regularly monitored by researchers since 2010. However, there are only six reported encounters for the rest of Saudi Arabia, which is likely due to the lack of local knowledge about the database (and not necessarily reflective of an absence of whale sharks).

The second highest number of encounters, 208, comes from Egypt. The remaining countries all have very low numbers of encounters recorded. The satellite tagging results of Berumen et al. (2014) suggest that the lack of records in *Wildbook* arises from a similar unawareness of the database and far fewer tourists in other areas. Egypt is a well-known Red Sea diving destination and many dive companies report sightings directly to *Wildbook*. The Red Sea Sharks Monitoring Programme (redseasharks.org), primarily operating at dive sites throughout the Egyptian Red Sea, maintains photo-identification databases for three species of shark, including oceanic whitetips (*Carcharhinus longimanus*), grey reef sharks (*Carcharhinus amblyrhynchos*), and silky sharks (*Carcharhinus falciformis*). The website also directs those interested in submitting whale shark and manta photos to Wildbook and Manta Matcher (mantamatcher.org), respectively. In addition to identifying >1000 individual sharks, the Red Sea Sharks Monitoring Programme has identified sightings of other species, such as scalloped hammerheads (*Sphyrna lewini*), pelagic threshers (*Alopias pelagicus*), and whitetip reef sharks (*Triaenodon obesus*).

A global genetic analysis also suggests regular connections of whale sharks between the Red Sea and the Indian Ocean. Very little genetic structure was detected within the Indo-Pacific, including samples from the Saudi Arabian aggregation (Vignaud et al. 2014). A follow-up study added additional locations by using DNA sequences obtained from copepod ectoparasites of whale sharks (*Pandarus rhincodonicus*), but found a similar genetic pattern (Meekan et al. 2017). Both studies show slight genetic structure between the Indo-Pacific and the Atlantic Ocean whale shark populations, and relative homogeneity within the Indo-Pacific.

Limited data is currently available on the identity of potential food sources that whale sharks target throughout the Red Sea, including at the Shib Habil aggregation site. Preliminary plankton tows collected next to feeding sharks have resulted in a near-monoculture of the sergestid shrimp (*Lucifer hanseni*) and in one case, copepods (*Acartia* spp.). These limited results suggest that, as described in Rohner et al. (2015), whale sharks most likely do not target one specific food source but rather target dense patches of prey without specific preferences. In 2016, 83% of the 53 encounters involved sharks feeding either at or just below the surface. This suggests that Shib Habil hosts a feeding aggregation, especially considering that the sharks are immature (based on size and clasper morphology in males) and breeding is therefore unlikely (Cochran et al. 2016). It is not clear what may drive the presumably high densities of prey that whale sharks feed upon in such a concentrated area (Hozumi 2015), although regional productivity may play a role (Racault et al. 2015). It also remains unclear if the mantas are targeting the same food source, or why the whale sharks venture so much farther from the site compared to the mantas. Understanding these drivers may become increasingly important if Saudi Arabia intends to develop marine ecotourism (see Sect. 8.5.3) in the near future; whale shark aggregations lend themselves to such initiatives, and can be sustainable if appropriate guidelines are adopted (e.g., Rowat and Engelhardt 2007; Catlin and Jones 2010).

8.4.3 Lessepsian Migrants

In addition to natural connectivity and movement patterns in the Red Sea, there is an important anthropogenically-induced connection in the far north of the Red Sea. The Suez Canal provides connectivity between the fauna of the Indo-West Pacific and Mediterranean biogeographical provinces (Por 1978). Since the opening of the canal in 1869, approximately 450 species of marine organisms (Bernardi et al. 2016), including 106 species of fishes (Rothman et al. 2016; Golani et al. 2017), have invaded the Mediterranean Sea from the Red Sea. The phenomenon, termed "Lessepsian migration" (named after the engineer Ferdinand de Lesseps, who supervised the construction of the canal), has been well-documented (Por 1978), particularly for fish taxa (Golani 1998; Golani and Appelbaum-Golani 2010; Azzurro et al. 2016). The canal has no locks or dams, providing little barrier to dispersal along the corridor. Two hypersaline lakes, known as the "Bitter Lakes", may have initially acted as an ecological barrier to dispersal. However, the salinity of these lakes has gradually equalized with the Red Sea over time (Edwards 1987); the large number of species successfully colonizing the Mediterranean is evidence of the ineffectiveness of the barrier. Despite the migratory pathway permitting bi-directional movement of marine fauna, only a few species have been confirmed as "reverse Lessepsian migrants" that immigrate from the Mediterranean and colonize the Red Sea (Ben-Tuvia 1971; Spanier and Galil 1991; Golani 1998, 1999). The largely unidirectional nature of Lessepsian migration may be attributed to the existence of unsaturated ecological niches in the Mediterranean and the competitive superiority and pre-adaptation of species originating in the highly diverse tropical Red Sea compared to those of a temperate origin (Golani 1999). Consequently, Lessepsian migrants are of significant ecological and economic concern, in some instances resulting in the displacement and local extirpation of native fish species in the Mediterranean (Galil et al. 2015). For example, the goldband goatfish, *Upeneus moluccensis*, a widespread Indo-Pacific species that invaded the Mediterranean via the Suez Canal, has largely replaced the native red mullet, *Mullus barbatus*, in Levantine fisheries (Goren and Galil 2005). Dramatic declines in biogenic habitat complexity, biodiversity, and biomass in the Levantine basin have also been attributed to Lessepsian invaders from the Red Sea. Research suggests the herbivorous invaders *Siganus luridus* and *Siganus rivulatus* are responsible for the rapid shift from well-developed native algal assemblages to "barrens" in the Mediterranean rocky infralittoral ecosystem (Sala et al. 2011). Some invasion events are relatively well-documented, and provide ideal opportunities to study the genetics associated with a rapid colonization of a new area. The bluespotted cornetfish (*Fistularia commersonii*) took 130 years to enter the Mediterranean, but only 4 years to expand as far as any other Lessepsian invaders had been recorded (Tenggardjaja et al. 2013). A new expansion of the Suez Canal was completed in late 2016, raising concerns of even further invasions to come (Galil et al. 2015).

8.4.4 Larval Ecology and Recruitment of Reef Fishes

The diversity of a larval fish pool, combined with species-specific distribution patterns, may provide useful information on spawning seasons and recruitment patterns of fish. Knowledge of such patterns facilitates efforts for ecosystem conservation and fisheries management, yet the research on the ecology of early life-history stages of fish in the Red Sea is still in its infancy. High species diversity and a paucity of diagnostic morphological characteristics for larval life stages of reef fishes have been among the major bottlenecks in larval ecological research in tropical waters (Leis 2014). These biological obstacles are further exacerbated by a lack of marine research opportunities and infrastructure in several Red Sea countries (see Sect. 8.2.2 above).

The primary sources of ichthyoplankton knowledge in the Red Sea are a few academic theses on larval fish taxonomy and ecology in the northern Red Sea, specifically from the Jordanian Gulf of Aqaba and Egyptian coastal waters (Abu El-Regal 1999, 2008; Froukh 2001). These studies identified larval stages at broad taxonomic levels (i.e., family level) and made predictions of potential fish spawning seasons.

The advancement of species identification through molecular techniques has boosted multi-species Red Sea ichthyoplankton studies (Isari et al. 2017a, b; Robitzch 2017; Kimmerling et al. 2018). Combining morphological characterization with DNA barcoding, Isari et al. (2017a) determined the larval fish diversity and assemblage variation throughout an annual cycle in coral reef waters of the central Saudi Arabian Red Sea using bongo net tows. Genetic analyses revealed high species richness in the area, and high water temperatures during the year appeared to be the main driver associated with the numerical increase of larvae in many families. Examination of coral reef fish recruitment patterns using light traps on coral reefs in the same area by Robitzch (2017) revealed a seasonal peak in the fall and early winter (i.e., October, November, and December) for most of the dominant families (e.g., Labridae and Gobiidae). Interestingly, other species appear to have spawning peaks during the cooler months of the year (e.g., *Amphiprion bicinctus* (Nanninga et al. 2015) and *Scarus niger* (Isari et al. 2017a)), which could likely reflect differentiation in reproductive thermal optima among species. Unfortunately, for many species, there is not even sufficient information to make an educated guess about the timing (or locations) of their spawning events.

Interestingly, larval fish collections by nets and light traps are now revealing previously unknown aspects of Red Sea fish biodiversity. Based on morphological criteria, new *Schindleria* records have been reported in the northern Red Sea (Abu El-Regal and Kon 2008; Fricke and Abu El-Regal 2017a, b). Genetic markers support a striking species richness of gobies in the central part of the basin (Isari et al. 2017a, 2017b), while high-throughput metabarcoding in ichthyoplankton collections from Gulf of Aqaba has been suggested as a promising tool in assessing the diversity of larval fish community at a species-level (Kimmerling et al. 2018).

Besides larval stages *per se*, studies on juveniles may also be informative regarding important ecological processes taking place during the larval phase. For instance, the duration of the pelagic larval phase and factors that may influence species recruitment across the Red Sea have been assessed on postlarval stages of pomacentrid species (Ben-Tzvi et al. 2007, 2008; Robitzch et al. 2016). These works showed a decrease in pelagic larval duration towards the southern Red Sea, mostly associated with the increase in food availability and water temperature (Racault et al. 2015; Robitzch et al. 2016), while increased downwelling current flow in the Gulf of Aqaba was associated with an enhancement of recruitment events (Ben-Tzvi et al. 2007). Otolith micro-chemistry analyses of newly-settled damselfishes at the Gulf of Aqaba have provided information on larval dispersal trajectories, showing heterogeneity in the dispersal routes that supply local populations (Ben-Tzvi et al. 2008; Ben-Tzvi et al. 2012). Coupling genetic analyses with biophysical dispersal models has verified a large scale of spatial dispersal of larval anemonefish in the central Red Sea (Nanninga et al. 2015; Raitsos et al. 2017; see also Sect. 8.3.2 above).

Despite the recent and growing interest in larval fish ecology in the Red Sea, thorough baseline data are missing. Much of the basic fish biology, larval biology and ecology, and other dynamics related to reproduction and recruitment processes remain unstudied or poorly known. Increased knowledge of early and late larval stages will improve our understanding of spawning, recruitment, and connectivity patterns, which are crucial components of effective management plans (McCook et al. 2009). Molecular techniques may be highly helpful in future studies to reveal not only the hidden diversity in Red Sea ichthyofauna (Kimmerling et al. 2018), but will also improve our knowledge of larval dispersal trajectories and their influence in population dynamics.

8.4.5 Particularly Understudied Areas

In terms of geography and depth of coverage in many topics, our knowledge of Red Sea fishes is in early stages. Nonetheless, there are some areas that are even less well understood ecologically, and some of these are noteworthy. This is not intended to be an exhaustive list, but we have highlighted some areas of potential interest that warrant future study.

8.4.5.1 Mesophotic Coral Ecosystems

In terms of reef habitats, depths greater than ~30 m are rarely the subject of thorough study, and only a portion of the limited studies address fishes inhabiting these depths (Hinderstein et al. 2010; Kahng et al. 2010). Such systems, termed "mesophotic coral ecosystems" (MCEs) are of increasing interest for several reasons, including the potential for reefs at these depths to serve as refugia from climate change and increasing temperatures in shallower reef systems. However, the technical challenge of accessing these depths (beyond the depth at which standard scuba diving can be conducted for any reasonable amount of time) remains a limiting factor. Often when deep-diving resources are available, such as remotely operated vehicles or manned submersibles, the target depths are deeper than the lower limit of MCEs. Only a handful of mesophotic reef studies have been conducted in the Red Sea. The Gulf of Aqaba was explored in seminal studies (Fricke and Schuhmacher 1983; Fricke and Hottinger 1983; Fricke and Knauer 1986), primarily with respect to the distributions of stony corals. More recent work has employed technical diving techniques and has focused on fishes in the Gulf of Aqaba (e.g., Brokovich et al. 2007, 2008, 2010). In terms of fishes, almost no other MCEs have been described in any detail in the Red Sea.

8.4.5.2 Al Wajh Lagoon Reefs

The Red Sea is often referred to as an extreme environment because it has summer sea surface temperatures regularly exceeding 30 °C and salinity often above 40 ppt (Ngugi et al. 2012; see also Chap. 1). Within the Red Sea basin, there are several coastal lagoon systems; these are often quite shallow and have limited water exchange with the broader Red Sea. These lagoons potentially experience even greater temperature and salinity ranges (due to reduced water exchange and increased evaporation) that could significantly influence benthic and fish communities. Most lagoons are small and support marginal reefs, however, there is one notable exception. The Al Wajh (sometimes transliterated from Arabic as "Wadj" or "Wahdj") lagoon system in the north-central region of the Red Sea (Fig. 8.3) is a distinct habitat that differs greatly from the adjacent deep, clear waters of the Red Sea basin. It is approximately 1500 km^2, consists of approximately 50 islands, and is contained within a barrier reef system with three very small channels providing limited hydrodynamic links to the broader Red Sea. Although tidal fluctuations are generally quite small (rarely more than 10s of cm, and often completely masked by wind-driven basin-wide shifts in sea level (Edwards 1987)), these narrow channels experience strong currents due to the volume of water in the lagoon. The lagoon is relatively shallow (mostly <30 m in depth) with a sandy substrate and shallow, patchy coral reefs.

While no temporal *in situ* environmental measurements have been reported from within the lagoon, SST satellite data (MODIS) reveals that temperature fluctuations are greater than the adjacent Red Sea basin with maximum summer temperatures up to 1 °C warmer and winter temperatures up to 3 °C cooler (Calder Atta, unpublished data). In January-February of 2016, several of the authors (MLB, THST, RSH, MDT, AK, and DJC) participated in an exploratory survey in the Al Wajh lagoon and experienced unexpectedly cold water temperatures, typically as low as 17–18 °C during dives at 10–15 m depth. It is conceivable that the lagoon may likewise reach peak temperatures well above 33 °C in summer. These extreme temperature ranges likely have an influence on fish communities, both directly and indirectly. Increased temperature ranges have been shown to directly influence metabolic rates, movement, and growth rates of fishes (Munday et al. 2008). In this regard, the Al Wajh lagoon may be more like the Arabian Gulf (see Sale et al. 2011), and only a subset of Red Sea fauna may be able to tolerate such large fluctuations in environmental conditions. Furthermore, the difference in the benthic reef communities (which have not yet been fully documented) may further influence the fish fauna, as indirect effects through changes in habitat are also expected to modify fish abundance and community structure (Wilson et al. 2006; Pratchett et al. 2008). The possibility of yet-undiscovered endemic species cannot be ruled out, as even the Gulf of Aqaba has endemic species (DiBattista et al. 2016b). While the Gulf of Aqaba is twice as large (~3100 km²), it has a much wider connection with the Red Sea (>5 km wide, compared to <1 km for Al Wajh). The Living Oceans Foundation included the Wajh lagoon in their habitat-mapping and groundtruthing of select areas of the Red Sea (Bruckner et al. 2011), but little data about the fish fauna from this unique habitat is available. This unique environment warrants future investigation to better understand how species present in this region adapt and cope in an extreme environment with implications to climate change within the region and globally.

8.5 Conservation Status and Future Challenges

The lack of historical data available on reef health (coral and fish communities) in the Red Sea presents challenges when assessing the current status of reefs, and, like many other regions, the Red Sea suffers from shifting baselines (Price et al. 2014). Nonetheless, consistent fish harvesting and recent disturbances suggest that this region is not immune to large-scale degradation and that it faces the same global threats (e.g., climate change, overfishing, coastal development, etc.) as reefs around the world. One notable exception is that terrestrial impacts (through fresh water input and nutrient runoff) are limited or inconsequential across many regions of the Red Sea due to limited rainfall and an absence of any permanent rivers entering the Red Sea. Nonetheless, inputs related to coastal development, fishing pressure, and increasing sea temperatures appear to be the main modern threats to reef-associated fishes of the Red Sea (Kotb et al. 2008; Wilkinson 2008; Furby et al. 2013; Spaet and Berumen 2015).

8.5.1 Bleaching and Thermal Stress

Historical information on coral bleaching in the Red Sea is limited, with some of the earliest reports of widespread bleaching documented during 1998 (in Egypt, Eritrea, Saudi Arabia, Sudan, and Yemen). This coincides with the global bleaching event at the time (see Hoegh-Guldberg 1999) and implies that while the Red Sea reefs experience higher water temperatures than other reef systems, they are not immune to the influences of global climate change (see also Cantin et al. 2010). Further bleaching has been reported in 2007 (Egypt), 2010 (Saudi Arabia), 2012 (Egypt), and more recently, large-scale coral loss was observed in 2015 throughout the southern reefs of Saudi Arabia (Osman et al. 2018; see also Chap. 3). Limited *in situ* data about flow-on effects restrict our understanding of how fish communities are influenced following disturbances, however, declines in coral cover and benthic structure are well-known to negatively affect many fish (Wilson et al. 2006; Pratchett et al. 2008).

In addition to coral loss, direct effects of climate change are predicted to have significant ramifications for fishes through increased water temperature and changes in ocean acidification (Munday et al. 2008). Fishes in the Red Sea are already existing in relatively high water temperatures, and several fishes may already be living beyond or at their thermal maxima for some periods of the year. It is unclear if fishes are already thermally stressed, or if fishes within the Red Sea have adapted to cope with greater temperature anomalies. Increased water temperatures can influence latitudinal distributions, depth structure, activity, growth, and metabolic processes (Booth et al. 2011; Johansen and Jones 2011; Nowicki et al. 2012). Latitudinal gradients in temperature, along with extreme regions like the Al Wajh lagoon (see Sect. 8.4.5.2), provide natural environments in the Red Sea to investigate the effects and adaption to future climate change scenarios.

8.5.2 Fisheries

The extraction of fishes by artisanal fisheries has historically been an integral component of food security in the Red Sea. Methods such as larger trawlers have recently been introduced in regions amendable to this method (e.g., southern Red Sea), however most fishing efforts employ more traditional methods, such as hook and line, gill nets, and traps (Tesfamichael and Pauly 2016). Accurate catch data in the Red Sea is difficult to source, particularly at a local scale (e.g., at the level of detail of individual fishing ports or landing sites) (Jin et al. 2012). As coastal populations increase, so will the demand for fish-based protein and associated catch rates, particularly in regions with large populations. Fishing pressure varies among countries (and among regions within countries) based on population, resources, and culture (Tesfamichael and Pauly 2016). Current estimates suggest that most targeted fishes are overfished in the Red Sea, with some groups, such as sharks, significantly reduced from historical numbers (Tesfamichael 2012; Spaet and Berumen 2015). Most fishers employ multi-gear, multi-species operations with no regional fisheries management organization oversight, and even bans on catching protected species are not enforced (Spaet et al. 2016). Some regions, such as Sudan, appear to experience lower levels of fishing pressure. A recent study comparing fish communities among comparable offshore reefs in south-central Saudi Arabia to reefs in Sudan revealed significantly lower abundance and biomass levels on Saudi Arabian reefs (Kattan et al. 2017). The cumulative evidence suggests that Saudi Arabian reefs generally experience heavy fishing pressure (e.g., Jin et al. 2012), however, this could be even higher in more populated regions (e.g., near Jeddah) and on reefs closer to shore. Data is lacking for catch rates, and is also deficient for more nuanced details such as the number of days at sea, discards, distance traveled, gear use, and targeted events (e.g., spawning aggregations). For example, in the southern islands of Saudi Arabia, longnose parrotfish (*Hipposcarus harid*, see Fig. 8.2h) are targeted in shallow waters during spawning aggregations (Gladstone 2006; Spaet 2013). These gaps in data need to be addressed if plans for sustainable fisheries are to be developed for future generations, in addition to simply maintaining the current level of associated goods and services that reef fisheries supply for Red Sea countries. The narrowness of the Red Sea (Morcos 1970) presents the potential for cross-basin connectivity through larval dispersal, specifically facilitated by periodic oceanographic features (Raitsos et al. 2017; see Sect. 8.3.2). This potential connectivity implies that regions of low fishing pressure (e.g., Sudanese reefs) could serve as a replenishment source for regions with depleted fish stocks (e.g., Saudi reefs) or for regions impacted by severe disturbances (e.g., recent mortality in the Farasan Banks due to bleaching, see Chap. 3).

While larval dispersal may provide some reason for optimism for reef fisheries, some highly mobile species (e.g., tunas or whale sharks, see Sect. 8.4.2) would benefit from formal management at the level of the entire Arabian Peninsula (e.g., Spaet et al. 2015). Fortunately, there does not currently appear to be a targeted fishery for mantas or whale sharks, two species of potential ecotourism value. *Mobula alfredi* is listed as vulnerable with on the IUCN Red List and *Rhincodon typus* is listed as endangered, with species population trends considered to be declining (Marshall et al. 2011; Pierce and Norman 2016). Surveys at the main Jeddah fish market revealed no manta or whale sharks (bimonthly surveys between 2011–2013), however, two species of mobulid ray were found (6 *Mobula thurstoni* and 1 *Mobula kuhlii*) (Spaet and Berumen 2015). The fishing fleet within the Al Lith area (near the whale shark aggregation site), like most of Saudi Arabia, is dominated by artisanal fishers using hand lines (e.g., Jin et al. 2012); mantas and whale sharks are not targeted. In 2011, one whale shark (previously tagged at the aggregation site) was accidentally captured in a gill net and died as a result (Cochran et al. 2016). Although it appears that bycatch in this form is rare, it is unclear if such instances would normally be reported. The nearshore location of Shib Habil and its proximity to the local port puts the mantas and sharks at risk from outboard motor strikes (Braun et al. 2015). Approximately half of all sharks encountered at the aggregation site have scars, with 15% of the scars apparently resulting from propeller trauma (Cochran et al. 2016). A manta aggregation in Dunganab Bay, Sudan, falls within a marine protected area that was declared a UNESCO World Heritage Site in 2016 (Kessel et al. 2017), affording the individuals at that location protection. Unfortunately, other elasmobranchs do not enjoy such reprieve and appear to be heavily impacted by fishing activies (Spaet and Berumen 2015; Spaet et al. 2016; see also Sect. 8.5.5).

8.5.3 Coastal Development, Ecotourism, and Saudi Arabia's Vision 2030

One Red Sea nation is poised to launch some of the most ambitious development projects ever undertaken. The Kingdom of Saudi Arabia has released and identified the nation's "Vison 2030", which outlines major economic goals for the country (details are available at http://vision2030.gov.sa). Among the many plans outlined, there are several coastal developments in northwestern Saudi Arabia that each have the potential to influence reefs in this area (ranging from the Al Wajh lagoon to Egyptian side of the Gulf of Aqaba). These coastal projects are described as "Giga-Projects" by the Saudi government's Public Investment Fund (www.pif.gov.sa/pifprograms/vrp_en). The NEOM project envisions a world-leading "smart city" occupying 438 km of coastline, sprawling into Egypt and Jordan. Among other lofty goals, the NEOM project has a plan to achieve a productive city with the highest per capita GDP in the world. A second coastal giga-project, known as "The Red Sea Project", is based in the Al Wajh lagoon area. This project focuses much more specifically on diversifying tourism activities in Saudi Arabia (projections include reaching 90,000 visitors to the Al Wajh lagoon's islands annually by 2022 and 1 million visitors annually by 2035). Eco-tourism and water sports are explicitly named among the attractions. The proposed scale and pace of development would set numerous records, especially considering the near-complete lack of infrastructure present in this region. These giga-projects will provide interesting case studies for years to come – hopefully they provide examples of 'successes' to serve as models for other regional developments.

Among the major goals of the new vision is an increased tourism sector, including the general introduction of tourism visas (reported to begin in April 2018). At this time, there is no mention of directly exploiting mantas or whale sharks. However, these species both readily lend themselves to ecotourism endeavors and are attractive targets in various locations worldwide. Access to the Al Lith whale shark aggregation site is relatively limited, despite the reef's proximity to shore, because there is currently only one dive operation in the Al Lith area with a limited number of vessels. Light ecotourism focused on the whale sharks has been ongoing since 2012. At present, there is no formal code of conduct for interactions with either species, which could lead to conflicts should tourism begin to increase. Whale sharks and mantas have the potential to play a role in sustainable development of the regional economy, but precautions must be taken to ensure the long-term viability and minimal risk to the animals. Some valuable lessons could be learned from Sudan. Sudanese reefs were brought to the world's attention following Jacques Cousteau's 1964 documentary, "World Without Sun", which documents the Cousteau team's adventures living underwater in the Conshelf 2 station. Today, the majority of international marine ecotourism in Sudan is centered around liveaboard dive boats and has grown rapidly (Chekchak 2013). In 2000, there were 8 liveaboard boats operating out of Port Sudan, but by 2017, there were 15 (with 7 boats from outside of Sudan). Between 2500–4000 divers visit annually (mostly from Europe), generating an estimated US$15–17 million per year in gross income, including tourism fees (Chekchak 2013). Tourism is the largest source of income for the Sanganeb Atoll Marine National Park (see Sect. 8.5.5). Nonetheless, marine tourism in Sudan can still be considered under-developed, and fortunately there seem to be minimal impacts on the conditions of the reefs or their resident fish communities. Many of the diving tourists to Sudan are attracted by the still-healthy populations of reef sharks (Hussey et al. 2013; Spaet et al. 2016).

8.5.4 Aquaculture

Al Lith is near Saudi Arabia's largest prawn farm, part of the National Aquaculture Group (NAQUA). The prawn farm pumps water into the initial stages of the farm and then uses a gravity-driven system to distribute the water. The effluent drains immediately adjacent to the sole marina available for visitor access to the region, and is directly inshore from the whale shark aggregation site. The prawn farm in Al Lith was established before focused study began on either the mantas or whale sharks, hindering a full understanding of the potential impacts (see also Hozumi 2015). NAQUA has recently introduced several sets of open-ocean fish cages growing barramundi (*Lates calcarifer*, a non-native species) approximately 15 km north of Shib Habil.

One aspect of Vision 2030 is continued and rapid development of aquaculture along the Saudi Arabian coast. There are at least two factors that will drive a major increase in demand for marine protein in the proposed plan for Saudi Arabia: a need to establish greater levels of food security (i.e., less reliance on imported foods) and an increase in international tourism and luxury seaside resorts, both of which can be expected to create demand for local seafood. A 2016 study identifying suitable potential sites for finfish cages along the Saudi Arabian coast suggested that the two southern-most sites in the study have the most potential (Salama et al. 2016). These locations were chosen due to their distance from industrial and residential areas (Salama et al. 2016), but they also align with nearshore reefs shown to be frequently used by *M. alfredi* (Braun et al. 2014). The tagging studies (see Sect. 8.4.2) can be used to inform development along the Saudi coastline, much like Kessel et al. (2017) focused on habitat use of the reef mantas in Sudan where development is being considered within the protected area.

8.5.5 Existing Protected Areas

Of all the countries bordering the Red Sea, Egypt and Sudan appear to have relative success in implementing and enforcing some forms of marine resource protection. Between 1983 and 2006, the Egyptian Environmental Affairs Agency (EEAA) declared the following areas as national parks or protected areas: Ras Mohamed, Nabq, and Abu Galum in the Sinai Peninsula, as well as Elba, Wadi El Gemal, and the Red Sea North Islands in the Red Sea Governorate (www.eeaa.gov.eg). These protected areas include both terrestrial and marine components and enjoy varying degrees of protection. While some tourism activities are permitted in each of these parks, entry usually requires special permits, and extractive activities (e.g., fishing) are prohibited. Outside the borders of protectorates, fishing regulations also prohibit the fishing of sharks and endangered species. Before the Arab Spring political uprising of 2011, the enforcement of protective regulations was carried out partially by rangers appointed by a branch of the EEAA and partially by the coast guard and the military. The current status of enforcement is unclear, although the same entities remain responsible. Anecdotal evidence from within the diving community in Sinai and Hurghada suggests possible higher levels of non-compliance by fishermen post-2011 as well as some potential positive impacts on coral reefs due to the reduction in tourism in recent years.

Sudan is home to some of the healthiest reefs in the Red Sea with relatively intact populations of sharks (Hussey et al. 2013; Spaet et al. 2016) and other top predators (Kattan et al. 2017). Currently, two marine protected areas exist in attempt to recognize and preserve the biodiversity and unique natural resources found along the coast of the Red Sea State: (1) Sanganeb Atoll Marine National Park was established in 1990, encompassing 22 km^2 around a prominent deepwater atoll, and (2) Dunganab Bay and Mukkawar Island National Park, a 2800 km^2 reserve established in 2004 that includes a mosaic of undisturbed coral reef, mangrove, seagrass, and intertidal mudflat habitats. These habitats collectively support regionally significant populations of endangered dugongs, sharks, manta rays, dolphins, nesting sea turtles, and birds (sudanmarineparks.info). Together these two sites were declared a World Heritage Site in July 2016. A management structure for these parks is in place, but faces three major challenges: (1) there is no broad community involvement, (2) it is missing a general facility for monitoring and enforcement, and (3) it lacks the capacity to absorb future growth in the region (Chekchak and Klaus 2013). Fortunately, hitherto underdeveloped levels of tourism (Chekchak and Klaus 2013) and fisheries (Tesfamichael and Pitcher 2006) have resulted in relatively minimal impacts to Sudanese marine resources. Some degree of self-policing by the local liveaboard dive boats creates a kind of *de facto* protection force, as the quality of the reefs is a driving factor in the success of the local ecotourism industry (see Sect. 8.5.3). With increasing interest in coastal development, fisheries, and tourism to the region, however, much effort is required to plan and coordinate for the long-term health of these fragile marine ecosystems (Chekchak 2013; Chekchak and Klaus 2013).

The Kingdom of Saudi Arabia, on the other hand, which controls most of the eastern coast of the Red Sea, has declared only two marine protected areas (MPAs), both of which currently appear to be little more than paper parks: the Farasan Islands and the island of Um Al-Qamari. The Farasan Islands ($3310 km^2$, see Fig. 8.3) were officially declared as protected in 1996 (Wood 2007). The islands are known to host a unique seasonal aggregation of the parrotfish *Hipposcarus harid* (Gladstone 1996; Spaet 2013). This MPA briefly enjoyed some success due to strong initial community involvement. However, its success was short-lived, as lack of long-term training and awareness programs for local rangers, combined with growing commercial fisheries in the area, led to a decline in the effectiveness of this MPA (Gladstone 2000). The island of Um Al-Qamari (located near Al-Qunfidhah in the Farasan Banks) was declared a protectorate in 1977, much earlier than the Farasan Islands, with an area of $2 km^2$ (Wood 2007). It was designated to protect a resident population of seabirds, and it is not clear whether any enforcement of protection currently takes place on the island or the surrounding waters. In addition to these declared MPAs, Saudi Arabia issued a royal decree in 2008 putting a total ban on the fishing of sharks (Spaet et al. 2016). However, little to no enforcement of this ban takes place; shark fishing occurs on a daily basis, and hundreds of sharks are landed in Saudi fish markets every month (Spaet and Berumen 2015).

8.5.6 Marine Invasive Species

The primary invasion threat Red Sea fish populations presently face appears to be limited to potential escapees from aquaculture operations. There are very few cases of invasive fishes colonizing the Red Sea (e.g., Por 1978). Planned rapid expansion of aquaculture efforts, particularly in Saudi Arabia, includes dramatic increases in open-sea cage farming of fishes (see Sect. 8.5.4). These operations have already commenced near Al Lith (see Fig. 8.3) and near Duba (north of Al Wajh). Adult barramundi (*Lates calcifer*) are routinely spotted at the Al Lith marina, apparently having escaped from the cages ~15 km to the north. Surveys of coastal reefs in the area, however, have yet to detect any barramundi between the marina and the fish farm (Alex Kattan, unpublished data). Barramundi may require estuarine or riverine areas for successful completion of some parts of the early life cycle (Copland and Grey 1987). The lack of these habitats in the Red Sea may preclude the establishment of a wild barramundi population, but large numbers of escaped barramundi (which are voracious predators) could still exert an impact on native reef fish populations.

As described in Sect. 8.4.3, the Red Sea appears to 'export' far more invasive species (into the Mediterranean) than it 'imports' (i.e., reverse Lessepsians are rare). It is possible that the relatively high temperatures and salinity levels may present physiological challenges for non-native species. If this is the mechanism reducing Mediterranean immigrants to the Red Sea, it may also be inhibiting potential invasive species that would otherwise arrive via traditional mechanisms (e.g., ship ballast water). These hypotheses remain to be formally tested.

Red Sea fishes have evolved in and adapted to some of the most challenging conditions in which modern coral reef ecosystems appear to be thriving. The opposite sides of the central Red Sea currently offer an interesting contrast that may reflect the impacts of anthropogenic pressures in recent decades. On one side, reef fish communities may be greatly altered by heavy fishing pressure and coastal development. On the other side, a lack of infrastructure and locally-initiated *de facto* protection may be preserving healthy reef communities, and may even be supplying important larval input to overfished populations across the basin. The anticipated additional future stressors (ranging from local to widespread) may create even more challenging conditions for Red Sea reefs, particularly planned 'giga-projects' with the potential to impact large portions of coastline. Responsible and sustainable management of Red Sea reef fish populations will require a more thorough understanding of the status of fisheries, the nuances of local ecology, and various aspects of connectivity. More than 250 years have passed since the first European natural historians began investigations into Red Sea fishes, yet we still have much to learn.

Acknowledgements Data acknowledgement: This research has made use of data and software tools provided by Wildbook for Whale Sharks, an online mark-recapture database operated by the non-profit scientific organization Wild Me with support from public donations and the Qatar Whale Shark Research Project.

We thank Malek Amr Gusti, Manal Bamashmos, and Prof. Khaled Salama for their assistance with Arabic translations.

We thank the staff of the KAUST Biosciences Core Laboratory for their assistance in the genetic analyses described in Sect. 8.3.3.

References

Abu El-Regal MA (1999) Some biological and ecological studies on the larvae of coral reef fishes in Sharm El-Sheikh (Gulf of Aqaba-Red Sea). MSc thesis. Suez Canal University, Egypt

Abu El-Regal MA (2008) Ecological studies on the ichthyoplankton of coral reef fishes in Hurghada, Red Sea, Egypt. PhD thesis. Suez Canal University, Egpyt

Abu El-Regal M, Kon T (2008) First record of the paedomorphic fish *Schindleria* (Gobioidei, Schindleriidae) from the Red Sea. J Fish Biol 72:1539–1543

Ackerman JL, Bellwood DR (2000) Reef fish assemblages: a re-evaluation using enclosed rotenone stations. Mar Ecol Prog Ser 206:227–237

Adams AJ, Dahlgren CP, Kellison GT, Kendall MS, Layman CA, Ley JA, Nagelkerken I, Serafy JE (2006) Nursery function of tropical back-reef systems. Mar Ecol Prog Ser 318:287–301

Aguilar-Perera A, Appeldoorn RS (2008) Spatial distribution of marine fishes along a cross-shelf gradient containing a continuum of mangrove–seagrass–coral reefs off southwestern Puerto Rico. Estuar Coast Shelf Sci 76:378–394

Almany GR, Planes S, Thorrold SR, Berumen ML, Bode M, Saenz-Agudelo P, Bonin MC, Frisch AJ, Harrison HB, Messmer V, Nanninga GB, Priest MA, Srinivasan M, Sinclair-Taylor T, Williamson DH, Jones GP (2017) Larval fish dispersal in a coral reef seascape. Nat Ecol Evol 1:148

Azzurro E, Maynou F, Belmaker J, Golani D, Crooks JA (2016) Lag times in Lessepsian fish invasion. Biol Invasions 18:2761–2772

Bearhop S, Adams CE, Waldron S, Fuller RA, MacLeod H (2004) Determining trophic niche width: a novel approach using stable isotope analysis. J Anim Ecol 73:1007–1012

Ben-Tuvia AA (1971) On the occurrence of the Mediterranean serranid fish *Dicentrarchus punctatus* (Bloch) in the Gulf of Suez. Am Soc Ichthyol Herpetol 1971:741–743

Ben-Tzvi O, Kiflawi M, Gildor H, Abelson A (2007) Possible effects of downwelling on the recruitment of coral reef fishes to the Eilat (Red Sea) coral reefs. Limnol Oceanogr 52:2618–2628

Ben-Tzvi O, Kiflawi M, Gaines SD, Al-Zibdah M, Sheehy MS, Paradis GL, Abelson A (2008) Tracking recruitment pathways of *Chromis viridis* in the Gulf of Aqaba using otolith chemistry. Mar Ecol Prog Ser 359:229–238

Ben-Tzvi O, Abelson A, Gaines SD, Bernardi G, Beldade R, Sheehy MS, Paradis GL, Kiflawi M (2012) Evidence for cohesive dispersal in the sea. PLoS One 7:e42672

Bernardi G, Azzurro E, Golani D, Miller MR (2016) Genomic signatures of rapid adaptive evolution in the bluespotted cornetfish, a Mediterranean Lessepsian invader. Mol Ecol 25:3384–3396

Berumen ML, Hoey AS, Bass WH, Bouwmeester J, Catania D, Cochran JE, Khalil MT, Miyake S, Mughal MR, Spät JL, Saenz-Agudelo P (2013) The status of coral reef ecology research in the Red Sea. Coral Reefs 32:737–748

Berumen ML, Braun CD, Cochran JE, Skomal GB, Thorrold SR (2014) Movement patterns of juvenile whale sharks tagged at an aggregation site in the Red Sea. PLoS One 9:e103536

Berumen ML, DiBattista JD, Rocha LA (2017) Introduction to virtual issue on Red Sea and Western Indian Ocean biogeography. J Biogeogr 44:1923–1926

Booth DJ, Bond N, Macreadie P (2011) Detecting range shifts among Australian fishes in response to climate change. Mar Freshw Res 62:1027–1042

Borman A, Wood TR, Black HC, Anderson EG, Oesterling MJ, Womack M, Rose WC (1946) The role of arginine in growth with some observations on the effects of argininic acid. J Biol Chem 166:585–594

Bouillon S, Raman AV, Dauby P, Dehairs F (2002) Carbon and nitrogen stable isotope ratios of subtidal benthic invertebrates in an estuarine mangrove ecosystem (Andhra Pradesh, India). Estuar Coast Shelf Sci 54:901–913

Bowen BW, Rocha LA, Toonen RJ, Karl SA, ToBo Laboratory (2013) The origins of tropical marine biodiversity. Trends Ecol Evol 28:359–366

Braun CD, Skomal GB, Thorrold SR, Berumen ML (2014) Diving behavior of the reef manta ray links coral reefs with adjacent deep pelagic habitats. PLoS One 9:e88170

Braun CD, Skomal GB, Thorrold SR, Berumen ML (2015) Movements of the reef manta ray (*Manta alfredi*) in the Red Sea using satellite and acoustic telemetry. Mar Biol 162:2351–2362

Brokovich E, Einbinder S, Kark S, Shashar N, Kiflawi M (2007) A deep nursery for juveniles of the zebra angelfish *Genicanthus caudovittatus*. Environ Biol Fish 80:1–6

Brokovich E, Einbinder S, Shashar N, Kiflawi M, Kark S (2008) Descending to the twilight-zone: changes in coral reef fish assemblages along a depth gradient down to 65 m. Mar Ecol Prog Ser 371:253–262

Brokovich E, Ayalon I, Einbinder S, Segev N, Shaked Y, Genin A, Kark S, Kiflawi M (2010) Grazing pressure on coral reefs decreases across a wide depth gradient in the Gulf of Aqaba, Red Sea. Mar Ecol Prog Ser 399:69–80

Bruckner A, Rowlands G, Riegl B, Purkis S, Williams A, Renaud P (2011) Khaled bin Sultan Living Oceans Foundation Atlas of Saudi Arabian Red Sea Marine Habitats. Panoramic Press, Phoenix

Campana SE, Thorrold SR (2001) Otoliths, increments, and elements: keys to a comprehensive understanding of fish populations? Can J Fish Aquat Sci 58:30–38

Cantin NE, Cohen AL, Karnauskas KB, Tarrant AM, McCorkle DC (2010) Ocean warming slows coral growth in the central Red Sea. Science 329:322–325

Catlin J, Jones R (2010) Whale shark tourism at Ningaloo Marine Park: a longitudinal study of wildlife tourism. Tour Manag 31:386–394

Chekchak T (2013) Toward a sustainable future for the Red Sea coast of Sudan. Part 2: socio-economic and governance survey. Cousteau Society, New York

Chekchak T, Klaus R (2013) Toward a sustainable future for the Red Sea coast of Sudan. Part 1: coastal and marine habitats survey. Cousteau Society, New York

Clements KD, German DP, Piché J, Tribollet A, Choat JH (2016) Integrating ecological roles and trophic diversification on coral reefs: multiple lines of evidence identify parrotfishes as microphages. Biol J Linnean Soc 120:729–751

Cocheret de la Morinière E, Ollux BJA, Nagelkerken I, van der Velde G (2002) Postsettlement life cycle migration patterns and habitat preference of coral reef fish that use seagrass and mangrove habitats as nurseries. Estuar Coast Shelf Sci 55:309–321

Cochran JEM, Hardenstine RS, Braun CD, Skomal GB, Thorrold SR, Xu K, Genton MG, Berumen ML (2016) Population structure of a whale shark *Rhincodon typus* aggregation in the Red Sea. J Fish Biol 89:1570–1582

Coker DJ, DiBattista JD, Sinclair-Taylor TH, Berumen ML (2018) Spatial patterns of cryptobenthic coral-reef fishes in the Red Sea. Coral Reefs 37:193–199

Coleman RR, Eble JA, DiBattista JD, Rocha LA, Randall JE, Berumen ML, Bowen BW (2016) Regal phylogeography: range-wide survey of the marine angelfish *Pygoplites diacanthus* reveals evolutionary partitions between the Red Sea, Indian Ocean, and Pacific Ocean. Mol Phylogenet Evol 100:243–253

Copland JW, Grey DL (1987) Management of wild and cultured sea bass / barramundi (*Lates calcarifer*): proceedings of an international workshop held at Darwin, N.T., Australia, 24–30 September 1986. ACIAR Proceedings No. 20, Australian Centre for International Agricultural Research, Canberra

Depczynski M, Bellwood DR (2003) The role of cryptobenthic reef fishes in coral reef trophodynamics. Mar Ecol Prog Ser 256:183–191

Depczynski M, Bellwood DR (2006) Extremes, plasticity, and invariance in vertebrate life history traits: insights from coral reef fishes. Ecology 87:3119–3127

Depczynski M, Fulton CJ, Marnane MJ, Bellwood DR (2007) Life history patterns shape energy allocation among fishes on coral reefs. Oecologia 153:111–120

Desalle R (2006) Species discovery versus species identification in DNA barcoding efforts: response to Rubinoff. Cons Biol 20:1545–1547

DiBattista JD, Berumen ML, Gaither MR, Rocha LA, Eble JA, Choat JH, Craig MT, Skillings DJ, Bowen BW (2013) After continents divide: comparative phylogeography of reef fishes from the Red Sea and Indian Ocean. J Biogeogr 40:1170–1181

DiBattista JD, Choat JH, Gaither MR, Hobbs JPA, Lozano-Cortés DF, Myers RF, Paulay G, Rocha LA, Toonen RJ, Westneat MW, Berumen ML (2016a) On the origin of endemic species in the Red Sea. J Biogeogr 43:13–30

DiBattista JD, Roberts MB, Bouwmeester J, Bowen BW, Coker DJ, Lozano-Cortés DF, Choat JH, Gaither MR, Hobbs JPA, Khalil MT, Kochzius M, Myers RF, Paulay G, Robitzch VSN, Saenz-Agudelo P, Salas E, Sinclair-Taylor T, Toonen RJ, Westneat MW, Williams ST, Berumen ML (2016b) A review of contemporary patterns of endemism in the Red Sea. J Biogeogr 43:423–439

DiBattista JD, Gaither MR, Hobbs J-PA, Saenz-Agudelo P, Piatek MJ, Bowen BW, Rocha LA, Choat JH, McIlwain JH, Priest MA, Sinclair-Taylor T, Berumen ML (2017) Comparative phylogeography of reef fishes from the Gulf of Aden to the Arabian Sea reveals two cryptic lineages. Coral Reefs 36:625–638

Docmac F, Araya M, Hinojosa IA, Dorador C, Harrod C (2017) Habitat coupling writ large: pelagic-derived materials fuel benthivorous macroalgal reef fishes in an upwelling zone. Ecology 98:2267–2272

Dromard CR, Bouchon-Navaro Y, Cordonnier S, Fontaine MF, Verlaque M, Harmelin-Vivien M, Bouchon C (2013) Resource use of two damselfishes, *Stegastes planifrons* and *Stegastes adustus*, on Guadeloupean reefs (Lesser Antilles): inference from stomach content and stable isotope analysis. J Exp Mar Biol Ecol 440:116–125

Edwards FJ (1987) Climate and oceanography. In: Edwards AJ, Head SM (eds) Red Sea – (Key environments). Pergamon Books Ltd, Exeter, pp 45–69

Elsdon TS, Wells BK, Campana SE, Gillanders BM, Jones CM, Limburg KE, Secor DH, Thorrold SR, Walther BD (2008) Otolith chemistry to describe movements and life-history parameters of fishes: hypotheses, assumptions, limitations and inferences. Oceanogr Mar Biol Annu Rev 46:297–330

Emms MA (2015) Broad-scale population genetics of the host sea anemone, *Heteractis magnifica*. MSc thesis. King Abdullah University of Science and Technology, Saudi Arabia

Fernandez-Silva I, Randall JE, Coleman RR, DiBattista JD, Rocha LA, Reimer JD, Meyer CG, Bowen BW (2015) Yellow tails in the Red Sea: phylogeography of the Indo-Pacific goatfish *Mulloidichthys flavolineatus* reveals isolation in peripheral provinces and cryptic evolutionary lineages. J Biogeogr 42:2402–2413

Fernandez-Silva I, Randall JE, Golani D, Bogorodsky SV (2016) *Mulloidichthys flavolineatusflavicaudus* Fernandez-Silva & Randall (Perciformes, Mullidae), a new subspecies of goatfish from the Red Sea and Arabian Sea. ZooKeys 605:131–157

Fricke R, Abu El-Regal M (2017a) *Schindleria elongata*, a new species of paedomorphic gobioid from the Red Sea (Teleostei: Schindleriidae). J Fish Biol 90:1883–1890

Fricke R, Abu El-Regal MA (2017b) *Schindleria nigropunctata*, a new species of paedomorphic gobioid fish from the Red Sea (Teleostei: Schindleriidae). Mar Biodivers:1–5. https://doi.org/10.1007/s12526-017-0831-z

Fricke HW, Hottinger L (1983) Coral bioherms below the euphotic zone in the Red Sea. Mar Ecol Prog Ser 11:113–117

Fricke HW, Knauer B (1986) Diversity and spatial pattern of coral communities in the Red Sea upper twilight zone. Oecologia 71:29–37

Fricke HW, Schuhmacher H (1983) The depth limits of Red Sea stony corals: an ecophysiological problem (a deep diving survey by submersible). Mar Ecol 4:163–194

Froukh TJ (2001) Studies on taxonomy and ecology of some fish larvae from the Gulf of Aqaba. MSc thesis. University of Jordan, Jordan

Furby KA, Bouwmeester J, Berumen ML (2013) Susceptibility of central Red Sea corals during a major bleaching event. Coral Reefs 32:505–513

Galil BS, Boero F, Campbell ML, Carlton JT, Cook E, Fraschetti S, Gollasch S, Hewitt CL, Jelmert A, Macpherson E, Marchini A, McKenzie C, Minchin D, Occhipinti-Ambrogi A, Ojaveer H, Olenin S, Piraino S, Ruiz GM (2015) "Double trouble": the expansion of the Suez Canal and marine bioinvasions in the Mediterranean Sea. Biol Invasions 17:973–976

Giles EC, Saenz-Agudelo P, Hussey NE, Ravasi T, Berumen ML (2015) Exploring seascape genetics and kinship in the reef sponge *Stylissa carteri* in the Red Sea. Ecol Evol 5:2487–2502

Gladstone W (1996) Unique annual aggregation of longnose parrotfish (*Hipposcarus harid*) at Farasan Island (Saudi Arabia, Red Sea). Copeia 1996:483–485

Gladstone W (2000) The ecological and social basis for management of a Red Sea marine-protected area. Ocean Coast Manag 43:1015–1032

Goatley CH, Brandl SJ (2017) Cryptobenthic reef fishes. Curr Biol 27:R452–R454

Golani D (1998) Impact of Red Sea fish migrants through the Suez Canal on the aquatic environment of the eastern Mediterranean. Yale F&ES Bull 103:375–387

Golani D (1999) The Gulf of Suez ichthyofauna-assemblage pool for Lessepsian migration into the Mediterranean. Isr J Zool 45:79–90

Golani D, Appelbaum-Golani B (2010) Fish invasions of the Mediterranean Sea: change and renewal. Coronet Books Incorporated, Philadelphia

Golani D, Bogorodsky S (2010) The fishes of the Red Sea – reappraisal and updated checklist. Zootaxa 2463:1–135

Golani D, Massutí E, Quignard J-P, Dulcic J, Azzurro E (2017) CIESM atlas of exotic fishes in the Mediterranean. http://www.ciesm.org/atlas/appendix 1.html

Goren M, Galil BS (2005) A review of changes in the fish assemblages of Levantine inland and marine ecosystems following the introduction of non-native fishes. J Appl Ichthyol 21:364–370

Green AL, Maypa AP, Almany GR, Rhodes KL, Weeks R, Abesamis RA, Gleason MG, Mumby PH, White AT (2015) Larval dispersal and movement patterns of coral reef fishes, and implications for marine reserve network design. Biol Rev 90:1215–1247

Gudger EW (1938) Four whale sharks rammed by steamers in the Red Sea region. Copeia 1938(4):170–173

Gudger EW (1940) Whale sharks rammed by ocean vessels: how these sluggish leviathans aid in their own destruction. New Engl Nat 7:1–10

Hansen T (1962) *Det Lykkelige Arabien*. Gyldendal, Copenhagen. Based on a 1964 translation published as: Arabia Felix: the Danish expedition of 1761–1767. Wm. Collins Sons & Co. Ltd, Glasgow

Hanson JM, Chouinard GA (2002) Diet of Atlantic cod in the southern gulf of St Lawrence as an index of ecosystem change, 1959–2000. J Fish Biol 60:902–922

Hare PE, Fogel ML, Stafford TW, Mitchell AD, Hoering TC (1991) The isotopic composition of carbon and nitrogen in individual amino-acids isolated from modern and fossil proteins. J Archaeol Sci 18:277–292

Harrison HB, Williamson DH, Evans RD, Almany GR, Thorrold SR, Russ GR, Feldheim KA, Van Herwerden L, Planes S, Srinivasan M, Berumen ML, Jones GP (2012) Larval export from marine reserves and the recruitment benefit for fish and fisheries. Curr Biol 22:1023–1028

Herler J (2007) Microhabitats and ecomorphology of coral-and coral rock-associated gobiid fish (Teleostei: Gobiidae) in the northern Red Sea. Mar Ecol 28:82–94

Herler J, Hilgers H (2007) A synopsis of coral and coral-rock associated gobies (Pisces: Gobiidae) from the Gulf of Aqaba, northern Red Sea. Aqua J Ichthyol Aquat Biol 10:103–132

Hernaman V, Munday P (2005) Life-history characteristics of coral reef gobies. I. Growth and life-span. Mar Ecol Prog Ser 290:207–221

Hernaman V, Munday P (2007) Evolution of mating systems in coral reef gobies and constraints on mating system plasticity. Coral Reefs 26:585–595

Hinderstein LM, Marr JCA, Martinez FA, Dowgiallo MJ, Puglise KA, Pyle RL, Zawada DG, Appeldoorn R (2010) Theme section on "Mesophotic coral ecosystems: characterization, ecology, and management". Coral Reefs 29:247–251

Hoegh-Guldberg O (1999) Climate change, coral bleaching and the future of the world's coral reefs. Mar Freshw Res 50:839–866

Hozumi A (2015) Environmental factors affecting the whale shark aggregation site in the south central Red Sea. PhD thesis. King Abdullah University of Science and Technology, Saudi Arabia

Hussey NE, Stroh N, Klaus R, Chekchak T, Kessel ST (2013) SCUBA diver observations and placard tags to monitor grey reef sharks, *Carcharhinus amblyrhynchos*, at Sha'ab Rumi, the Sudan: assessment and future directions. J Mar Biol Assoc UK 93:299–308

Isari S, Pearman JK, Casas L, Michell CT, Curdia J, Berumen ML, Irigoien X (2017a) Exploring the larval fish community of the central Red Sea with an integrated morphological and molecular approach. PLoS One 12:e0182503

Isari S, Pearman JK, Casas L, Michell CT, Curdia J, Berumen ML, Irigoien X (2017b) Integrating morphology and genetics to study the larval community of gobies in the central Arabian Red Sea. Abstracts from the 10th Indo-Pacific Fish Conference, Tahiti, French Polynesia. p83

Jin D, Kite-Powell H, Hoagland P, Solow A (2012) A bioeconomic analysis of traditional fisheries in the Red Sea. Mar Resour Econ 27:137–148

Johansen J, Jones G (2011) Increasing ocean temperature reduces the metabolic performance and swimming ability of coral reef damselfishes. Glob Chang Biol 17:2971–2979

Kahng SE, Garcia-Sais JR, Spalding HL, Brokovich E, Wagner D, Weil E, Hinderstein L, Toonen RJ (2010) Community ecology of mesophotic coral reef ecosystems. Coral Reefs 29:255–275

Kalish JM (1991) $\delta^{13}C$ and $\delta^{18}O$ isotopic disequilibria in fish otoliths: metabolic and kinetic effects. Mar Ecol Prog Ser 75:191–203

Kattan A, Coker DJ, Berumen ML (2017) Reef fish communities in the central Red Sea show evidence of asymmetrical fishing pressure. Mar Biodivers 47:1227–1238

Kennedy BP, Klaue A, Blum JD, Folt CL, Nislow KH (2002) Reconstructing the lives of fish using Sr isotopes in otoliths. Can J Fish Aquat Sci 59:925–929

Kessel ST, Elamin NA, Yurkowski DJ, Chekchak T, Walter RP, Klaus R, Hill G, Hussey NE (2017) Conservation of reef manta rays (*Manta alfredi*) in a UNESCO world heritage site: large-scale island development or sustainable tourism? PLoS One 12:e0185419

Khalaf MA, Kochzius M (1992) Community structure and biogeography of shore fishes in the Gulf of Aqaba, Red Sea. Helgol Mar Res 55:252–284

Khalil MT, Bouwmeester J, Berumen ML (2017) Spatial variation in coral reef fish and benthic communities in the central Saudi Arabian Red Sea. PeerJ 5:e3410

Kieckbusch DK, Koch MS, Serafy JE, Anderson WT (2004) Trophic linkages among primary producers and consumers in fringing mangroves of subtropical lagoons. Bull Mar Sci 74:271–285

Kimirei IA, Nagelkerken I, Trommelen M, Blankers P, Van Hoytema N, Hoeijmakers D, Huijbers CM, Mgaya YD, Rypel AL (2013) What drives ontogenetic niche shifts of fishes in coral reef ecosystems? Ecosystems 16:783–796

Kimmerling N, Zuqert O, Amitai G, Gurevich T, Armoza-Zvuloni R, Kolesnikov I, Berenshtein I, Melamed S, Gilad S, Benjamin S, Rivlin A, Moti O, Paris CB, Holzman R, Kiflawi M, Sorek R (2018) Quantitative species-level ecology of reef fish larvae via metabarcoding. Nat Ecol Evol 2:306–316

Kotb MM, Hanafy MH, Rirache H, Matsumura S, Al-Sofyani A, Ahmed A, Bawazir G, Al Horani F (2008) Status of coral reefs in the Red Sea and Gulf of Aden region. In: Wilkinson C (ed) Status of Coral Reefs of the World: 2008. Australian Institute of Marine Science, Townsville, pp 67–78

Kristensen E, Lee SY, Mangion P, Quintana CO, Valdemarsen T (2017) Trophic discrimination of stable isotopes and potential food source partitioning by leaf-eating crabs in mangrove environments. Limnol Oceanogr 62:2097–2112

Larsen T, Taylor DL, Leigh MB, O'Brien DM (2009) Stable isotope fingerprinting: a novel method for identifying plant, fungal, or bacterial origins of amino acids. Ecology 90:3526–3535

Larsen T, Wooller MJ, Fogel ML, O'Brien DM (2012) Can amino acid carbon isotope ratios distinguish primary producers in a mangrove ecosystem? Rapid Commun Mass Spectrom 26:1541–1548

Larsen T, Ventura M, Andersen N, O'Brien DM, Piatkowski U, McCarthy MD (2013) Tracing carbon sources through aquatic and terrestrial food webs using amino acid stable isotope fingerprinting. PLoS One 8:e73441

Layman CA (2007) What can stable isotope ratios reveal about mangroves as fish habitat? Bull Mar Sci 80:513–527

Letourneur Y, De Loma TL, Richard P, Harmelin-Vivien ML, Cresson P, Banaru D, Fontaine MF, Gref T, Planes S (2013) Identifying carbon sources and trophic position of coral reef fishes using diet and stable isotope ($\delta^{15}N$ and $\delta^{13}C$) analyses in two contrasted bays in Moorea, French Polynesia. Coral Reefs 32:1091–1102

Lieske E, Myers RT (2004) Coral Reef Guide: Red Sea. HarperCollins Publishers Ltd, London

Malcolm H, Jordan A, Smith SA (2010) Biogeographical and cross-shelf patterns of reef fish assemblages in a transition zone. Mar Biodivers 40:181–193

Marguillier S, van der Velde G, Dehairs F, Hemminga MA, Rajagopal S (1997) Trophic relationships in an interlinked mangrove-seagrass ecosystem as traced by $\delta^{13}C$ and $\delta^{15}N$. Mar Ecol Prog Ser 151:115–121

Marshall A, Kashiwagi T, Bennett MB, Deakos M, Stevens G, McGregor F, Clark T, Ishihara H, Sato K (2011) *Manta alfredi*. The IUCN red list of threatened species 2011: e.T195459A8969079

McCook LJ, Almany GR, Berumen ML, Day JC, Green AL, Jones GP, Leis JM, Planes S, Russ GR, Sale PF, Thorrold SR (2009) Management under uncertainty: guide-lines for incorporating connectivity into the protection of coral reefs. Coral Reefs 28:353–366

McMahon KW, Fogel ML, Johnson BJ, Houghton LA, Thorrold SR (2011a) A new method to reconstruct fish diet and movement patterns from $\delta^{13}C$ values in otolith amino acids. Can J Fish Aquat Sci 68:1330–1340

McMahon KW, Berumen ML, Mateo I, Elsdon TS, Thorrold SR (2011b) Carbon isotopes in otolith amino acids identify residency of juvenile snapper (Family: Lutjanidae) in coastal nurseries. Coral Reefs 30:1135–1145

McMahon KW, Berumen ML, Thorrold SR (2012) Linking habitat mosaics and connectivity in a coral reef seascape. Proc Natl Acad Sci USA 109:15372–15376

McMahon KW, Thorrold SR, Houghton LA, Berumen ML (2016) Tracing carbon flow through coral reef food webs using a compound-specific stable isotope approach. Oecologia 180:809–821

Meekan M, Austin CM, Tan MH, Wei NWV, Miller A, Pierce SJ, Rowat D, Stevens G, Davies TK, Ponzo A, Gan HM (2017) iDNA at sea: recovery of whale shark (*Rhincodon typus*) mitochondrial DNA sequences from the whale shark copepod (*Pandarus rhincodonicus*) confirms global population structure. Front Mar Sci 4:420

Meeker ND, Hutchinson SA, Ho L, Trede NS (2007) Method for isolation of PCR-ready genomic DNA from zebrafish tissues. BioTechniques 43(5):610–614

Morcos SA (1970) Physical and chemical oceanography of the Red Sea. Oceanogr Mar Biol Annu Rev 8:73–202

Mulcahy SA, Killingley JS, Phleger CF, Berger WH (1979) Isotopic composition of otoliths from a benthopelagic fish, *Coryphaenoides acrolepis*, Macrouridae: Gadiformes. Oceanol Acta 2:423–427

Mumby PJ, Edwards AJ, Arias-González JE, Lindeman KC, Blackwell PG, Gall A, Gorczynska MI, Harborne AR, Pescod CL, Renken H, Wabnitz CC (2004) Mangroves enhance the biomass of coral reef fish communities in the Caribbean. Nature 427:533–536

Munday PL, Jones GP, Pratchett MS, Williams AJ (2008) Climate change and the future for coral reef fishes. Fish Fish 9:261–285

Nagelkerken I, Dorenbosch M, Verberk WCEP, Cocheret de la Moriniére E, van der Velde G (2000) Importance of shallow-water biotopes of a Caribbean bay for juvenile coral reef fishes: patterns in biotope association, community structure and spatial distribution. Mar Ecol Prog Ser 202:175–192

Nagelkerken I, Blaber SJM, Bouillon S, Green P, Haywood M, Kirton LG, Meynecke JO, Pawlik J, Penrose HM, Sasekumar A, Somerfield PJ (2008) The habitat function of mangroves for terrestrial and marine fauna: a review. Aquat Bot 89:155–185

Nanninga GB (2013) Merging approaches to explore connectivity in the anemonefish, *Amphiprion bicinctus*, along the Saudi Arabian coast of the Red Sea. Ph.D. Thesis. King Abdullah University of Science and Technology, Saudi Arabia

Nanninga GB, Saenz-Agudelo P, Manica A, Berumen ML (2014) Environmental gradients predict the genetic population structure of a coral reef fish in the Red Sea. Mol Ecol 23:591–602

Nanninga GB, Saenz-Agudelo P, Zhan P, Hoteit I, Berumen ML (2015) Not finding Nemo: limited reef-scale retention in a coral reef fish. Coral Reefs 34:383–392

Ngugi DK, Antunes A, Brune A, Stingl U (2012) Biogeography of pelagic bacterioplankton across an antagonistic temperature-salinity gradient in the Red Sea. Mol Ecol 21:388–405

Nowicki JP, Miller GM, Munday PL (2012) Interactive effects of elevated temperature and CO_2 on foraging behavior of juvenile coral reef fish. J Exp Mar Biol Ecol 412:46–51

Osman EO, Smith DJ, Ziegler M, Kürten B, Conrad C, El-Haddad KM, Voolstra CR, Suggett DJ (2018) Thermal refugia against coral bleaching throughout the northern Red Sea. Glob Chang Biol 24:e474–e484

Pierce SJ, Norman B (2016) *Rhincodon typus*. The IUCN red list of threatened species 2016: e.T19488A2365291

Por FD (1978) Lessepsian migration. The influx of Red Sea biota into the Mediterranean by way of the Suez Canal. Springer Verlag, Berlin

Post DM (2002) Using stable isotopes to estimate trophic position: models, methods, and assumptions. Ecology 83:703–718

Pratchett MS, Munday P, Wilson SK (2008) Effects of climate-induced coral bleaching on coral-reef fishes - ecological and economical consequences. Oceanogr Mar Biol Annu Rev 46:251–296

Price A, Ghazi S, Tkaczynski P, Venkatachalam A, Santillan A, Pancho T, Metcalfe R, Saunders J (2014) Shifting environmental baselines in the Red Sea. Mar Pollut Bull 78:96–101

Priest MA, DiBattista JD, McIlwain JL, Taylor BM, Hussey NE, Berumen ML (2016) A bridge too far: dispersal barriers and cryptic speciation in an Arabian Peninsula grouper (*Cephalopholis hemistiktos*). J Biogeogr 43:820–832

Racault MF, Raitsos DE, Berumen ML, Brewin RJW, Platt T, Sathyendranath S, Hoteit I (2015) Phytoplankton phenology indices in coral reef ecosystems: application to ocean-colour observations in the Red Sea. Remote Sens Environ 160:222–234

Raitsos DE, Pradhan Y, Brewin RJ, Stenchikov G, Hoteit I (2013) Remote sensing the phytoplankton seasonal succession of the Red Sea. PLoS One 8:e64909

Raitsos DE, Brewin RJW, Zhan P, Dreano D, Pradhan Y, Nanninga GB, Hoteit I (2017) Sensing coral reef connectivity pathways from space. Sci Rep 7:9338

Reeds PJ (2000) Dispensable and indispensable amino acids for humans. J Nutr 130:1835S–1840S

ReFuGe 2020 Consortium (2015) The ReFuGe 2020 consortium—using "omics" approaches to explore the adaptability and resilience of coral holobionts to environmental change. Front Mar Sci 2:68

Roberts CM, Shepherd ARD, Ormond RFG (1992) Large-scale variation in assemblage structure of Red Sea butterflyfishes and angelfishes. J Biogeogr 19:239–250

Roberts MB, Jones GP, McCormick MI, Munday PL, Neale S, Thorrold S, Robitzch VSN, Berumen ML (2016) Homogeneity of coral reef communities across 8 degrees of latitude in the Saudi Arabian Red Sea. Mar Pollut Bull 105:558–565

Robinson DP, Jaidah MY, Bach SS, Rohner CA, Jabado RW, Ormond R, Pierce SJ (2017) Some like it hot: repeat migration and residency of whale sharks within an extreme natural environment. PLoS One 12:e0185360

Robitzch VSN (2017) The assessment of current biogeographic patterns of coral reef fishes in the Red Sea by incorporating their evolutionary and ecological background. Ph.D. Thesis. King Abdullah University of Science and Technology, Saudi Arabia

Robitzch V, Banguera-Hinestroza E, Sawall Y, Al-Sofyani A, Voolstra CR (2015) Absence of genetic differentiation in the coral *Pocillopora verrucosa* along environmental gradients of the Saudi Arabian Red Sea. Front Mar Sci 2:5

Robitzch VSN, Lozano-Cortés D, Kandler NM, Salas E, Berumen ML (2016) Productivity and sea surface temperature are correlated with the pelagic larval duration of damselfishes in the Red Sea. Mar Pollut Bull 105:566–574

Rohner CA, Armstrong AJ, Pierce SJ, Prebble CE, Cagua EF, Cochran JE, Berumen ML, Richardson AJ (2015) Whale sharks target dense prey patches of sergestid shrimp off Tanzania. J Plankton Res 37:352–362

Rothman SBS, Stern N, Goren M (2016) First record of the Indo-Pacific areolate grouper *Epinephelus areolatus* (Forsskål, 1775) (Perciformes: Epinephelidae) in the Mediterranean Sea. Zootaxa 4067:479–483

Rowat D, Engelhardt U (2007) Seychelles: a case study of community involvement in the development of whale shark ecotourism and its socioeconomic impact. Fish Res 84:109–113

Rowat D, Meekan MG, Engelhardt U, Pardigon B, Vely M (2007) Aggregations of juvenile whale sharks (*Rhincodon typus*) in the Gulf of Tadjoura, Djibouti. Environ Biol Fish 80:465–472

Rubinoff D (2006) Utility of mitochondrial DNA barcodes in species conservation. Consver Biol 20:1026–1033

Saenz-Agudelo P, DiBattista JD, Piatek MJ, Gaither MR, Harrison HB, Nanninga GB, Berumen ML (2015) Seascape genetics along environmental gradients in the Arabian Peninsula: insights from ddRAD sequencing of anemonefishes. Mol Ecol 24:6241–6255

Sala E, Kizilkaya Z, Yildirim D, Ballesteros E (2011) Alien marine fishes deplete algal biomass in the Eastern Mediterranean. PLoS One 6:e17356

Salama AJ, Satheesh S, Balqadi AA, Kitto MR (2016) Identifying suitable fin fish cage farming sites in the eastern Red Sea Coast, Saudi Arabia. Thalassas Int J Mar Sci 32:1–9

Salas De la Fuente EM (2016) Reef fish population genomics and hybridization using RADSeq: a case study with *Dascyllus trimaculatus*. PhD thesis. University of Santa Cruz

Sale PF, Feary DA, Burt JA, Bauman AG, Cavalcante GH, Drouillard KG, Kjerfve B, Marquis E, Trick CG, Usseglio P, Van Lavieren H (2011) The growing need for sustainable ecological management of marine communities of the Persian Gulf. Ambio 40:4–17

Sawall Y, Kürten B, Hoang BX, Sommer U, Wahl M, Al-Sofyani A, Al-Aidaroos AM, Marimuthu N, Khomayis HS, Gharbawi WY (2014) Coral communities, in contrast to fish communities, maintain a high assembly similarity along the large latitudinal gradient along the Saudi Red Sea coast. J Ecosyst Ecography S4:003

Schwarcz HP, Gao Y, Campana S, Browne D, Knyf M, Brand U (1998) Stable carbon isotope variations in otoliths of Atlantic cod (*Gadus morhua*). Can J Fish Aquat Sci 55:1798–1806

Sheppard C, Sheppard AS (1991) Corals and coral communities of Arabia. Fauna Arab 12:3–170

Sheppard C, Price A, Roberts C (1992) Marine ecology of the Arabian region - patterns and processes in extreme tropical Environments. Academic, London

Spaet J (2013) Predictable annual aggregation of longnose parrotfish (*Hipposcarus harid*) in the Red Sea. Mar Biodivers 43:179–180

Spaet JLY, Berumen ML (2015) Fish market surveys indicate unsustainable elasmobranch fisheries in the Saudi Arabian Red Sea. Fish Res 161:356–364

Spaet JLY, Thorrold SR, Berumen ML (2012) A review of elasmobranch research in the Red Sea. J Fish Biol 80:952–965

Spaet JLY, Jabado RW, Henderson AC, Moore ABM, Berumen ML (2015) Population genetics of four heavily exploited shark species around the Arabian Peninsula. Ecol Evol 5:2317–2332

Spaet JLY, Nanninga GB, Berumen ML (2016) Ongoing decline of shark populations in the Eastern Red Sea. Biol Conserv 201:20–28

Spalding MD, Fox HE, Allen GR, Davidson N, Ferdaña ZA, Finlayson M, Halpern BS, Jorge MA, Lombana AL, Lourie SA, Martin KD, McManus E, Molnar J, Recchia CA, Robertson J (2007) Marine ecoregions of the world: a bioregionalization of coastal and shelf areas. Bioscience 57:573–583

Spanier E, Galil BS (1991) Lessepsian migration: a continuous biogeographical process. Endeavour 15:102–106

Stephenson PC, Edmonds JS, Moran MJ, Caputi N (2001) Analysis of stable isotope ratios to investigate stock structure of red emperor and Rankin cod in northern Western Australia. J Fish Biol 58:126–144

Tamura K, Stecher G, Peterson D, Filipski A, Kumar S (2013) MEGA6: molecular evolutionary genetics analysis version 6.0. Mol Biol Evol 30:2725–2729

Tenggardjaja K, Jackson A, Leon F, Azzurro E, Golani D, Bernardi G (2013) Genetics of a Lessepsian sprinter: the bluespotted cornetfish, *Fistularia commersonii*. Isr J Ecol Evol 59:181–185

Tesfamichael D (2012) Assessment of the Red Sea ecosystem with emphasis on fisheries. PhD thesis. University of British Columbia

Tesfamichael D, Pauly D (2016) The Red Sea ecosystem and fisheries. Springer Netherlands, Dordrecht

Tesfamichael D, Pitcher TJ (2006) Multidisciplinary evaluation of the sustainability of Red Sea fisheries using Rapfish. Fish Res 78:227–235

Thorrold SR, Latkoczy C, Swart PK, Jones CM (2001) Natal homing in a marine fish metapopulation. Science 291:297–299

Thorrold SR, Zacherl DC, Levin LA (2007) Population connectivity and larval dispersal using geochemical signatures in calcified structures. Oceanography 20:80–89

Tietbohl MD (2016) Assessing the functional diversity of herbivorous reef fishes using a compound-specific stable isotope approach. MSc thesis. King Abdullah University of Science and Technology, Saudi Arabia

Tornabene L, Ahmadia GN, Berumen ML, Smith DJ, Jompa J, Pezold F (2013) Evolution of microhabitat association and morphology in a diverse group of cryptobenthic coral reef fishes (Teleostei: Gobiidae: *Eviota*). Mol Phylogenet Evol 66:391–400

Troyer E (2017) Microhabit association of cryptobenthic fishes (Family Gobiidae) in the central Red Sea. MSc thesis. King Abdullah University of Science and Technology, Saudi Arabia

Troyer EM, Coker DJ, Berumen ML (2018) Comparison of cryptobenthic reef fish communities among microhabitats in the Red Sea. PeerJ 6:e5014

Truett G, Heeger P, Mynatt R, Truett A, Walker J, Warman M (2000) Preparation of PCR-quality mouse genomic DNA with hot sodium hydroxide and tris (HotSHOT). BioTechniques 29:52–54

Vignaud TM, Maynard JA, Leblois R, Meekan MG, Vázquez-Juárez R, Ramírez-Macías D, Pierce SJ, Rowat D, Berumen ML, Beeravolu C, Baksay S (2014) Genetic structure of populations of whale sharks among ocean basins and evidence for their historic rise and recent decline. Mol Ecol 23:2590–2601

Waldrop E, Hobbs JPA, Randall JE, DiBattista JD, Rocha LA, Kosaki RK, Berumen ML, Bowen BW (2016) Phylogeography, population structure and evolution of coral-eating butterflyfishes (Family Chaetodontidae, genus *Chaetodon*, subgenus *Corallochaetodon*). J Biogeogr 43:1116–1129

Walter RP, Kessel ST, Alhasan N, Fisk AT, Heath DD, Chekchak T, Klaus R, Younis M, Hill G, Jones B, Braun CD (2014) First record of living *Manta alfredi* × *Manta birostris* hybrid. Mar Biodivers 44:2016–2001

Ward RD, Zemlak TS, Innes BH, Last PR, Hebert PD (2005) DNA barcoding Australia's fish species. Philos Trans Roy Soc B Biol Sci 360:1847–1857

White WT, Corrigan S, Yang L, Henderson AC, Bazinet AL, Swofford DL, Naylor GJP (2017) Phylogeny of the manta and devilrays (Chondrichthyes: mobulidae), with an updated taxonomic arrangement for the family. Zool J Linnean Soc 182:50–75

Wilkinson C (2008) Status of Coral Reefs of the World: 2008. Global Coral Reef Monitoring Network and Reef and Rainforest Research Centre, Townsville

Wilson SK, Graham NAJ, Pratchett MS, Jones GP, Polunin NVC (2006) Multiple disturbances and the global degradation of coral reefs: are reef fishes at risk or resilient? Glob Chang Biol 12:2220–2234

Wood LJ (2007) MPA global: a database of the world's marine protected areas. Sea Around Us Project, UNEP-WCMC & WWF. http://www.mpaglobal.org

Wyatt ASJ, Waite AM, Humphries S (2012) Stable isotope analysis reveals community-level variation in fish trophodynamics across a fringing coral reef. Coral Reefs 31:1029–1044

Xu W, Ruch J, Jónsson S (2015) Birth of two volcanic islands in the southern Red Sea. Nat Commun 6:7104

Yao F, Hoteit I, Pratt LJ, Bower AS, Zhai P, Köhl A, Gopalakrishnan G (2014) Seasonal overturning circulation in the Red Sea: 1. Model validation and summer circulation. J Geophys Res 119:2238–2262

Zhan P, Subramanian AC, Yao F, Hoteit I (2014) Eddies in the Red Sea: a statistical and dynamical study. J Geophys Res – Oceans 119:3909–3925

Printed by Printforce, United Kingdom